THE
FACTS ON FILE
DICTIONARY OF
NUMERICAL ALLUSIONS

THE
FACTS ON FILE
DICTIONARY OF
NUMERICAL ALLUSIONS

by
Laurence Urdang

Facts On File Publications
New York, New York • Oxford, England

The Facts On File Dictionary of Numerical Allusions

Copyright © 1986 by Laurence Urdang Inc.

First Published in the United States of America by Facts On File, Inc.
460 Park Avenue South, New York, New York 10016.

Library of Congress Cataloging-in-Publication Data
Urdang, Laurence.
 The Facts on File dictionary of numerical allusions.

 1. Numbers, natural—Dictionaries. 2. Symbolism of
numbers—Dictionaries. 3. English language—Glossaries,
vocabularies, etc. I. Facts on File, Inc. II. Title.
III. Title: Dictionary of numerical allusions.
AG5.U73 1986   031   86-22873
ISBN 0-8160-1300-4

Printed in the United States of America
10 9 8 7 6 5 4 3 2 1

Table of Contents

Foreword

Although it has to do entirely with numbers, this book has very little to do with mathematics. There are a few oddments—*pi*, the *Archimedean* and the *Kepler-Poinsot solids*—but for the most part, there is little here that would be of interest to mathematicians and numerologists only. Consequently, those who abhor mathematics or otherwise find the subject difficult, annoying, or formidable need have no fear.

Numbers surround us, more today than ever before. Everyone is beset by social security numbers, telephone numbers, house numbers, checking or savings account numbers, automobile registration numbers, ZIP codes, credit card numbers, and so on. One who despises such numbers as depersonalizing and thinks of them as a manifestation of many of the ills of modern society and its current obsession with computers should be aware that the notion of numbers has always been with us: as far back as we can trace mankind, numbers have played an important role in society. Notions of unity, duality, and trinity pervade the earliest traces of civilization; reflexes of three, seven, and nine (in particular) appear throughout the classical period; down through history, wars have been identified by their duration; in modern times, numbers are used to identify all sorts of things—political groups, laws, and so on. Their variety covers the entire spectrum of human experience.

Not everything encountered in the research for this book has been included; as may well be imagined, an enormous amount of information found to be of remote interest, boring, or insignificant has been omitted. Those who seek in these pages numerological hocus-pocus will find it only occasionally, and then mentioned sparingly: other sources should be consulted for that sort of thing. At the start, it was thought that song, book, film, and other titles—*Three Little Words, Tea for Two, 2001, Two on a Mattress, Cheaper by the Dozen, Six are Enough*, etc.—ought to be included, but the policy was changed when their quantity became apparent; besides, merely listing them would be of little use.

The task, though pleasurable, was at times frustrating, for, as a moment's reflection will show, information is not usually arranged after the fashion of this book, and finding appropriate entries could be done only by reading a huge volume of writings in history, mythology, literature, and other areas. In a few instances, entries were culled from sources which could not be readily corroborated: such material is accompanied by a reference to the secondary source where it was found and is shown in square brackets at the end of the (recast) definition. The selection is eclectic, reflective of personal bias.

For all of the excellent suggestions she made and, generally, for the immense help she has been to me throughout the preparation of this book, I am especially grateful to Jacquelyn Goodwin. Frank R. Abate helped, particularly, with sports entries and classical references. William A. Kunstler, Ronald Kuby, and Aidan Baker provided valuable material on entries like the *Chicago Seven, Catonsville Nine,* and *Shrewsbury Two.* John Ayto contributed some entries, as well, for England. And I am indebted to John Thornton, my editor at Facts On File, who supported this work—often on faith—for, in the early stages, he wasn't entirely sure what would emerge.

The editor would sincerely appreciate suggestions from readers, browsers, and users of *Numerical Allusions* regarding ways in which the book could be improved.

Laurence Urdang

Essex, Connecticut
August, 1986

Organization of Information

The organization of *Numerical Allusions* is so simple as to be almost self-evident. The main divisions are in numerical order; the numbers, ranked from the smallest to the largest, are set in the outside margin of each page; numbers occurring at the beginnings of headwords are always spelt out (e.g., **14th** or **XIVth Amendment** appears as **Fourteenth Amendment**). Within each numerical category (where there is more than one entry), entries in **bold type** are in alphabetical order. Some entries have variants that differ essentially in spelling, and those are shown following the entry word, separated by the word *or*. If the variant is a different word or name, it appears at the end of the entry following the words "Also called . . . . " Because the entries do not appear throughout the text in a single alphabetical set, they are so gathered together in an Index at the back where they appear in **bold type**.

There are many subentries in this book. For example, if you look up **Muse**, under **9**, you will find all nine Muses listed. Their names also appear in the Index, interfiled with the headwords and their variants. Titles of books, plays, films, and longer poems are shown in *italic* type; those of short stories, poems, and such are shown in double quotation marks; individual glosses (e.g., 'deceitful' 2: two-faced) are shown in single quotation marks. In some cases, references to longer headwords have been abbreviated, most often by repeating only the words preceding the first comma but occasionally by replacing several of the words at the end by "...". Such abbreviations should not interfere with the identification of the headword; in the few instances where ambiguity does arise, the selection of the proper entry is readily made by glancing at the text.

In the Appendix, starting on page 227, there is a list of the words for the set of numbers from 1 to 21 and on up to 100, 1,000, and 1,000,000 for French, Spanish, Italian, German, and Latin. It was thought that such lists might be useful. Both ordinals and cardinals are shown.

THE
FACTS ON FILE
DICTIONARY OF
NUMERICAL ALLUSIONS

-273

absolute zero, the theoretical temperature ($-273°$ Celsius) at which all molecular movement associated with heat ceases.

-3

double eagle, (in golf) the holing of a ball in three strokes less than par for the hole. See also -1: **birdie,** -2: **eagle.**

-2

eagle, (in golf) the holing of a ball in two strokes less than par for the hole. See also -1: **birdie,** -3: **double eagle.**

$\sqrt{-1}$, $1^{-\frac{1}{2}}$

i, or **i,** (in mathematics) the symbol for the square root of minus one.

-1

birdie, (in golf) the holing of a ball in one stroke less than par for the hole. See also -2: **eagle,** -3: **double eagle.**

0

bit, See 2: **bit.**

double or nothing *or* **double or quits,** See 2: **double or nothing.**

duck's egg, *British.* (in cricket) the zero that marks in the score the fact that a player or side makes nothing; a score of nothing, as *to win a duck's egg.* Compare 0: **goose-egg.**

goose-egg *U. S. informal.* zero; nothing. Compare 0: **duck's egg.**

ground zero, the precise point where an explosion is detonated.

nil, nothing; zero, used especially in Britain in giving scores of sporting events: *West Bromwich 3, Wolverhampton nil.*

nothing, nought; zero.

null, nothing; zero; cipher.

round o, zero.

single space, to leave no line of space between lines of text, especially in typing.

Zero, a Japanese fighter plane of World War II.

1

0

zero-dimensional, noting the magnitude of a point, which has position but no length, breadth, or thickness.

zero in on, focus on, usually used metaphorically, as *After a general introduction, the speaker zeroed in on the problem of unemployment in the shoreline area.*

Zeroth Law of Thermodynamics, a commutative law stating that if system A has the same temperature as system B, and if B has the same temperature as system C, then the temperatures of system A and C are the same. Compare **1: First Law of Thermodynamics, 2: Second Law of Thermodynamics, 3: Third Law of Thermodynamics.**

$\frac{1}{1000}$

mil, **1.** a unit of measure equal to 0.001 of an inch.
2. a money of account in the U. S., equivalent to one thousandth of a dollar, used chiefly in the calculation of tax rates.

millesimal, a thousandth; consisting of thousandth parts, as a *millesimal fraction.*

milli, a word element, taken from Latin, meaning 'one thousandth,' as in *millisecond,* or 'one thousand,' as in *millipede.*

milliare, (in the metric system) a unit of area equal to one thousandth of an are, or 155 square inches.

$\frac{1}{600}$

decimal minute (of an angle) one six-hundredth of a right angle.

$\frac{1}{360}$

chronic degree, one three-hundred-sixtieth part of a tropical year.

local degree, one three-hundred-sixtieth part of the zodiac.

$\frac{1}{240}$

decimal degree (of an arc) one two-hundred-fortieth of a circumference.

$\frac{1}{100}$

centesimal, **1.** a hundredth.
2. hundredth.

centesimation, the punishment of one man in a hundred, as in cases of mutiny or desertion from the army. Compare **10: decimate.**

1/100

centiare, one square meter, being the hundredth part of an are, equal to 1.19 square yards.

centigram, *or* **centigramme,** in the metric system, a measure of weight one hundredth of a gram, or 0.15432 grain troy.

centiliter, *or* **centilitre,** a measure of liquid in the metric system, one hundredth of a liter, or 0.33814 fluid ounce.

centimeter, *or* **centimetre,** (in the metric system) the hundredth part of a meter, or 0.3937+ inch.

decimal minute (of time) one hundredth of an hour; 0.6 of a minute.

penny, a coin of the United States, worth one hundredth of a dollar. Also called **cent.**

1/86

denarius, a Roman weight, the 86th or 94th of a Roman pound.

1/72

point, (in the United States) a unit of measure for the bodies of printing types, 0.0138 inches, or $1/72$ of an inch, or 0.0351+ centimeters. See also **0.0376: point.**

1/64

double demisemiquaver, (in music) a sixty-fourth note.

sixty-fourth note, (in musical notation) a hemidemisemiquaver, being equivalent in time-value to one half of a thirty-second-note.

1/60

decimal degree (of an angle) one sixtieth of a right angle.

second, 1. the sixtieth part of a minute of time.
2. the sixtieth part of a degree.

1/50

two-cent piece, a former coin of the United States, worth two cents.

3/100

three-cent piece, a former coin of the United States, worth three cents.

3

1/32

demisemiquaver, (in music) half of a demiquaver; a thirty-second note. Also called **hemidemisemiquaver.**

0.0376

point, (in France and Germany) a unit of measure for the bodies of printing types, 0.0376+ centimeters. Also called **Didot point.** See also **1/72: point.**

1/24

karob, a measure of weight formerly used by goldsmiths, one twenty-fourth of a grain.

1/20

nickel, a coin of the United States, worth one twentieth of a dollar.

1/16

demiquaver, (in music) a sixteenth note. Also called **semiquaver.**

fusella, (in medieval music) a sixteenth-note.

sixpence, a former American silver coin, worth one sixteenth of a dollar.

sixteenth-note, (in musical notation) a semiquaver, being equivalent in time-value to one half of an eighth-note.

1/10

deci-, a prefix or word element meaning 'tenth.'

deciare, a unit of land measure, the tenth part of an are.

decigram, *or* **decigramme,** a weight measure equal to one tenth of a gram.

decil, (in astrology) an aspect or position of two planets when they are a tenth part of the zodiac (36°) distant from each other.

deciliter, *or* **decilitre,** a volume measure equal to one tenth of a liter.

decima, 1. an organ-stop whose pipes sound a tenth above the keys struck.
2. (in Spanish money) the tenth of a real vellon.
3. a tenth part; a tithe, or tax of one tenth.

decimal degree, (of time) one tenth of an hour, or six minutes.

decimate, *Obsolete.* to take the tenth part of or from; tithe.

$\frac{1}{10}$

decime, a French coin, the tenth of a franc.

decimeter, *or* **decimetre,** (in the metric system) a measure of length equal to the tenth part of a meter, or 3.937 inches.

decimo, 1. (in Spain) **(a)** the tenth part of a peso or dollar. **(b)** the tenth part of an oncia, or ounce.
2. (in La Plata) a copper coin, the tenth part of a medio.

decinormal, having one tenth of the normal or usual strength.

decistere, (in the metric system) a cubic measure equal to the tenth part of a stere, 100,000cc, or 3.531 cubic feet.

deem, *or* **deeme,** *Obsolete.* a tithe; a tenth. [var. of *dime, disme*]

dime, 1. *Obsolete.* tithe. See **10: dime, 2.**

dinero, a silver coin of Peru, the tenth of a sol.

Saladin's tenth, a tax laid in England and France, in 1188, by Pope Innocent III for the crusade of Richard I and Philip Augustus against Saladin, Sultan of Egypt, then going to besiege Jerusalem. Gibbon (*Decline and Fall of the Roman Empire*) considered it the foundation of the ecclesiastical tithing system.

tithe, a tenth part of anything, hence a tax levied by a civil or ecclesiastical body on the people for some use.

$\frac{1}{9}$

nonage, a ninth part, especially of movables, which in former times was paid to the English clergy on the death of persons in their parish and claimed on pretense of being devoted to pious uses.

$\frac{1}{8}$

doit, a small copper coin, formerly of the Netherlands, worth one eighth of a stiver.

eighth-note, (in musical notation) a note having half the time-value of a quarter-note; a quaver.

fusa, (in medieval music) an eighth-note, or quaver.

octave, a small cask of wine containing one eighth of a pipe.

octoroon, *or* **octaroon,** an offspring of a quadroon and a white person; a person having one eighth negro blood.

quintroon, the offspring of a white person and one who is one fourth part negro, or a quadroon; a person with one eighth negro blood.

shilling, a former American silver coin, worth one eighth of a dollar.

⅙

sextans, a bronze coin of the ancient Roman republic, equivalent to one sixth of an as.

sextant, 1. one sixth part of a circle.
2. a device used in navigation and surveying for measuring the angular distance of two heavenly bodies or other objects, or the altitude of a heavenly body above the horizon by causing images of the two points of reference to coincide when transmitted through an arrangement of mirrors. [So called because the arc through which the measurement is made is one sixth of a circle, or 60°.]

⅕

cinquième, a coin of Louis XV of France, the fifth part of an écu.

fifth, (in the U. S.) a standard size of container for alcoholic spirits, being equal to one fifth of a gallon, or four fifths of a quart.

quintile, (in astrology) noting the aspect or position of two planets that are distant from each other the fifth part of the zodiac, or 72°.

¼

demi-semi, half of a half; applied contemptuously to a person or thing of little account or consequence.

farthing, a former English coin, worth one fourth of a penny.

firkin, a measure of capacity, formerly usually the fourth part of a barrel.

grand quarter, (in heraldry) 1. one of the four primary divisions of a shield, namely, dexter chief, sinister chief, dexter base, and sinister base.
2. an ordinary occupying one fourth of the field, placed in the dexter chief unless otherwise specified. Compare **9: point.**

pice, (formerly) a coin of India under British rule, worth one quarter of an anna. [From Hindi *paisā* < Sanskrit *padī* 'quarter']

quadrans, in the ancient Roman monetary system, a bronze coin which was equal to one fourth of an as.

quadrant, a quarter of a circle.

quadroon, the offspring of a mulatto and a white person; a person having one quarter negro blood.

quart, a quarter of a gallon.

quarter, one of four equal or equivalent parts into which anything can be divided, as *a quarter of an hour* (15 minutes), *a quarter (of a dollar)* (twenty-five cents), the fourth part of the moon's period, one of the four

6

$^1/_4$

cardinal points of the compass, the fourth part of the school year, the fourth part (three months) of a year, etc.

quarter-horse, (in the U. S.) a horse used for racing a distance of a quarter of a mile.

quarter-point, (in navigation) the fourth part of the distance between points on a compass-card, being one fourth of 11°15′, or 2°48′45″.

quarter-waiter, *or* **quarterly waiter,** an officer or gentleman usher at the English court who is in attendance a quarter of a year at a time.

quarto, a size of book one fourth that of a size of paper accepted as standard, namely: cap quarto (7″ x 8½″); demy quarto (8″ x 10½″); folio-post quarto (8½″ x 11″); medium quarto (9″ x 12″); royal quarto (10″ x 13″).

$^1/_3$

riding, one of the three administrative districts (originally) of Yorkshire, namely: the East Riding; the North Riding; and the West Riding. The term *Riding* was later applied in other areas in Great Britain and in her colonies abroad. [From a misanalysis of *North Thriding* 'north third(ing),' with assimilation of the two *-th-* sounds.]

trigon, (in astrology) **1.** one of the four divisions, consisting of three signs each, into which the zodiac is divided, namely: the watery trigon (Cancer, Scorpio, Pisces); the earthy trigon (Taurus, Virgo, Capricornus); the airy trigon (Gemini, Libra, Aquarius); and the fiery trigon (Aries, Leo, Sagittarius).
2. an aspect of two planets 120° distant from one another; trine.

trine, (in astrology) noting the aspects or position of two planets that are distant from each other the third part of the Zodiac, or 120°.

.44

forty-four, the caliber of the pistol used by Frankie to shoot Johnnie, in the ballad, *Frankie and Johnnie.*

.45

forty-five, *Informal.* a .45-caliber pistol.

$^1/_2$

better half, *Facetious.* one's spouse, usually one's wife.

demi-, a prefix denoting 'half.'

demi-cadence, (in music) a half cadence, usually denoting the progression from tonic to dominant.

demicarlino, a coin equal in value to half a carlino.

7

¹⁄₂

demi-column, a column with about half the shaft engaged in the wall.

demifarthing, a coin of Ceylon, worth half a farthing.

demigalonier, *Obsolete.* a vessel for table use with a capacity of half a gallon.

demi-jambe, a piece of armor covering only the front of the leg. Also called **demi-grevière.**

demi-kindred, persons related by either maternal or paternal blood.

demi-landau, a half-landau; a landaulet.

demi-metope, (in architecture) a half metope, sometimes found at the angles of a Doric frieze in Roman, Renaissance, or other debased examples.

demi-pique, *Obsolete.* a saddle with a pique about half as high as that of a military saddle.

demisang, (in law) a person of half-blood.

demi-volt, (in manège) a half turn in the air with the forelegs raised.

demi-wolf, a mongrel, half wolf and half dog.

demy, 1. half, as applied to a particular size of paper.
2. a size of writing paper in America, 16″ x 21″, and of printing paper in Great Britain, 17½″ x 22″; English writing demy is 15″ x 20″. Double-demy is 26″ x 38½″.

dichotomy, subdivision into halves or pairs.

dimidiate, (in heraldry) to cut in half, showing only one half.

flat, (in music) 1. a tone one half-step below a given tone.
2. to depress a tone a half-step.

half-and-half, 1. a mixture of half milk and half light cream, used for coffee.
2. (in Britain) a mixture (usually) of half porter and half ale.

half-assed, *or Brit.* **half-arsed,** *Slang.* inept; bungling.

half-baked, *Informal.* 1. unenthusiastic, as *Anne's response to the invitation was pretty half-baked.*
2. wishy-washy, noncommittal, as *The senator gave a half-baked answer when asked about his campaign promise to improve housing.*

half-breed, a person who is the offspring of parents of different races, usually applied to the offspring of an American Indian and a white.

half-dollar, a coin of the United States, worth half a dollar.

half model, a model of half the longitudinal hull of a vessel, usually of a sailboat, especially mounted on a polished board for display.

halfpenny, *or ha'penny,* **1.** equal to one cronebane, circulated in Ireland toward the end of the 18th century.
2. a former coin of England, worth half a penny.

half seas over, drunk.

hemiobolion, a coin of ancient Athens, equal in value to half an obol.

hemisome, half of an animal's body.

hemisphere, half a sphere, especially half of the terrestrial globe, which may be divided into the northern hemisphere, which includes Europe, Asia, and North America, and the southern hemisphere, which includes South America, Africa, Australia, and Oceania, or into the eastern hemisphere, which includes Asia west to the Ural Mountains, Australia, and Oceania, and the western hemisphere, which includes Europe, Africa, and North and South America.

mezzo-rilievo, (in sculpture) relief higher than bas relief, but lower than high relief; demi-relief.

quarter-watch, (in nautical terminology) one half of the watch on deck.

semitone, (in music) an interval approximately equal to half of a tone; a demitone.

decimal minute (of time) one hundredth of an hour; six tenths of a minute.

dodrans, three fourths, especially three fourths of a Roman foot (8.73 inches).

three-quarter, **1.** (formerly) a portrait painted on a canvas 30 inches by 25 inches, or about three quarters the size of a kitcat (36 inches by 28 inches).
2. a portrait showing about three quarters of the figure.
3. three-quarter face, a portrait, especially a photograph, showing the face midway in position between a profile and a frontal view.

three-quarter time, (in music) the meter of a musical composition having a time signature of $\frac{3}{4}$ and three quarter notes or their equivalents in each measure. Also called **waltz time.**

1

A, a, **1.** the first letter and first vowel of the English alphabet.
2. the first item in a system where A = 1, B = 2, C = 3, etc.

aleph, the first letter of the Hebrew alphabet, corresponding to the Greek *alpha* and the English *A, a*.

alif, the first letter of the Arabic alphabet.

All Fools' day, the first of April, on which it has long been customary to mock the unwary by sending them on some bootless errand or by making them the object of a good-humored practical joke.

alpha, the first letter of the Greek alphabet, corresponding to the English *a*.

America firster, a person who refuses to consider the rights of any other country before those of the U. S., who denies aid to any other nation till every person in the U. S. has been given like consideration, and who generally maintains an isolationist point of view.

Ash Wednesday, the first day of Lent.

at first blush, upon the initial presentation or viewing of, as *At first blush, I thought it a good idea.*

at first hand, personally; not through an intermediary.

B, b, the first consonant of the English alphabet.

bird in the hand is worth two in the bush, a, a proverb meaning that it is more worthwhile to take advantage of immediate opportunities than to wait for more speculative ones.

bogey, (in golf) the holing of a ball in one stroke more than par for the hole. See also **2: double bogey, 3: triple bogey.**

burn one, *U. S.* **1.** (in a bar) an order to serve up one beer.
2. (in a short-order restaurant) **(a)** an order to serve up one malted milk or the equivalent. **(b)** an order to cook thoroughly the food ordered.

calends, the, *or* **kalends, the,** the first day of the month in the ancient Roman calendar. Compare **13: ides, 9: nones.**

committee of one, a person who appoints himself as the sole judge or authority, without recourse to the opinions of others.

die, *pl.* **dice.** a small cube marked on its faces with spots numbering from one to six, used in gaming and gambling.

dollar-a-year man, *U. S.* a person appointed to a governmental post who is paid only a token salary, though not literally a dollar a year.

double space, to leave an empty line of space between lines of text, especially in typing.

draw one, (in a short-order restaurant) to serve one cup of coffee.

dyadic syntheme, (in Greek prosody) a dyadic disyntheme in which each element occurs only once. Compare **2: dyadic disyntheme.**

enthymeme of the first order, (in logic) a syllogism with only the major premise expressed.

first, **1.** the highest rank in an examination for honors, as *He took a first in science.* Compare **2: double-first.**
2. the first, or lowest gear used to move a motor vehicle forward. Also called **low, low gear, first gear,** *Brit.* **first speed.**

first aid, emergency medical help given at the scene of an accident or the like till the patient can be treated by a fully equipped medical facility.

first base, (in baseball) **1.** the first of the bases from the home plate, in a counterclockwise direction.
2. the player positioned at that base.

First Cause, the, that which is not caused by anything else, i.e., God.

first-class, **1.** best-equipped, most luxurious, and most expensive, especially used in reference to travel and hotel accommodations.
2. *Informal.* first-rate; excellent in every way.
3. in a first-class style or manner, as *She always travels first-class.*

first class mail, mail that consists chiefly of private letters or postcards carried at a high rate for prompt delivery.

first come, first served, 'service will be rendered in the sequence of arrival of those who are to be attended to.'

first covenant, a promise made by God to Noah (Genesis 9:9) that never again would a flood destroy the earth. The rainbow is the token of that covenant. See also **2: second covenant.**

First-day, Sunday, the first day of the week, so called by the Society of Friends.

first-day cover, (in philately) a cover, or envelope postmarked to indicate that it was mailed on the first day of a stamp's issue from a city where it was issued.

first-degree burn, a relatively slight burn in which the surface of the skin is reddened.

first difference, (in logic) the most fundamental difference.

first digit, **1.** the innermost digit of a five-digit limb; the thumb or great toe in man.
2. the index finger.

1

First Empire, the empire established by Napoleon Bonaparte in 1804, replacing the First Republic and lasting until 1814.

first family, *U. S.* **1.** the family of the president of the United States or of the governor of a state.
2. a family ranking high in social standing in a given community.
3. a family descended from one of the first or earliest settlers of a given area.

first floor, 1. (in the U. S.) the floor in a building nearest the ground and above the basement; the ground floor.
2. (in Britain) the next floor up.

first-foot, (in Scotland) **1.** the person who first enters a dwelling in the New Year. **2.** the first person or object met on setting out on an important journey or undertaking.

first fruits, the first yield or result of anything.

first good, (in ethics) something that is desirable for itself; the ultimate end.

first-hand, 1. new; unused.
2. directly from the original source. Compare **2: second-hand.**

first-in, first-out, *or* **FIFO,** a commercial and financial plan under which merchandise first received is delivered first, with the value of all inventory set at the cost of the most recent purchases, making inventory values proportionate to any price rise. See also **1: last-in, first-out.**

first intention, (in logic) a general conception obtained by abstraction from the ideas or images of sensible objects. Compare **2: second intention.**

first lady, *U. S.* **1.** the wife of the president of the United States or of a governor of a state.
2. the wife of the leader of any country.
3. one of the most prominent women in any field.

First Law of Thermodynamics, the change in the internal energy of a system is equal to the heat it absorbs less the work it does. Compare **0: Zeroth Law of Thermodynamics, 2: Second Law of Thermodynamics, 3: Third Law of Thermodynamics.**

first lieutenant, a lieutenant ranking above a second lieutenant and below a captain.

first light, dawn; daybreak.

firstling, anything that comes first, as a first-born child.

first mate, (on a commercial vessel) the officer ranking next below the captain.

first mortgage, the senior mortgage, which takes precedence over any other mortgages, liens, or encumbrances and gives the mortgagee priority to payment in the event of default. Compare **2: second mortgage.**

first name, given name; Christian name.

first nighter, a person who habitually attends plays, ballets, and other repeated performances on the opening night.

first offender, a person convicted of any offense, usually a felony, for the first time.

first papers, *U. S.* the documents associated with the first stage of application for U. S. citizenship by an alien.

first person, (in grammar) a category of pronouns and of verb forms referring to the speaker or speakers or to the speaker and his companions.

first position, (in ballet) a position in which the feet have the heels touching and the toes pointing in opposite directions to the left and right.

first quarter, (in astronomy) the appearance of the half moon approximately one week after the new moon.

first-rate, of the highest excellence; preeminent in quality or estimation; first-class.

First Reich, (in Germany) the medieval Holy Roman Empire until its dissolution in 1806. Compare **2: Second Reich, 3: Third Reich.**

First Republic, the French republic established in 1792, as a result of the French Revolution, replaced in 1804 by the First Empire.

first-run, 1. (of a motion picture) not shown previously; in its initial run. 2. (of a motion-picture theater) showing only, or chiefly, first-run motion pictures.

first sergeant, the senior noncommissioned officer in the U. S. Army who is responsible for personnel and administration of a military unit.

first set, (in whaling) the first thrust of the harpoon. Also called **first lance.**

first-string, 1. consisting of the main team players; not including substitute players, second-string players, etc. 2. of the best or most highly respected group.

first things first, 'everything should be taken in proper order.'

First Triumvirate, the coalition (60 BC) between Pompey, Julius Caesar, and Crassus. See also **2: Second Triumvirate.**

first watch, (in nautical use) the watch stood from 8 P.M. (2000 hours) to midnight (2400 hours).

1

first water, 1. (of diamonds) of the finest quality.
2. (of anything) of the best quality.

get to first base, *U. S. informal.* 1. to succeed in completing the initial step in a process, activity, or series.
2. to succeed in the initial stages of seduction, usually used by males in referring to females and usually used in negative contexts, as, *He never could get to first base with Marie.* [From baseball, in which a batter must get to first base in order to get a chance to score at home.]

healing by first intention, (in medicine) the union of the edges of a wound, without granulation. Compare **2: healing by second intention.**

Hecatombaeon, the first month of the Attic year, corresponding to the latter half of July and the first half of August.

hole in one, (in golf) a dropping of the ball into the hole as the result of a single stroke from the tee.

I, *or* **i,** the Roman numeral for one.

ichiban, number one; the first; the best. [From Japanese *ichi* 'one' + *ban,* a suffix indicating number.]

in the country of the blind the one-eyed is king, a saying meaning that it requires little talent or intelligence to surpass those possessed of neither.

J, *or* **j,** the roman numeral I, especially when it appears at the end of a series, as in *vj* 'six,' *viij* 'eight,' etc.

January, the first month of the year in the modern calendar.

last-in, first-out, *or* **LIFO,** a commercial and financial plan under which merchandise received last is delivered first, with inventory valued at the market price prevailing at time of delivery.

lever of the first class *or* **kind,** a straight lever in which the power is at one end and the weight at the other, with the fulcrum in between.

life, house of, (in astrology) the first house. See **12: house.**

Lone Star State, a nickname of Texas.

magpie, to see only one magpie is unlucky. [Brewer, *Reader's Handbook,* 1889.] See also **2: magpies.**

masculine rime, (in prosody) a rime between only one stressed syllable in each word, as *despite/contrite/all-night, consign/resign/assign/align.* Compare **2: female rime, 2: feminine rime, 3: feminine rime.**

1

May day, the first of May, an international date observed mainly in celebrating labor, or left-wing or communistic sympathies; formerly a day for celebrating the first spring planting.

milliarium, a Roman milestone, a column set at intervals of a thousand steps (**mille passus**), equivalent to five thousand Roman feet, along military roads.

M. O. 1, (in Britain) one of several units of the Military Operations Directorate, established in 1904; it was concerned with imperial defense and strategic distribution of the army. Other units were: M. O. 2, which dealt with foreign intelligence; M. O. 3, with administration and special duties; and M. O. 4, with topography. M. O. 5 was established in 1916 to deal with counterintelligence. M. I. 1(a) dealt with operational intelligence, M. I. 1(b) with censorship and propaganda. Various other departments have appeared from time to time, for example: M. I. 9, which dealt (1940–45) with escapees from Germany; M. I. 3, which may have served (1940–45) as a repository of expertise about activities in Germany; and M. I. 19, which dealt with interrogation.

M-1A, *U. S.* a measurement of the amount of money in circulation, defined by the Federal Reserve as consisting of currency, and checking account deposits (also called demand deposits) in commercial banks. See also **1: M-1B; 2: M-2; 3: M-3.**

M-1B, *U. S.* a measurement of the amount of money in circulation, defined by the Federal Reserve as consisting of M-1A plus other interest-bearing check-like deposits in all other depository institutions, such as NOW accounts, automatic transfer accounts, credit union share drafts, and demand deposits at mutual savings banks and thrift institutions. See also **1: M-1A; 2: M-2; 3: M-3.**

monandrous, having or taking only one husband at a time. Compare **1: monogynous.**

monism, (in philosophy) the view that seeks to explain all sorts of phenomena by the assumption that there is only one absolute substance or principal. Compare **2: dualism.**

mono-, a prefix of Greek origin meaning 'one, single.'

monocerous, having one horn, as a unicorn.

monochromatic, *or* **monochroic,** consisting of only one color or of one color and white; in painting, it may refer to the use of many shades of a single hue.

monochrome, consisting of only one color and white, usually of black and white.

monocle, a single eyeglass lens worn by fitting it between the eyebrow and the upper cheek.

1

monocracy, government or rule by one person; autocracy.

monogamy, 1. the principle or practice of marrying only one, especially one that forbids remarriage after the death of a spouse.
2. the condition of being married to only one person at a time.

monoglot, speaking or using only one language.

monograph, a treatise on a single subject or a single set of things or subjects.

monogynous, having only one wife; monogamous. Compare **1: monandrous.**

monolatry, the idolatrous worship of only one divinity.

monolith, a single stone; a single structure or object formed of a single piece, as an obelisk or column, or seemingly so, as a large company with a more or less vertical administrative structure.

mononym, a name, especially a scientific name, consisting of one term. Compare **2: binomial, 2: dionym, 3: trionym.**

monophobia, a morbid dread of being left alone.

monophthong, a single vowel sound, as that of the *a* in *father*. Compare **2: diphthong, 3: triphthong.**

Monophysitism, the doctrine that in the person of Christ there is one nature only. Compare **2: Dyothelitism.**

monorhine, having only one nasal passage, instead of the usual two, as lampreys and hags. Compare **2: double-nostriled.**

monosyllable, 1. a word of one syllable, as *good*.
2. **the monosyllable,** usually, the word "yes" or "no."

Monothelitism, *or* **Monotheletism,** *or* **Monothelism,** the doctrine that in the person of Christ there are but one will and one energy or operation. Compare **2: Dyothelitism.**

monoxylon, something formed from a single piece of wood, as a sculpture.

monozygotic, coming from a single ovum, as (identical) *monozygotic* twins, triplets, etc. Compare **2: dizygotic.**

murder in the first degree, (in U. S. law) usually, premeditated murder or murder incidental to a planned felony. Also called **first-degree murder,** *Informal,* **murder one.**

Newton's First Law of Motion, one of three laws describing the relationships between forces acting on a body and the motion of the center of its mass, applicable only relative to an inertial reference frame (i.e., one based on the positions of the fixed stars). It states that if a body is at rest or moving at a constant speed in a straight line, it will continue to remain so unless acted upon by a force. Compare **2: Newton's Second Law of Motion, 3: Newton's Third Law of Motion.**

New-Year's day, the first day of the year in the modern calendar; the first of January.

number one, 1. a child's euphemism for urination. Compare **2: number two.**
2. numero uno, oneself.

olfactory nerve, the nerve involved in the sense of smell. Also called **first cranial nerve.**

once, one time; occurring one time.

once bitten, twice shy, a saying meaning that a person who has had a bad experience will avoid the circumstances that may lead to its repetition.

once in a blue moon, very rarely, if ever.

one, 1. being a single unit or individual.
2. unity; the first whole number or the symbol representing it.

"One, two, buckle my shoe . . . ," from an old nursery counting rime, said to have gone up to thirty (*Oxford Dictionary of Nursery Rhymes,* Iona and Peter Opie, O.U.P., 1951 and 1975):
>One, two, buckle my shoe;
>Three, four, open the door;
>Five, six, pick up sticks;
>Seven, eight, lay them straight;
>Nine, ten, a big, fat hen;
>Eleven, twelve, dig and delve;
>Thirteen, fourteen, maids a-courting;
>Fifteen, sixteen, maids in the kitchen;
>Seventeen, eighteen, maids in waiting;
>Nineteen, twenty, my plate's empty.

one, two, three a-leery, a child's rime recited while bouncing a rubber ball and performing various acrobatic actions.

one-a-cat, *or* **one-o'-cat,** *or* **one old cat,** a simplified form of baseball, formerly played by children, in which there is one base and one batter on each side who must run to the base and back to home plate before being put out. Compare **2: two-a-cat, 3: three-a-cat, 4: four-a-cat.**

one-alarm fire, a fire that can be controlled by fire-fighting equipment from the first call issued by a fire department. See also **2: two-alarm fire.**

1

one-armed bandit, *Informal.* a gambling device into which a player inserts money and pulls a lever, causing three or more wheels bearing pictures on their periphery to spin, a different amount of money being paid to the player depending on the pattern of identical pictures that appear in a row, with most patterns paying nothing at all.

one-cent sale, (in merchandising) a temporary promotion attempting to create traffic in a retail shop by offering to sell a customer two selected identical items in exchange for one cent more than the price of one such item.

one-dimensional, noting the magnitude of a line, which has only length, but no breadth or thickness.

one for the road, a drink before leaving; a stirrup cup.

one-horse town, *U. S. informal.* a small, insignificant village; a backwater so provincial that there is but one horse in it.

"One if by land, and two if by sea, . . . " the signals agreed on by Paul Revere to warn of the arrival of the British forces (1775). In *"Paul Revere's Ride,"* by Henry Wadsworth Longfellow (1807–82):
> One if by land, and two if by sea;
> And I on the opposite shore will be.

According to a letter to Dr. Jeremy Belknap (1775), Paul Revere (1735–1818) arranged with Colonel Conant of the Charlestown Committee of Safety:
> If the British went out by water, to show two lanterns in the North Church steeple; and if by land, one as a signal, for we were apprehensive it would be difficult to cross the Charles River or get over Boston Neck. [*Bartlett's Familiar Quotations,* 15th ed., 1980.]

"One I love . . . ," from a child's rime recited when plucking petals from a daisy:
> One I love, two I love,
> Three I love, I say.
> Four I love with all my heart,
> Five I cart away.
> Six he loves, seven she loves, eight both love.
> Nine he comes, ten he tarries,
> Eleven he courts, twelve he marries.

one-man dog, a dog that, either by temperament or training, obeys or tolerates the presence only of its master.

oneness, the quality of being one; unity.

one-night stand, 1. (originally) a stopover of only one night by an itinerant group of performers, a carnival, etc.
2. *U. S. slang.* any activity that is characterized by its short-lived or tem-

porary nature, especially a sexual encounter involving no permanent relationship, used literally and figuratively.

one-upmanship, a facetious term referring to being "one-up" on one's fellows or to having an advantage over them by a clever ploy that actually puts them down. [*The Theory and Practice of Gamesmanship*, by Stephen Potter, 1962.]

only, 1. single as regards number, class, or kind; one and no more or other.
2. alone; nothing or nobody else than.

Palm Sunday, the first day of Holy Week.

penny post, a postal system that carries and delivers letters, etc., for a penny. The original penny post was set up in London about 1680 by William Dockwra and Robert Murray for delivery to all parts of London and its suburbs of letters and packets weighing less than one pound for the sum of one penny each.

Polyphemus, a cyclops, so named, from whom Odysseus and his men escaped by blinding him and then concealing themselves under the bellies of his sheep when they were put out to graze.

prima ballerina, the principal female dancer in a ballet or ballet company.

prima buffa, the principal female singer in a comic opera or comic opera company.

primacy, the condition of being first in order, power, or importance.

prima donna, the principal female singer in an opera or opera company.

primal, first in time, order, or importance.

primal scream, the first, or infant birth cry.

primal urge, the first, or most important force in man; the sexual urge.

primary, first in order, dignity, or importance; original.

primary deviation, (in ophthalmology) the deviation of the weaker eye from that position which would make its visual line pass through the object-point of the healthy eye.

primary education, the first level of schooling at which students are taught basic skills, like reading, writing, and arithmetic. See also **2: secondary education.**

Primates, the first or highest order of *Mammalia,* including man, monkeys, and lemurs.

1

prime, 1. a canonical hour. Compare **3: terce, 6: sext, 9: none.**
2. the first of eight parries or guards against thrusts in fencing. Compare **2: seconde, 3: tierce, 4: quarte, 5: quinte, 6: sixte, 7: septime, 8: octave.**

primigenial, *or* **primogenial,** first-born; primary.

primigravida, a woman who is pregnant for the first time.

primipara, a woman who bears a child for the first time.

primo, (in music) the first or principal part, as in duets and trios.

primogeniture, the state of being the first-born among children of the same parents, hence, according to some laws and customs, legally entitled to succeed to the father's real estate in preference to and to the absolute exclusion of the younger sons and daughters.

protagonist, (in ancient Greek drama) the first actor, or hero. Originally, the drama was performed by the chorus only; legend has it that Thespis introduced the novelty of a single actor speaking separately, and the protagonist came to be. See also **2: deuteragonist, 3: tritagonist.**

Sabbath, (in the Christian calendar) the first day of the week, Sunday, or the Lord's Day, called First-day by the Friends.

single, one; consisting of one; unmarried.

single-cut file, a file with a single series, or course, of teeth.

single file, an arrangement of a single line of persons or objects, one behind the other. Also called **Indian file.**

singleton, (in certain card games, as bridge) a card that is the only one of that suit in a hand.

singular, 1. being one; (in grammar) pertaining to or noting one person or thing.
2. (in grammar) a form used when noting one person or thing; in English and many other modern languages, contrasting with *plural;* in some, especially older languages, contrasting with *dual* and *plural.*

square one, back to, start all over again. [From the rules of various board games.]

St. David's day, a festival observed by the Welsh on March 1st in honor of St. David, patron saint of Wales, bishop of St. David's in Pembrokeshire, who flourished in the fifth and sixth centuries, and is said to have lived to the age of 110.

Sunday, the first day of the week.

1

unicorn, a fabled animal, depicted with the body of a horse and, sometimes, the tail of a lion, having a single, long pointed horn projecting from its forehead. Also called **monoceros.**

unicuspid, having only one cusp, or such a tooth, as an incisor or canine. See also **2: bicuspid.**

unicycle, a velocipede consisting of a single rubber-tired wheel set in a frame with a seat and no handlebars, used chiefly by trick cyclists.

unidactyl, *or* **unidactyle,** *or* **unidactylous,** having only one (functional) digit, as a horse. Also **monodactyl, unidigitate.**

unilateral, one-sided.

uninominal, *or* **uninomial,** (of a scientific name) consisting of only one element or word. Compare **2: binomial, 3: trinomial.**

uniocular, having only one eye. Also called **monocular.**

unipara, a woman who has borne only one child.

uniparous, giving birth to only one offspring at a time.

uniped, having only one foot.

unique, having no like or equal; single; only.

unit, a single thing or person.

unity, the state of being only one; oneness.

univalent, (in chemistry) having a valence of one.

univalve, 1. having only one valve, as a snail.
2. such a mollusk. Compare **2: bivalve.**

War of the First Coalition, a war (1793–97) between France and an alliance of European monarchies led by Austria, in which Napoleon Bonaparte was the victor.

Washington, George, (1721–99) the first president of the United States (1789–97), described, by Henry "Light-Horse Harry" Lee (1756–1818), on the occasion of Washington's death, as " . . . first in war, first in peace, and first in the hearts of the citizens." "Citizens" was later changed to "countrymen."

wine of one ear, (in Rabelais) good wine.

word, (in computer technology) a set of characters composed of bits that occupies one storage location and is treated in processing as one unit of data. See also **2: bit.**

1⅛

sesquioctaval, in the ratio of 9 to 8, or 1⅛ to 1.

$1\frac{1}{7}$

sesquiseptimal, in the ratio of 8 to 7, or $1\frac{1}{7}$ to 1.

$1\frac{1}{6}$

sesquisextal, in the ratio of 7 to 6, or $1\frac{1}{6}$ to 1.

$1\frac{1}{5}$

sesquiquintal, in the ratio of 6 to 5, or $1\frac{1}{5}$ to 1.

$1\frac{1}{4}$

sesquiquartal, in the ratio of 5 to 4, or $1\frac{1}{4}$ to 1.

$1\frac{1}{3}$

sesquitertian, *or* **sesquitertial,** *or* **sesquitertianal,** in the ratio of 4 to 3, or $1\frac{1}{3}$ to 1.

$1\frac{1}{2}$

picnostyle, (in architecture) intercolumniation of $1\frac{1}{2}$ diameters, in the classification of Vitruvius. See also **2: intercolumniation.**

sesqui-, a Latin prefix meaning 'one half more,' that is, one and a half times some unit.

sesquialteral, *or* **sesquialterate,** *or* **sesquialterous,** one half more; one and a half times.

2

à deux, for two, as *an intimate dinner à deux.*

ai, the two-toed sloth. See also **2: unau, 3: three toed-sloth.**

ambivalent, offering two choices or unable to choose between two alternatives.

at second hand, not personally; through an intermediary.

B, b, 1. the second letter of the English alphabet.
2. the second item in a system where A = 1, B = 2, C = 3, etc.

Berrigan Brothers, the, Daniel and Philip F. Berrigan, Roman Catholic priests and militant political and anti-Viet Nam war activists during the 1960s and '70s. They were convicted for destroying draft records in Catonsville, Maryland, in 1967. See also **9: Catonsville Nine.**

beta, the second letter of the Greek alphabet, corresponding to the English *b*.

beth, the second letter of the Hebrew alphabet, corresponding to the Greek *beta* and the English *b, v.*

between two fires, faced by equally unattractive alternatives.

bi-, a prefix of Latin origin, cognate with *di-* and *two*, meaning 'two, twice, double, twofold.'

biarchy, dual governments or sovereignty.

biathlon, a winter sports event combining the skills of cross-country skiing and marksmanship.

bicameral, having or consisting of two legislative bodies.

bicorn, a hat, popular in the eighteenth and early nineteenth centuries, with a wide brim turned up and fastened on two sides to the crown so as to form two pointed projections which point either sideways or front and back. Compare **3: tricorn.**

biennial, of or pertaining to a period of two years.
2. a commemoration or celebration of an event that occurred two years earlier; a second anniversary.
3. lasting for two years, as *a biennial plant*; occurring once every three years.

bifarious, twofold; double; binary; divided in two parts.

bifid, forked, as a snake's tongue; cleft; divided in two parts halfway down its length. Compare **3: trifid.**

bifocal, having two foci, especially in reference to spectacles that allow for corrected vision of close and distant objects by the combination of two lenses in the same frame.

bifrons, having two faces, as the Roman god Janus.

bifurcate, *or* **bifurcated,** *or* **bifurcous,** forked.

biga, (in Roman antiquity) a chariot drawn by two horses abreast. Compare **3: triga, troika; 4: quadriga.**

bigamy, the state or condition of being married to two spouses at the same time. Compare **2: deuterogamist, digamy, 3: trigamist.**

bilateral, having two sides; involving both of two sides.

bilingual, involving or using two languages, as speech, an inscription, etc.

biliteral, containing two letters, as a root, word, or syllable.

bimanous, having two hands. Compare **4: quadrumanous.**

bimensal, bimonthly; occurring once every two months; bimestrial.

2

bimestrial, bimonthly; lasting for two months.

bimetallism, a monetary system employing two metals with relative values established by legislative enactment, especially a system permitting the unlimited coinage of silver and gold at a fixed relative value.

binary, twofold; dual; double; twin; dimerous.

binary name, a taxonomic name consisting of two elements, as *Felis leo*; a binomial.

binary number, a quantity represented in the binary system, in which 2 is the base, or radix. Figures used are 0 and 1. The quantity represented by 140 in the decimal system (base 10) would be shown as 10001100 in the binary system: $(1 \times 2^7) + (0 \times 2^6) + (0 \times 2^5) + (0 \times 2^4) + (1 \times 2^3) + (1 \times 2^2) + (0 \times 2^1) + (0 \times 2^0)$. Compare **8: octal number, 16: hexadecimal.**

binary star, two stars that orbit around each other, kept together by the force of gravity so that they are unable to escape from each other. Astronomers believe that most stars are binaries. Also called **double star**.

binary system, any system composed of only two elements, as 0 and 1, *yes* and *no*, *true* and *false*. Also called **dyadic system.** See also **2: bit, 2: binary number.**

binaural, 1. having or involving two (or both) ears.
2. Also called **simulcast.** (of broadcasting) simultaneously broadcast on AM and on FM so that simultaneous reception on the two wavelengths simulates a stereophonic effect.

binocular, *or* **binoculate, 1.** having or involving two (or both) eyes.
2. suited for the use of both eyes, as *binocular fieldglasses,* a *binocular microscope,* especially to afford stereoscopic views.

binomial, *or* **binominal,** of, pertaining to, or involving two names or elements; binary (name).

biped, having two feet, as human beings.

bipennis, an ancient two-headed ax.

bipod, any support with two legs, as may be used for an automatic rifle. Compare **2: fork-rest, 3: tripod.**

birefringent, (of a lens) doubly refractive.

bireme, an ancient galley with two banks of oars on each side.

bisexual, 1. having the organs of both sexes in one individual; hermaphrodite.
2. (of a person) erotically inclined to both sexes.

bisulcate, 1. having two grooves or furrows.
2. having cloven hooves, as an ox, pig, or goat.

bit, (in computer technology)) a unit of data in the binary numbering system, in which all numerical quantities are represented by using two numerals, 0 and 1. The bit is the fundamental unit of information in computerized data. A bit can be represented by the presence or absence of electrical charge in a circuit or storage medium, hence is the key to electronic data storage and manipulation. [Blend of *binary* dig*it*.] See also **1: word, 2: binary number, 4: nibble, 8: byte.**

bivalve, 1. having two valves, as a clam or oyster.
2. such a mollusk. Compare **1: univalve.**

Books of Discipline, two documents known as the *First* and *Second Book of Discipline,* constituting the original standards of government for the Church of Scotland. The former, adopted by an assemblage of reformers led by John Knox in January 1561, dealt only with the government of individual churches or congregations; the latter, adopted by the General Assembly in April, 1578, abolished episcopacy and regulated the organization and functions of the various governing bodies or ecclesiastical courts of the church. Neither was ratified by the state authorities, but both were generally accepted and served as the groundwork of the ultimate constitution of the church.

brace, two.

Brumaire, the second month of the French revolutionary calendar, beginning about October 22nd and ending about November 20th.

C, c, the second consonant of the English alphabet.

catamaran, a light-displacement vessel, usually a sailboat, having two parallel hulls. Compare **3: trimaran.**

causeuse, a small sofa or settee for two persons.

checkering file, a file formed of two files riveted together, used in checkerwork, as in decorating gunstocks.

complementary color, one of two primary or secondary colors contrasted on a color wheel or other scale to the other member of the set, as orange and blue, violet and yellow, and green and red.

couple, two of anything, especially a husband and wife.

D, (in music) 1. the second tone, or *re*, of the scale of C.
2. a note which represents this tone.
3. the keynote of the key of two sharps.
4. (on the piano keyboard) the white note between each group of two black keys.

2

daily double, (in horseracing) a scheme that allows a bettor to make a single wager on the winners of two specified races in any one day's racing schedule.

découplé, (in heraldry) parted into two; uncoupled, used especially of a chevron when two rafters are separated by a slight space.

degorder, the pair of numbers signifying the degree and order of any mathematical form.

Delian problem, the problem of the duplication of the cube, that is, of finding a cube double the volume of a given cube. [So called because the oracle at Delos told the Athenians that a pestilence would cease when they had doubled the altar of Apollo, which was cubical.]

demi-bateau, one of two small boats which when united form a pontoon-boat. Also called **demi-pontoon.**

demi-season, intermediate between two seasons, as in style, weight, etc.

dennet, a light, open, two-wheeled pleasure carriage suspended on two half-elliptical springs and one cross-spring; a gig.

depas amphikypellon, (in Greek archaeology) a double cup; a cup having two handles or ears, or one divided into two parts by a partition; sometimes, a vessel consisting of two bowls joined at their bottoms, so that either can serve as a foot for the other, as in a double egg-cup.

derodidymus, *pl.* **derodidymi.** a double-headed monster.

dessert-spoonful, as much as a dessert-spoon will contain; about two drams.

desultor, a bareback rider in the Roman circus who rode two (or more) horses at once, leaping from one to another.

deuces wild, (in certain card games, as poker) a rule by which every card with a value of two can count for any value, except in ties, when a natural card outranks it.

deuteragonist, (in ancient Greek drama) the second actor, who has dialogue with the protagonist. The deuteragonist was probably introduced by Æschylus early in the 5th century BC. See also **1: protagonist, 3: tritagonist.**

deuteranopia, a form of red-green color blindness.

deuterium, an isotope of hydrogen, having a nucleus (called a deuteron) double the mass of the nucleus of hydrogen. Also called **heavy hydrogen.** Compare **3: tritium.**

deutero-, an element in words of Greek origin, meaning 'second.'

deuterocanonical, forming or belonging to a second canon.

deuterocanonical books, those books of the Bible as received by the Roman Catholic Church which are regarded as constituting a second canon, accepted later than the first but of equal authority.

deuterogamist, a person who married for the second time. Compare **2: bigamy, digamy, 3: trigamist.**

deuterogenic, of secondary origin; specifically (in geology) applied to those rocks which have been derived from protogenic rocks by mechanical action.

deutero-Nicene, pertaining to the second Nicene council.

Deuteronomy, the second law, or the second statement of the law: the name given to the fifth book of the Pentateuch, consisting of three addresses purported to have been made by Moses to Israel shortly before his death. [From Greek 'second law,' an apparent translation in the Greek Septuagint of Hebrew *mishneh hat-Torah* 'copy of the law.'] See also **5: Deuteronomy.**

deuteropathy, *or* **deuteropathia,** a secondary infection, the result of another, earlier infection, as retinitis from nephritis.

deuteroscopy, *Rare.* **1.** second sight.
2. a second view, or that which is seen upon a second view.

deuto-, a prefix used in chemistry to denote strictly the second term in an order or a series. It is often used as equivalent to *bi-* or *di-* with reference to the constitution of compounds, distinguishing them from *mono-* or *proto-* compounds.

deutocerebrum, *pl.* **deutocerebra,** *or* **deuterencephalon.** the second lobe of an insect's brain.

deutomala, *pl.* **deutomalae.** the second pair of jaws, or mouth-appendages, of the *Myriapoda.*

deutopsyche, the diencephalon.

deutotergite, (in entomology) the second dorsal segment of the abdomen of insects.

Deuxième Bureau, an intelligence bureau of the French government.

deux-temps, **1.** a rapid form of the waltz, containing six steps to every two of the trois-temps, or regular waltz.
2. the music for this dance.

devil on two sticks, a toy consisting of a hollow and well-balanced piece of wood turned in the form of an hourglass. It is first placed upon a cord loosely hanging from two sticks held in the hands, and upon being made to rotate by the movement of the sticks it exhibits effects similar to those of a top. Also called **diabolo.**

2

di-, a prefix of Greek origin, cognate with *bi-*, and meaning 'two,' 'two-fold,' 'double.' In chemistry, it denotes that a chemical compound contains two units of the element or radical to which *di-* is prefixed, as manganese *di*oxide, MnO_2, a compound of one atom of manganese and two of oxygen.

diactine, *or* **diactinal,** having two rays.

diadelphic, being one of a set of two.

diaderm, a collective term for both the ectoderm and the entoderm.

diagyíos, (in ancient prosody) consisting of two members, specifically, the paeon or paeonic foot commonly known as the Cretic. See also **2: paeon diagyíos, paeon epibatus.**

diandrous, 1. (in botany) having two stamens.
2. (in ornithology) having two male mates.

diapsid, having two temporal arches.

diarchy, *or* **dyarchy,** *or* **duarchy,** a form of government in which the executive power is vested in two persons, as in the case of William and Mary of England.

diatonic, (in chemistry) consisting of two atoms.

diaulos, *pl.* **diauli.** 1. an ancient Greek musical instrument, consisting of two single flutes, either similar or different, so joined at the mouthpiece that they could be played simultaneously.
2. (in ancient Greek games) a double course, in which the racers passed around a goal at the end of the course and returned to the starting-place.

diaxon, 1. having two axes, as a sponge-spicule.
2. a sponge-spicule with two axes.
3. a neuron having two axis-cylinder processes.

dibrach, *or* **dibrachys,** (in ancient prosody) a foot consisting of two short syllables; a pyrrhic.

dibranchiate, having two gills.

dicatalexis, (in prosody) incompleteness of both a middle and a final foot in a line.

dicellate, two-pronged, as a sponge-spicule.

dicephalus, a monster having two heads on one body.

dicerion, a candlestick with two lights, representing the two natures of Christ, used in the Greek Orthodox Church by bishops in blessing the people.

dicerous, (in entomology) having a pair of developed antennae or horns.

2

dichasium, *pl.* **dichasia.** (in botany) a cyme having two main axes.

dichlamydeous, (in botany) having a double perianth, consisting of both calyx and corolla.

dicho-, a first element in scientific terms, meaning 'in two parts,' 'in pairs.'

dichogamism, hermaphroditism, or the presence of the accessory reproductive organs as well as the gonads of both sexes.

dichord, an ancient musical instrument of the lute or harp class, having two strings.

dichotic, involving the use of both ears for the simultaneous hearing of tones of different pitch. Compare **2: diotic.**

dichotomy, *pl.* **dichotomies.** a cutting in two; division into two parts or into twos.

dichroism, the property of exhibiting different colors when viewed from different directions, possessed by many doubly refracting crystals. Compare **3: trichroism.**

dichromat, *or* **dichromate,** (in ophthalmology) a person who possesses only two of the three normal color-sensations; a red-blind or green-blind person.

dichromatism, the state or condition of normally presenting two different colors or systems of coloration.

dichromic, relating to or embracing two colors only.

dichromism, color-blindness in which only two of the three primary colors are experienced.

dichronous, 1. (in ancient prosody) **(a)** having two times or quantities; sometimes long and sometimes short. **(b)** consisting of two normal short times, or morae; lasting for two morae; disemic.
2. (in botany) having two periods of growth in one year.

dicoccous, (in botany) formed of two cocci: applied to fruits having two separate lobes.

dicoelous, (in anatomy) having two cavities; bilocular.

dicolon, *pl.* **dicola.** (in prosody) a verse or period consisting of two cola, or members.

dicondylian, having two occipital condyles, as the skull of a mammal or an amphibian.

dicotyledon, *pl.* **dicotyledons** *or* **dicotyledones.** (in botany) a plant that produces an embryo having two cotyledons. Also, *Informal,* **dicot.**

2

dicrotic, *or* **dicrotous,** double-beating; noting the pulse when for one heartbeat there are two arterial pulses.

dicycle, a bicycle having the two wheels parallel on the same axis instead of in line.

dicynodont, 1. having two tusk-like canine teeth.
2. an animal with such teeth, especially a suborder of extinct reptiles.

didactyl, *or* **didactyle,** *or* **didactylous,** 1. having only two toes or fingers, as arthropods with limbs ending in a forceps or chela.
2. an animal with two digits on each foot, as the two-toed sloth.

didelphian, *or* **didelphic,** *or* **didelphoid,** having a double womb, as marsupials.

di-diurnal, twice a day.

didrachm, *or* **didrachma,** a silver coin of ancient Greece, worth two drachmae.

didymium, *Obsolete.* an element that was thought to be a twin to the element lanthanum with which it was found in minerals and from which it could be separated only with difficulty. It was later shown to be a mixture of neodymium and praseodymium.

didymous, 1. (in botany) twofold; twin; growing double.
2. (in zoology) twain; paired; forming a pair touching each other.

didynamous, *or* **didynamian,** *or* **didynamic,** (in botany) in two unequal pairs; having four stamens in two unequal pairs.

dietheroscope, an apparatus for measuring the refractions that occur in a mirage.

digamma, a character of the ancient Greek alphabet, in pronunciation similar to that of modern English *w* and in form yielding modern *F*. So named because it resembles two Greek gammas, Γ, set one above the other.

digamy, second marriage, after the death of the first spouse. Compare 2: **bigamy, deuterogamist.**

digastric, (in anatomy) having two fleshy bellies, or lobes, as a muscle.

digenesis, (in biology) successive generation by two different processes, as sexual and asexual; parthenogenesis alternating with ordinary sexual reproduction.

digenous, bisexual; of or pertaining to both sexes; syngenetic.

diglossia, the condition of having two tongues.

diglot, *or* **diglottist,** 1. using, speaking, or written in two languages; bilingual.
2. a person who speaks or uses two languages, especially equally well; a bilingual.

diglottism, the use of two languages among a people, or of words derived from two languages.

diglyph, (in architecture) an ornament consisting essentially of two associated cuts or channels. Compare **3: triglyph.**

digoneutic, (in entomology) having two broods during a single year; double-brooded.

digonous, (in botany) having two angles, as a stem.

digraph, 1. two letters used to represent one sound, as *ea* in *head*, *th* in *path*, *th* in *this*, *sh* in *should*, *ng* in *sing*, etc.
2. two letters written together, as *ae, oe.*

digraphic, 1. pertaining to or represented by a digraph.
2. written in two distinct alphabets or characters, as in longhand and shorthand or in Hebrew and Greek.

digyn, (in botany) a plant having two styles, or carpels.

dihedral angle, the angle at which two plane surfaces intersect.

dihedron, *pl.* **dihedra, dihedrons.** a figure with two sides or surfaces, as a crystal.

diiamb, *or* **diiambus,** *pl.* **diiambi.** (in ancient prosody) two iambs, or an iambic dipody regarded as a single compound foot.

dilambdodont, having oblong molar teeth with two V-shaped ridges.

dilemma, a form of argument in which it is shown that whoever maintains a certain proposition must accept one or the other of two alternative conclusions, each of which involves the denial of the proposition. See also **3: trilemma.**

dilogy, (in rhetoric) 1. the use of a word or words twice in the same context; repetition, especially for the sake of emphasis.
2. intentional use of an ambiguous expression, or the word or expression so used.

dimastigate, (in zoology and botany) having two flagellae; biflagellate.

dimerous, 1. consisting of or divided into two parts; bipartite.
2. (in botany) (of flowers) having two members in each whorl.
3. (in entomology) (of plant-lice) having two-jointed tarsi.

dimetallic, (in chemistry) containing two atoms of a metallic element.

2

dimeter, (in prosody) **1.** consisting of two measures; divisible into two feet, or dipodies.
2. a verse or period consisting of two feet, or dipodies, as *Ionic dimeter, iambic dimeter.*

dimidiate, 1. to divide into two equal parts; to halve.
2. divided into two equal parts.

dimorphism, 1. the property of assuming or of existing in two distinct forms, as carbon, which is diamond in one form and graphite in another; flowers, which may occur in two forms on the same plant; animals, which exhibit sexual dimorphism between individuals of the same species, etc.
2. the existence of a word under two or more forms, called doublets, as *dent* and *dint, fat* and *vat*, etc.

dimyarian, *or* **dimyary,** having two muscles; double-muscled, as certain bivalves.

diobely, (in Greek antiquity) an allowance of two obols a day to each citizen present during the Athenian festivals, to pay for seats in the theater.

diobol, a silver coin of ancient Greece, worth two obols.

diode, an electronic device, especially a vacuum tube, consisting of an anode and a cathode, each with different characteristics of voltage and amperage.

diodont, 1. having two teeth.
2. an animal, especially a certain family of fishes, having only two teeth.

dioecious, *or* **diecious, 1.** (in botany) having the male and female flowers borne on separate plants.
2. (in zoology) sexually distinct; having two sexes in different individuals. Also **dioecian, diecian.**

dioestrum, *pl.* **dioestra,** a brief interval (from 6 to 14 days) between two periods of sexual desire in animals.

dionym, a name, especially a scientific name, consisting of two terms. Compare **1: mononym, 2: binomial, 3: trionym.**

diota, *pl.* **diotae,** a Greek vase with two ears or handles, similar to an amphora.

Diothelism, *or* **Dyothelism,** (in theology) the doctrine that Christ during his earthly life possessed two wills, a human and a divine. Compare **1: Monothelitism.**

diotic, involving the use of both ears in the perception of a single tone. Compare **2: dichotic.**

dioxide, *or* **dioxid,** an oxide consisting of one atom of a metal and two atoms of oxygen.

dipenthemimeres, (in ancient prosody) a verse consisting of two penthemimeres, or groups of five half-feet (two and a half feet) each, as a line composed of a dactylic pentameter and an iambic monometer hypercatalectic.

dipetalous, (in botany) having only two petals.

diphthong, a coalition of two vowels pronounced in one syllable, as *oi* in *boil, ou* in *bound,* etc.

diphthongia, (in pathology) a condition in which two sounds of different pitch are produced simultaneously by the voice. Also called **diplophonia.**

diphyletic, (in zoology) having two sets of ancestors; derived from two distinct groups of animals.

diphyllous, (in botany) having two leaves, as a calyx formed of two sepals.

diphyodont, 1. having two sets of teeth.
2. a mammal which has two sets of teeth.

Diphysitism, *or* **Dyophysitism,** (in theology) the doctrine of two distinct natures in Christ, a divine and a human. Compare **1: Monophysitism, 2: Dyothelitism.**

diplacusis, (in otology) double hearing; usually, the hearing of the same tone in a different pitch by the two ears; less often, the arousing of two tonal sensations in the same ear by a single stimulus.

diplanar, pertaining to or of two planes.

diplasiasmus, 1. a figure of orthography consisting in the doubling of a letter that is usually written single, as, in American spelling, *panelling* for *paneling.*
2. (in rhetoric) repetition of a word or name for the sake of emphasis, as *O, Jerusalem, Jerusalem, thou that killest the prophets,* Mat. xxii, 37.

diplasic, 1. double; twofold.
2. (in ancient prosody) constituting the proportion of two to one.

diplegia, (in pathology) paralysis of corresponding parts on the two sides of the body, as of the two arms.

dipleidoscope, an instrument for indicating the passage of the sun or a star over the meridian by the coincidence of two images of the object, one formed by single and the other by double reflection.

diplex, double, as the method of transmitting two messages in the same direction at the same time over a single transmission line.

2

diploglossate, (in zoology) having a double tongue formed so that the anterior portion telescopes into the posterior.

diploneural, having a double nerve supply, each from a separate source, as a muscle.

diplopia, *or* **diplopy,** (in ophthalmology) a defect of vision in which a single object appears double.

diplosome, (in cytology) a double centrosome, or one that has divided into two daughter-chromosomes which have not yet moved apart to form the poles of a karyokinetic spindle.

diplospire, a double spire.

dipneumonous, having two lungs or respiratory organs, as a spider, lungfish, or holothurian; double-lunged.

dipnoous, having both gills and lungs, as the lungfish.

dipode, 1. having only two feet; biped.
2. an animal appearing to have only two feet because the forelimbs are rudimentary.

dipody, *pl.* **dipodies,** (in prosody) a group of two like feet; a double foot, especially a pair of feet constituting a single measure.

dipolar, having two poles, as polarized light.

dipole, 1. two electrical points or magnetic poles of opposite charge but equal magnitude set close together.
2. a radio or television antenna consisting of two poles projecting in opposite directions from their connectors.

diprotodont, 1. having only two lower front teeth, as herbivorous marsupials.
2. a marsupial with such dentition.

dipteral, 1. (in entomology) having only two wings; dipterous.
2. (in architecture) (of a portico) consisting of or furnished with a double range of columns.

dipterous, 1. (in entomology) See **2: dipteral.**
2. (in botany) (of stems, fruits, seeds, etc.) having two wing-like membranous appendages; bialate.

dipterygian, having two fins only.

diptych, 1. (in ancient times) a hinged, two-leaved tablet of wood, ivory, or metal, having the inner surfaces waxed and used for writing on with a stylus.
2. similar, ornamental tablets of wood, ivory, or metal used for ceremonial purposes in ancient Rome and Byzantium.
3. (in the early church) **(a)** the tablets on which were written the names of

those who were to be specially commemorated at the celebration of the Eucharist. Removal of a name implied an accusation of heresy. **(b)** the list so recorded. **(c)** the intercessions in the course of which these names were introduced.
4. (in art) a pair of pictures or carvings on two panels hinged together. See also **3: triptych.**

disepalous, (in botany) having two sepals.

disomatous, having two bodies; double-bodied.

dispermous, *or* **dispermatous,** (in botany) (of fruits and their cells) containing only two cells.

dispermy, the entrance of two spermatozoa into a single egg.

dispireme, (in cytology) the stage of karyokinesis during which the chromosomes unite to form two skeins, one for each of the daughter-nuclei.

dispondee, (in prosody) a double spondee; two spondees regarded as forming one compound foot.

disquiparancy, *or* **disquiparance,** *Rare.* the denotation of two objects, as being related, by different names, as *father and son, master and servant, brother and sister.*

disquiparant, **1.** an object so related to a second that the latter is in a different relation to it, as a husband or wife is a disquiparant, a spouse an equiparant.
2. of or pertaining to disquiparancy.

dissyllable, a word consisting of two syllables only, as *paper, into, defer.*

dissymmetry, symmetry between two objects, with respect to a plane of symmetry, as between the right and left hands.

distich, **1.** (in prosody) a group or system of two lines or verses.
2. See **2: distichous.**

distichous, *or* **distich,** disposed in two rows; biserial; bifarious; dichotomous.

distigmatic, (in botany) having two stigmas.

distomatous, having two mouth-like structures, as the flukes.

distylous, (in botany) having two styles.

ditesseral, having two small squares or cubes.

ditetragonal, (in crystallography) twice four-angled or twice four-sided, as an eight-faced pyramid (or double pyramid.)

divalent, (in chemistry) having a valence of two.

2

dizoic, producing two young, as a spore.

dizygotic, coming from two ova, as (fraternal) *dizygotic* twins, triplets, etc. Compare **1: monozygotic.**

DNA, deoxyribonucleic (or desoxyribonucleic) acid, called the double helix in reference to the configuration of its structure.

dobhash, (in the East Indies) a person who speaks two (or more) languages; an interpreter.

doblon, a gold coin of Spain; doubloon, the double escudo.

Doctrine of the Two Swords, the politico-religious tenet that guided government in the Middle Ages, namely, that of the secular sword of worldly empire and that of the spiritual sword toward salvation.

dogger, a two-masted Dutch fishing-vessel, formerly used in the North Sea, resembling a ketch.

doppio, when modifying a feminine noun, **doppia.** (in music) double, as *doppio tempo* 'double time,' *lira doppia* 'a double, or bass lyre.'

double, 1. consisting of two in a set together; being a pair; coupled; two-fold.
2. twice as large or as much; multiplied by two.
3. of twice the weight, size, or strength (or relatively so), as a *double bed.*
4. (of musical instruments) producing a tone an octave lower, as a *double bassoon.*
5. a twofold quantity or size; a number, sum, value, or measure twice as great as one taken as standard.
6. See **5: grandsire.**
7. (in ecclesiastics) a feast on which the antiphon is doubled.
8. **(a)** (in bridge) a doubling of the value of trick points. **(b)** a call against the latest bidding opponent, challenging the ability of the opposing team to fulfill the contract bid. Also called **business double.** Compare **2: informatory double.**
9. **doubles.** a game of tennis, badminton, etc., with two players on each side.
10. (in tennis) two successive faults in serving; double-fault.
11. (in baseball) a ball batted so as to enable the batter to reach second base. Also called **two-bagger, two-base hit.**
12. (in astronomy) two stars which seem one to the unaided eye but which can be distinguished through the telescope.
13. to call a double in bridge.
14. to hit a double in baseball.

double agent, an agent who spies for two rival countries, companies, etc.

double ax, a double-headed ax, used in Hittite hieroglyphics and recurring as the special symbol of the Cretan Zeus; labrys.

double bar, (in music) two parallel vertical lines in a score indicating the end of a composition or of a part of it.

double-barreled, 1. having two barrels, as a gun.
2. *Jocular.* having two parts, as a name: *Smyth-Jones.*
3. serving a double purpose or producing a double result.

double bed, a bed wide enough to accommodate two adults side by side, usually now called **full size.**

double-bitt, (in nautical use) to pass a line, cable, etc., around two bitts or around the same bitt twice, for greater security.

double-blind, noting an experimental procedure in which neither the experimenters nor the subjects are given any information about the products or procedures being tested.

double-bodied, 1. having two bodies; disomatous.
2. (in astrology) (of signs) one of the four zodiacal signs: Gemini, Virgo, Sagittarius, Pisces.

double bogey, (in golf) the holing of a ball in two strokes more than par for the hole. See also **1: bogey, 3: triple bogey.**

double boiler, a cooking pot consisting of a lower pot, containing water, and an upper pot that fits into it, containing food, used to control the temperature at which the food is heated and to ensure its not warming to more than the boiling point of water. Also called, *Brit.,* **double saucepan.**

double-breasted, made alike on both sides, as a coat, jacket, etc., that has two rows of buttons and (originally) buttonholes, enabling it to be buttoned on either side.

double-brooded, (in entomology) digoneutic.

double chin, a fold of flesh under the chin, giving the appearance of two chins.

double consonant, a character representing two consonant sounds, as x = [ks], Ψ = [ps], etc.

doublecross, 1. to promise (a person) to do something and then not do it, often doing the opposite, which usually brings harm.
2. the act of doublecrossing.

double-cut file, a file with two series of straight cuts crossing each other.

double dagger, (in typography) a reference mark ‡ used in text next in order after the dagger, or obelisk; diesis.

double date, *U. S. informal.* a social meeting in which two couples go out together.

2

double-dealer, a person guilty of duplicity.

double-dealing, duplicity.

double-decker, 1. a ship with two decks above the waterline.
2. a serving of ice cream, as in a cone, with two scoops, one atop the other.
3. a sandwich consisting of three slices of bread with fillings between; same as a triple-decker.
4. a bus or train carriage with two sets of seats, arranged in compartments one above the other.

double dies, dies which cut two blanks, one small and one large, side by side, or a small hole and a blank next to it.

double dip, 1. to work at a government job while receiving a government pension from a previous job; the term connotes unethical behavior, especially if one is receiving a disability pension while working at another job.
2. a serving of two scoops of ice cream, usually in a cone.

double-distilled, *Informal.* absolutely pure; thorough-going; out-and-out, as *a double-distilled blackguard.*

double down, (in blackjack) to turn up the card dealt face down if it is of the same value as that dealt face up and to double the bet on the first card, allowing for one or more cards to be dealt, face down, on each of the doubled cards.

double dummy, a game of whist in which two play, each with two hands, his own and another exposed.

double Dutch, 1. unintelligible gibberish; **double-talk.**
2. a child's rope-jumping game in which two ropes are used.

double duty, twice the function expected, as a car used for both business and pleasure.

double-dyed, 1. twice dyed.
2. deeply imbued, as with guilt; thorough; complete, as a *double-dyed villain.*

double-eagle, 1. a former gold coin of the United States, worth two eagles, or $20.
2. the heraldic representation of an eagle with two heads, as in the national arms of Austria and Poland, and, formerly, of Russia. It is an ancient emblem of the Byzantine and Holy Roman Empires.

double-edged, 1. (of a sword or other weapon) having two edges.
2. cutting both ways, as an argument that works both for and against the person employing it.

2

double-ender, anything with two ends alike, as a boat with a stern similar in configuration to its bow.

double entendre, a word or phrase with two meanings or allowing of two interpretations, one of which is usually obscure or indelicate.

double-entry bookkeeping, a system, originated in the Middle Ages, for keeping the financial records of a business. It is based on the formula that the sum of the assets is equal to the sum of the claims against the assets plus the ownership equity (if any). Put differently, this amounts to the equity being equal to the assets less the claims against the assets. In practice, it means that every increase in assets (entered on the left side of the ledger) is balanced by an increase in the liabilities or proprietorship (entered on the right side).

double exposure, (in photography) a fault, sometimes deliberately achieved for effect, in which one image is exposed on a film directly over another.

double feature, a motion-picture show in which two films are exhibited to the same audience for a single admission fee.

double fever, intermittent fever in which there are two paroxysms in each cycle.

double figures, (in sports) any statistical total of 10 to 99, as in basketball, where scoring in double figures by a player in a single game is considered a minimum standard of adequate performance.

double-first, (at Oxford University) **1.** one who gains the highest place in the examinations in both classics and mathematics.
2. the degree itself, as *he took a double-first.* Compare **2: double-man.**

double fitché, (in heraldry) terminating in two points, as the arms of a cross.

double flat, (in music) a tone two half-steps lower than a given tone.

double-gild, to gild with double coatings of gold; hence, to cover up by flattery and cajolement.

double glazing, (in Britain) storm windows or double panes of glass usually set into one frame in order to better insulate the windows of a building, especially a home.

double-half-round file, a file with two curved sides and convex edges of different angles, used for dressing, or crossing out, balance-wheels. Also called **cross-file.**

doubleheader, a sporting event in which two games, between the same teams or different teams, are presented in succession for a single admission fee.

2

double-hung, noting a window with two vertically sliding sashes, the outer for closing the upper part, the inner for closing the lower.

double indemnity, a provision in a life insurance policy allowing for the payment of twice the face value of the policy if death occurs under certain specified conditions, usually before the age of 60 and attributable to an accident as set forth in the policy.

double jeopardy, the Constitution of the United States provides (Fifth Amendment): " . . . nor shall any person be subject for the same offence to be twice put in jeopardy of life or limb." Jeopardy is defined as initiating when a jury is sworn in or, if no jury is sworn, when the first witness is sworn in.

double knit, a stitch used to weft-knit fabric that produces a tightly constructed material resistant to wrinkles.

double life, a mode of existence in which a person leads essentially two different lives, the people associated with him in one knowing nothing about the other; a dual existence.

double line, (between lanes of traffic on a road) a pair of parallel continuous lines that should not be crossed by a vehicle.

double-lunged, dipneumonous.

double-man, (at Cambridge University) a student proficient both in classics and in mathematics. Compare **2: double-first.**

double-meaning, 1. having or conveying two meanings; misleading; deceitful.
2. ambiguous.

double-minded, indecisive; unstable; undetermined.

double negative, a construction in English, considered nonstandard in modern English, employing two negative words to express a single idea of negation, as *He don't have none, She didn't go nowhere, They don't visit no more.*

double-nostriled, having two nasal passages; amphirhine, as all skulled vertebrates except the lampreys and hags. Compare **1: monorhine.**

double or nothing *or* **double or quits,** (in gambling) said about the stake that is to be double an understood amount or nothing, according to the favorable or unfavorable outcome of a certain chance.

double play, (in baseball) a play in which a fielder, either alone or with assistance, puts out two runners before the ball is returned to the pitcher for delivery.

double pneumonia, pneumonia affecting both lungs.

double possessive, a construction in English in which possession is expressed both by the preposition *of* and by the possessive form of the word following, as *He is a friend of mine, She is an employee of my mother's.*

double quotes, two pairs of quotation marks, " and ", as appear around quoted text in U. S. usage and, in British usage, usually as secondarily quoted text within single quotation marks.

double quotidian fever, intermittent fever in which two paroxysms occur within twenty-four hours.

double room, a room that can accommodate two persons in a hotel, inn, etc.

double scull, a racing shell accommodating two rowers, one behind the other.

double solitaire, solitaire played by two people in competition, the winner being the player who finishes first.

double standard, a set of principles providing differently for one group of people from another, especially that under which women are treated differently from men. Compare **2: ethical dualism.**

double-stop, to stop two strings of a violin simultaneously with the fingers to produce two-part harmony.

double-struck, (of a coin) struck twice with the same design owing to an accidental shift of the blank in the die.

doublet, **1.** a duplicate form of a word in a later stage of a language traceable to a single form at an earlier stage, as *dent* and *dint, discreet* and *discrete*, etc. Compare **2: dimorphism, 2.**
2. Usually, **doublets.** one of a pair of dice that turn up the same number of spots.
3. an outer garment as worn by men from the end of the 15th to the middle of the 17th century. [So called because it was lined.]

double take, (chiefly in comedic acting) a delayed reaction of amazement, feigned or genuine, in response to a remark or situation whose significance was not first grasped.

double-talk, speech that is deliberately made unintelligible in order to confuse the listener, either by the use of confusing syntax, the insertion of extra syllables in words, the use of nonwords, or all of these.

double tertian fever, intermittent fever with two paroxysms having features distinct from each other, as in severity or distance from the last paroxysm, in one cycle of forty-eight hours. [Called "tertian" because, in counting forty-eight hours as two days, the first, second, and last days are counted as three.]

2

doubleton, (in bridge, etc.) a holding of only two cards of a suit in a hand.

double-tongue, (in playing the flute and certain brass instruments) to apply the tongue rapidly to the teeth and hard palate alternately, so as to ensure a brilliant execution of a staccato passage.

double-topsail, noting a rig in which the square topsail is replaced by two smaller sails and yards, in order to lessen the labor of the crew and enable them to reduce sail more rapidly.

double up, to share with another person, as accommodation.

double-whammy, a hex or enchantment placed by one who fixes the victim with an eye-bulging stare. [From *Li'l Abner,* by Al Capp, U. S. comic strip.]

doubloon, a former gold coin of Spain and the Spanish-American states, originally double the value of a pistole.

dual, 1. relating to two.
2. a grammatical form of a noun, adjective, or verb in certain languages that denotes the number two. It exists, for example, in classical Greek along with the singular (denoting one) and the plural (denoting three or more). Relics of it occur in English *both, between,* etc.
3. (in chess) a problem which has two solutions, that is, one in which the mate can be given either by one or by two pieces, or by one piece on two or more different squares.

Dual Alliance, 1. the alliance between Russia and France, from 1893 to 1917.
2. the secret alliance between Austria-Hungary and Germany against Russia, from 1879 to 1918.

dual carriageway, (in Britain, etc.) a divided roadway, usually with two lanes of traffic in each direction.

dual citizenship, the state of having citizenship in two countries.

dualism, 1. division into two parts; a twofold division; duality.
2. (in philosophy) the view that seeks to explain all sorts of phenomena by the assumption of two radically independent and absolute elements. Compare **1: monism.**
3. (in theology) **(a)** the doctrine that there are two independent divine beings or principles, one good and the other evil. **(b)** the (heretical) view of the personality of Christ which regards him as consisting of two personalities.

duality, the state of being two or of being divided into two parts; twofold division or character; twoness.

dualize, to make or regard as two.

2

Dual Monarchy, the Austro-Hungarian monarchy from 1867 to 1918.

dual personality, (in psychopathology) two dissociated personalities manifest in one individual.

due corde, (in music) a direction to play the same note simultaneously on two strings of an instrument of the violin class.

duel, 1. a premeditated and prearranged combat between two persons, usually in the presence of two witnesses, called seconds, for the purpose of deciding a quarrel, avenging an insult, or clearing the honor of one of the combatants or of some third party whose cause he champions.
2. to fight a duel.

duello, 1. a duel.
2. the art or practice of dueling or the code of laws that regulate it.

duet, a musical composition either for two voices or for two instruments, or for two performers on one instrument, with or without accompaniment.

duettino, a short, unpretentious duet.

duettist, one who takes part in a duet.

duetto, a duet.

due volte, (in music) a direction to play a passage twice.

dufoil, (in heraldry) 1. a head of two leaves growing out of a stem.
2. having only two leaves.

Duija, ('twice-born') those caste groups in India that are entitled to wear the sacred thread and to perform the initiation rite of *upanayana.*

duo, 1. a duet.
2. any two people considered together.

duodrama, a dramatic piece for two performers only.

duogravure, a method of photoengraving requiring two plates for the production of prints of one color.

duole, (in music) a group of two notes to be performed in the time normally occupied by three.

duologue, a dialogue.

duoparental, of or from two parents or sexual elements; amphigonic.

dupion, *or* **doupion,** 1. a double cocoon formed by two silkworms spinning together.
2. the coarse silk produced by such a double cocoon.

duplation, multiplication by two; doubling.

2

duple ratio, a ratio such as that of 2 to 1, 8 to 4, etc. Compare **2: subduple ratio.**

duple rhythm, (in music) a rhythm characterized by two beats or pulses to the measure; double time.

duplex, 1. double; twofold.
2. a doubling or duplicating.
3. an apartment dwelling on two floors, one above the other. Also called **duplex apartment.** Compare **2: duplex house.**

duplex house, a two-family dwelling. Compare **2: duplex, 3.**

duplex telegraphy, a system in which two messages are transmitted simultaneously over a single transmission medium; it includes both diplex and contraplex.

duplicate, 1. to double; repeat; replicate; copy.
2. (in physiology) to divide into two by natural growth or spontaneous division.
3. to become double.
4. (in ecclesiastics) to celebrate Holy Communion twice in the same day.
5. double; twofold; consisting of or relating to a pair or pairs or to two corresponding parts.
6. exactly like or corresponding to something made or done before; repeating an original; matched, as *a duplicate copy, a duplicate action.*
7. one of two or more things corresponding in every respect to each other; an additional original.

duplicate bridge, a form of bridge used in competitions in which different teams of players are dealt the same hands, the team with the best over-all score being declared the winners. Also called **duplicate.**

duplicature, a doubling; a fold or folding.

duplicity, the acting or speaking differently in relation to the same thing at different times to the same person or to different persons, with intention to deceive; double-dealing.

duplo-, a prefix meaning 'twofold' or 'twice as much.'

dupondius, *pl.* **dupondii.** a Roman bronze coin, first issued by Augustus, having the value of two asses; called *second brass* by numismatists to distinguish it from the sestertius, or *first brass.*

Dutch door, a door divided in two horizontally, with each part separately hinged, allowing either to be kept open or shut.

duumvir, *pl.* **duumvirs, duumviri,** *or* **duovir,** *pl.* **duovirs, duoviri,** (in ancient Rome) two magistrates (*duum perduellionis*), priests (*duumviri sacris faciundus*), naval officers (*duumviri navales*), etc., who administered justice and, respectively, religious, military, and other activities. The *duumviri sacris faciundus*, who originally had responsibility of the books of the Sibylline oracles, were increased to ten (See **10: decemvir**) in 367 BC, then to fifteen (See **15: quindecemvir**) by Sulla (82–70 BC).

dyad, 1. two units treated as one.
2. pertaining or relating to the number two or to a dyad; dyadic.
3. (in prosody) a group of two lines having different rhythms.
4. noting an axis of twofold symmetry. Also, **duad.**

dyadic, 1. dyad.
2. (in Greek prosody) **(a)** comprising two different rhythms or meters. **(b)** consisting of groups of systems each of which contains two unlike systems, or pericopes, as *a dyadic poem.*

dyadic disyntheme, (in Greek prosody) any combination of dyads, with or without repetition, in which each element occurs twice and no oftener. Compare **1: dyadic syntheme.**

dyaster, *or* **diaster,** the double-star figure occurring in or resulting from karyokinesis.

Dynamic Duo, (in U. S. comics) the crime-fighting team of Batman and Robin.

dyophysitic, having two natures.

Dyothelitism, the doctrine that Christ had two wills. Compare **1: Monothelitism.**

E, the second vowel in the English alphabet.

enthymeme of the second order, (in logic) a syllogism with only the minor premise expressed.

epanadiplosis, (in rhetoric) a figure in which the same word is duplicated at the beginning and end of a sentence, as *Rejoice in the Lord alway: and again I say, Rejoice.* [Phil. iv.4.]

epanastrophe, (in rhetoric) a figure in which the same word is duplicated at the end of a phrase or sentence and at the beginning of the next, as "And if children, then heirs; heirs of God . . . " [Rom. viii.17.] Also called **anadiplosis.**

epiparados, (in ancient Greek tragedy) a second or additional parados, or entrance of the chorus.

ethical dualism, 1. a moral system that demands one kind of conduct toward fellow-members of one's own social group, and the opposite kind of conduct toward all others. Compare **2: double-standard.**
2. (in primitive society) the recognition of one set of duties toward fellow-clansmen and of opposite duties toward strangers not of the blood-bond.

even, divisible by two with no remainder.

2

fall between two stools, miss being appropriate to either of two categories; miss both of two alternatives.

February, the second month of the modern calendar.

feel like two cents, to be very embarrassed; to have been made to feel very cheap or petty.

female rime, (in prosody) a double rime, as *notion, motion,* consisting of two-syllable words that end in an unaccented syllable, distinguished from feminine rime in that the latter may be of more than two syllables, with the concluding syllables unaccented, as *merrily, verily.* Compare **2: feminine rime, 3: feminine rime, 1: masculine rime.**

feminine number, an even number, that is, a number divisible by two with no remainder. See also **Misc.: masculine number.**

feminine rime, (in prosody) a rime of two syllables in which the first syllable is stressed, as *hugger/mugger, cater/later.* Also called **double rime.** Compare **1: masculine rime, 2: female rime, 3: feminine rime.**

feminine sign, (in astrology) one of the even signs of the zodiac, that is, the second, fourth, sixth, etc. See also **Misc.: masculine sign.**

feriae, (in Roman antiquity) two classes of holidays during which free Romans suspended their political transactions and lawsuits and slaves enjoyed a cessation of labor: feriae publicae were days on which public religious festivals were observed; feriae privatae were observed by single families or individuals in commemoration of some particular event of consequence to themselves or their ancestors.

figure of Lissajous, a curve that illustrates the combination of two simple harmonic motions.

Filioque, the clause of the Nicene creed in its western form that asserts that the Holy Ghost proceeds both from the Father and from the Son ("qui ex Patria Filioque procedit"); the doctrine of the double procession. The Greek Church maintains the original form of the Nicene Creed, in accordance with John xv.26, "the Spirit of truth, which proceedeth from the Father," and the controversy, called the Filioque controversy, was one of the causes of the schism between the two churches.

fork-rest, formerly, a two-pronged pole carried by a soldier as a rest in aiming a heavy firearm. Compare **2: bipod.**

fraternal twin, one of a pair of twins developed simultaneously from different ova. Compare **2: identical twin.**

gemel, a twin.

geminate, 1. to double or become double.
2. doubled; twin; binate.

Gemini, a sign of the zodiac, representing the **Twins,** Castor and Pollux. See also **2: Twins.**

gimmer, a ewe that is two years old.

Goody Two-Shoes, 1. (originally) a little girl who delighted in having and wearing her two shoes. [From *The History of Goody Two-Shoes,* 1766.]
2. (in modern use) a person too good, sweet, and innocent to be true; a goody-goody.

Harpies, Aello and Ocypete (according to Hesiod). See also **3: Harpies.**

have two left feet, be awkward, ungainly, or uncoordinated.

have two strings to one's bow, have more than one way of accomplishing something; have talent, skill, or knowledge in more than one area.

healing by second intention, (in medicine) the closing of a wound the edges of which do not unite, resulting in the development of granulation tissue and a somewhat larger scar than that seen in healing by first intention. Compare **1: healing by first intention.**

Hermanas, Sala de las dos, ('Hall of the Two Sisters') a chamber in the Alhambra, an ancient Moorish palace and fortress at Granada, in southern Spain, so called after two large white marble slabs laid into its pavement.

hoots (in hell), not care (*or* **give) two,** not care at all, as *I don't care two hoots if you go out with Suzanne.*

horns of the dilemma, the alternatives offered by a dilemma, traditionally two in number. Compare **3: trilemma, 4: tetralemma.**

iamb, *or* **iambus,** (in prosody) a foot of two syllables, the first short, or unaccented, the second long, or accented, as the normal pronunciation stress pattern of *delight, begin, confer,* etc.

identical twin, one of a pair of twins developed simultaneously from the same ovum. Compare **2: fraternal twin.**

informatory double, (in bridge) a call of double, not to challenge the latest bid of an opponent but to impart information about the bidder's hand to his partner. Compare **2: double, 8(b).**

interamnian, situated between two rivers, applied specifically to Mesopotamia. [Literally equivalent to *mesopotamian.*]

intercolumniation, (in architecture) the space between two columns, measured at the lower part of their shafts, usually from center to center. See also **1½: picnostyle; 2: systyle; 3: diastyle; 4: areostyle; 2¼: enstyle.**

2

It takes two to tango, a truism carrying with it the implication that a misdeed is seldom performed without an accomplice.

Janus, a primitive Italic solar deity, regarded among the Romans as the doorkeeper of heaven and the patron of the beginning and ending of all undertakings. As the god of the sun's rising and setting, he had two faces, one looking to the east, the other to the west.

Janus-faced, two-faced, like the Roman household god, Janus; hence, double-dealing; deceitful.

jereboam, a bottle, usually of champagne, containing two magnums, or about 100 fluid ounces. Also called **double magnum.**

jumelle, 1. twin; forming a pair or couple, as a jumelle operaglass (one having two tubes).
2. **jumelles,** a pair of operaglasses.

kill two birds with one stone, accomplish two ends while expending the effort required to accomplish only one.

lever of the second class *or* **kind,** a straight lever in which the fulcrum is at one end and the power at the other, with the weight in between.

little casino, *or* **small casino,** (in the card game of casino) the two of spades.

Louis Napoleon, Louis Napoleon was associated with the number two in the following ways: December 2, 1851 was the *coup d'état*; he was made emperor on December 2, 1852; the Franco-Prussian War began on August 2, 1870, at Saarbrücken; he surrendered to William of Prussia on September 2, 1870. See also **2: Napoleon.**

love seat, a small sofa or settee for two persons. See also **2, 4: tête-à-tête.**

magpies, to see two magpies denotes merriment or a marriage, or, according to a Lancashire superstition, bad luck. [Brewer, *Reader's Handbook*, 1889.] See also **3: magpies.**

make a double, (in shooting) to kill two birds or beasts in succession, one with each barrel of a double-barreled gun.

make the pot with two ears, *Obsolete.* put the arms akimbo.

manille, (in ombre or quadrille) the two of clubs or spades, the second highest card.

mark twain, (formerly) a call given by a crew member to the pilot of a river boat, especially on the Mississippi, to indicate that the depth of water is two fathoms.

mittelhand, (in skat) the second player on the first trick.

Monday, the second day of the week.

monomachy, single combat; a duel.

M-2, *U. S.* a measurement of the amount of money in circulation, defined by the Federal Reserve as consisting of M-1B plus small denomination deposits in savings accounts at all depository institutions. See also **1:M-1A; 1:M-1B; 3:M-3.**

murder in the second degree, (in U. S. law) usually, unpremeditated or accidental murder, as incidental to an unplanned felony or owing to an accident; manslaughter. Also called **second-degree murder,** *Informal,* **murder two.**

Napoleon, Napoleon Bonaparte was associated with the number two in being crowned on December 2, 1804; also the victory at Austerlitz was December 2, 1805. See also **2: Louis Napoleon.**

Newton's Second Law of Motion, the time rate of change of the directed speed or acceleration of a body is directly proportional to the force acting upon the body and inversely proportional to the mass of the body. Compare **1: Newton's First Law of Motion, 3: Newton's Third Law of Motion.**

no two ways about it, there is no alternative.

number two, a child's euphemism for a bowel movement. Compare **1: number one.**

on the double, **1.** a double-quick military marching tempo, or 165 steps to the minute, each 33 inches long.
2. with all haste; quickly; at a run; double-quick.

optic nerve, either of two nerves involved in the sense of sight. Also called **second cranial nerve.**

other, the second or alternative between two.

paeon diagyíos, the ordinary cretic, a paeonic foot of two semeia, or divisions. Compare **paeon epibatus.**

paeon epibatus, a compound paeonic foot, divided into four parts, double the magnitude of the paeon diagyíos. See also preceding entry.

pair, **1.** two of anything, especially when forming a matched set.
2. a single thing that consists of two pieces or parts, not necessarily separable, as *a pair of spectacles, trousers, scissors,* etc.
3. (in poker) two cards in a hand that are of the same denomination, as *a pair of kings, eights, etc.*

pair of colors, the two flags carried by an infantry regiment, one of which is the national ensign, the other bearing mottoes, devices, etc., that identify the regiment.

2

Passion Sunday, the second Sunday before Easter Day. See also **5: Passion Sunday.**

Pillars of Hercules, the classical name for the Rock of Gibraltar, (classical Calpe) and Jebel Musa (classical Abyla), at Centa, in Morocco, the promontories flanking the Strait of Gibraltar. Legend has it that they were originally a single unit, torn apart by Hercules.

play second fiddle, to take a subordinate position behind or below another.

pricket's sister, a two-year-old fallow doe.

quadratic, (in algebra) **1.** noting any equation or expression in which the unknown quantity is of no higher power than 2; of the second degree, as $a^2 = b^2$.
2. noting an equation of the form $A^2 + 2Bx + C = 0$.

quadratic equation, (in mathematics) an equation of the second degree, of the form, $A^2 + 2Bx + C = 0$.

quadric, (in algebra and geometry) noting an equation, expression, or a plane or solid figure in which there is more than one variable.

quantic, (in mathematics) a rational integral homogeneous function of two or more variables, classified according to their dimensions as quadric, cubic, quartic, quintic, etc., denoting quantics of the second, third, fourth, fifth, etc. degrees. They are further distinguished as binary, ternary, quaternary, etc., according to whether they contain two, three, four, or more variables.

riches, house of, (in astrology) the second house, a succedent house. See **12: house.**

rule of two, (in mathematics) the procedure for finding a third number from two given numbers, as by addition, subtraction, multiplication, or division.

sacrifice double, (in bridge) a call of double in order to prevent the opponents from bidding a game.

second, 1. next after the first in order.
2. (in baseball) second base.
3. one who assists another in a duel, boxing match, etc.

secondary deviation, (in ophthalmology) the deviation of the healthier eye from the position that would make its visual line pass through the object-point of the weaker eye.

secondary education, the second level of schooling, between primary school and college or university education, as given at a secondary school. See also **1: primary education.**

secondary planet, one of the many satellites that orbit the planets. Compare **9: primary planet.**

second childhood, the state of senility or dotage, sometimes reached by an adult who reverts to childish or childlike behavior.

second class mail, 1. (in the U. S.) mail consisting of periodicals carried at a preferential rate.
2. (in the U. K.) mail carried at a rate cheaper than that charged for first class and with a delayed delivery.

Second Coming, the, (in Christianity) the belief that Christ will come in glory at the end of the world and judge all men. The Book of Revelation also mentions a thousand-year reign of Christ and the righteous. Also called **the Parousia.**

second covenant, a promise made by God to Israel (Jeremiah 31:31-34; Hebrews 8:8-13) that He would forgive all their sins. Christians believe Jesus is the fulfillment of this covenant. See also **1: first covenant.**

Second-day, Monday, the second day of the week, so called by the Society of Friends.

second death, (in theology) the state of lost souls after physical death; eternal punishment.

second-degree burn, a moderately severe burn in which the skin is blistered.

seconde, the second of eight parries or guards against thrusts in fencing. Compare **1: prime, 3: tierce, 4: quarte, 5: quinte, 6: sixte, 7: septime, 8: octave.**

second-hand, used; not purchased new, but acquired from another person rather than the maker, retailer, etc.; hand-me-down.

second intention, (in logic) a general conception obtained by reflection and abstraction applied to first intentions as objects. Compare **1: first intention.**

Second Law of Thermodynamics, heat always flows from a higher to a lower temperature. Compare **0: Zeroth Law of Thermodynamics, 1: First Law of Thermodynamics, 3: Third Law of Thermodynamics.**

Second Order of St. Francis, an order of nuns, founded by St. Clare early in the thirteenth century. Also called **Clarisses, Poor Clares, Poor Ladies.** Compare **3: Third Order of St. Francis.**

second person, (in grammar) a category of pronouns and of verb forms referring to the person or persons addressed.

2

second position, (in ballet) a position in which the feet are parallel and slightly apart, the toes pointing in opposite directions to the left and right.

Second Reich, the German Empire from 1871 to 1918. Compare **1: First Reich, 3: Third Reich.**

seconds, one or more additional helpings of food.

second sight, a supposed power that enables its possessor to see into the future; clairvoyance.

second string, (especially in sports) one or more players who are not as skilled as those in the first string.

Second Triumvirate, the coalition in Rome (43 BC) between Mark Antony, Octavian (Augustus), and Lepidus.

second wind, a revival of energy after an initial feeling of tiredness, often experienced by those engaged in vigorous, sustained physical exercise.

secundigravida, a female who bears young for the second time.

secundipara, a woman who is giving birth for the second time.

secundogeniture, the right of inheritance pertaining to a second son. Compare **1: primogeniture.**

see double, to see two images of the same object, as from drunkenness, rage, etc.

serve two masters, accommodate two, possibly conflicting, obligations, usually used negatively, as *You can't serve two masters.*

show (someone) a clean (*or* fair) pair of heels, to run away (from); to escape (from) successfully.

Shrewsbury Two, the, Eric Tomlinson and Dennis Warren, building workers jailed in 1973 for their part in violence on sites in and around Shrewsbury, England, during a strike the previous year. They were sentenced, respectively, to two and three years' imprisonment. The inclusion of a conspiracy charge (which provided for such severe penalties) in the indictments led to their being regarded as labor martyrs.

Sicilies, the Two, the former legal designation of the kingdom of Naples, founded by the Norman dynasty in the Middle Ages and comprising Sicily, the southeastern provinces of Italy, and a number of nearby islands.

snake-eyes, *Slang.* a pair of dice showing a total score of two.

span, (in the U. S.) a pair of horses or mules harnessed together, especially a pair of horses matched for driving or work.

2

stand on one's own two feet, to be able to conduct oneself and one's affairs without the help of anyone else.

subduple ratio, a ratio such as that of 1 to 2, 4 to 8, etc. Compare **2: duple ratio.**

systyle, (in architecture) intercolumniation of 2 diameters, in the classification of Vitruvius. See also **2: intercolumniation.**

take-out double, (in bridge) a call of double in an attempt to prevent the opponents from bidding a game.

tête-à-tête, 1. a meeting, conversation, etc., between two people only.
2. of or pertaining to such a meeting.
3. alone together; with only two present.
4. a seat for two persons, S-shaped in plan, allowing them to face one another, as for intimate conversation, a **love seat.**

think twice, consider carefully before arriving at a decision.

triple space, to leave two empty lines of space between lines of text, especially in typing.

trochee, (in prosody) a foot of two syllables, the first long or accented, the second short or unaccented, as in the words *shortly, feeble.*

twibill, a double-bladed battle-ax.

twicetold, stale; hackneyed; trite.

twill, a textile fabric in which the weft-threads pass over one warp-thread and under two, over one and under three, over one and under eight or ten, depending on the kind. [From Low German *twillen* 'make double.']

twin, one of a pair of offspring delivered at the same birth. See also **2: fraternal twin, 2: identical twin.**

twine, a coarse cord made of two strands twisted together. [From Middle English *twyne* 'double thread.']

Twins, the, Castor and Pollux, twin sons of Zeus. Castor is a horse-tamer, Pollux (Greek Polydeuces) the master of the art of boxing. In Homer, they are represented as the sons of Leda and Tyndareos. Also called **Dioscuri, Gemini, Tyndaridae.** See also **2: Gemini.**

two, the number two, or the dyad, symbolizes diversity, the evil principle. [Pythagoras, *On the Pentad.*]

two-a-cat, *or* **two-o'-cat,** *or* **two old cat,** a simplified form of baseball, formerly played by children, in which there are two bases and two players. Compare **1: one-a-cat, 3: three-a-cat, 4: four-a-cat.**

two-alarm fire, the second call the fire department issues for further help in fighting a fire. See also **1: one-alarm fire.**

2

two-bit, *Slang.* insignificant. [After a *bit,* a Spanish or Mexican silver real worth 12½ cents, in former use in parts of the U. S.]

two-by-four, **1.** a long piece of lumber, typically measuring 1⅝ x 3⅝ inches.
2. *U. S. slang.* tiny; cramped, as living quarters, an office, etc.

two can play at that game, (of a questionable, unfair, or nefarious act or action) the perpetrator is not the only person who can behave badly (implying that another, often the speaker, can and might behave just as badly).

two cents, **1.** **put in one's two cents,** *U. S.* add one's comments to a discussion already under way.
2. **give two cents for (something)** *U. S.* (*usually negative*) deprecate something by implying that it is not worth more than two cents, as *I wouldn't give two cents for this year's model of their computer.*

two cents plain, *or* **2 cents plain,** a glass of seltzer costing two cents, especially as formerly served at soda fountains, chiefly in New York City.

two-dimensional, noting the magnitude of a surface, which has length and breadth, but no thickness.

two-dog night, (especially in arctic conditions) a very cold night [From the number of sled dogs one has to sleep with in order to keep warm.] Compare **3: three-dog night.**

two-faced, deceitful; insincere.

twofer, *Informal.* a coupon enabling a person to purchase two tickets to a theatrical production for the price of one.

two-fisted, aggressive; vigorous.

two heads are better than one, a saying meaning that the intelligence of two people can accomplish more than that of one.

two-legged mare, the, a gallows.

two minds, be of, be undecided, as *She was of two minds whether to marry him or just live with him.*

twopence, **1.** (formerly) a coin of Great Britain worth one sixth of a shilling. Also called **half-groat.**
2. an allowance for each day of service paid to soldiers and noncommissioned officers in the British army on discharge or death. Also called **deferred pay.**

twopenny, of little value; almost worthless.

twopenny damn, not worth a, worthless.

twopenny nail, **1.** a nail that, before 1500, cost twopence a hundred. **2.** (in modern use) a common nail 1 inch long.

two-platoon system, (in American football) a system under which each team consists of one set of players trained and skilled in offensive strategy and another set in defensive strategy, with the replacement of all or most players in accordance with the condition of the game.

two's company, three's a crowd, the presence of a third person can dampen the ardor of two who are bent on a tête-à-tête, sexual activity, etc. Also, *Brit.* **two's company, three's not.**

two shakes of a lamb's tail, in, at once; anon; immediately; very soon; very quickly, as *I'll be with you in two shakes of a lamb's tail.* Often shortened to **in two shakes.**

"two sides to every question," attributed (in *Bartlett's Familiar Quotations*, 15th ed., 1980) to Protagoras (c485–c410 BC).

two-time (someone), **1.** double-cross or deceive (someone). **2.** be unfaithful to or break a trust with (a spouse, friend, or other associate or someone who trusts one).

"Two-Ton" Tony, a common reference to any man considered overweight, an allusion to Tony Galento, an American heavyweight boxer (1910–1979) known for his adiposity.

two-way mirror, a half-silvered mirror, usually mounted in a wall between two rooms, which serves as a normal mirror when viewed from the front but which can be seen through from the back; used for watching activity in a room without the watcher being seen.

Two Years Before the Mast, a personal narrative (1840), by Richard Henry Dana, that recounts his experiences as an ordinary seaman after his graduation from Harvard University.

unau, the South American two-toed sloth. See also **2: ai, 3: three-toed sloth.**

unlucky number, the regents of Great Britain have been unlucky with the number 2: **1,** Ethelred II abdicated; **2,** Harold II died at the Battle of Hastings; **3,** William II was shot; **4,** Henry II had to fight for his crown, which was usurped by Stephen; **5,** Edward II was murdered; **6,** Richard II was deposed; **7,** Charles II was exiled; **8,** James II abdicated; **9,** George II lost at Fontenoy and Lawfeld, was disgraced by General Braddock and Admiral Byng, and was harassed by Charles Edward, the Young Pretender.

War of the Second Coalition, a war (1798–1801) between France and an alliance of European monarchies led by Austria, in which Napoleon Bonaparte was the victor.

2

wear two hats, be engaged in two major activities, as *I wear two hats in the company: I am the sales manager and the publicity director.*

zygodactyl, (of a bird's foot) having two toes facing forward and two back; pair-toed; yoke-toed.

2¼

enstyle, (in architecture) intercolumniation of 2¼ diameters, in the classification of Vitruvius. See also **2: intercolumniation.**

2½

sesquiduplicate, two and a half times; in the ratio of 5 to 2, or 2½ to 1.

sestertius, a coin of ancient Rome, originally worth two and a half asses.

twopenny-halfpenny, a disparaging term for anything of little value.

3

Agni, (in Hinduism) the three forms of Agni, the ancient Aryan god are: Grhapati (as god of fire or the hearth); Surya (as the sun); and Trita (as lightning). The three forms are called Tryambaka 'three-mothered.' From Jobes, *Dictionary of Mythology, Folklore and Symbols*.] See also **8: Agni.**

Alliance of the Three Kings, a short-lived association between Prussia (king Frederick William IV), Hanover (elector Frederick William I), and Saxony (king Frederick Augustus II), organized by General Joseph Maria von Radowitz, on May 26, 1849, to aid in the unification of Germany. It was effectively dissolved less than a year later.

Andrews Sisters, the, a close-harmony singing group popular on the stage and in motion-picture musicals in the 1940s, namely: Laverne; Maxine; and Patty.

Animalia, the first of the three kingdoms of organized beings, including those that are true animals and not true plants (*Vegetabilia*) or neither (*Primalia*).

Apocrypha, one of the three classes into which Biblical literature is divided: the canonical and inspired (the *Bible*); the noncanonical and uninspired, but on account of their character, worthy of use in the services of the church (the *Apocrypha*); and those which, though Biblical in form, so vary from the Biblical writings in spirit that they are not deemed worthy of any place in religious use (the *Pseudepigrapha*).

Apostolic Hours of Prayer, (early Christianity) in accordance with the psalm, "In the evening and at morning and noon day will I pray," the earliest Christians chose to pray at nine o'clock, when Christ was nailed to the cross; twelve o'clock, when the darkness came; and three o'clock, when Christ died. See also **7: Canonical Hours.**

aum, **1.** (in Buddhism) a monosyllable symbolizing the triratna, namely: Buddha (intelligence and soul); dharma (body and matter); and shangha (the union of Buddha and dharma).
2. (in Hinduism) a monosyllable symbolizing the trinity, namely: Aditi; Varuna; and Mitra.

Big Three, the, **1.** the representatives of France, Great Britain, and the United States at the Paris Peace Conference in 1919, namely: Georges Clemenceau; David Lloyd George; and Woodrow Wilson.
2. the heads of government of Great Britain, the United States, and the Union of Soviet Socialist Republics during World War II, namely: Winston Churchill; Franklin Delano Roosevelt; and Josef Stalin.
3. the athletic teams of Harvard; Princeton; and Yale.

Boëdromion, the third month of the ancient Greek year, corresponding to the latter part of September and the early part of October.

brethren, house of, (in astrology) the third house, a cadent house. See **12: house.**

Brontë, three sisters, Acton (1820–49), Currer (1816–55), and Ellis Bell (1818–48), who wrote novels in the nineteenth century under the noms de plume of Anne, Charlotte, and Emily Brontë, respectively.

C, c, **1.** the third letter of the English alphabet.
2. the third item in a system where A = 1, B = 2, C = 3.

Cerberus, (in Greek mythology) the three-headed watchdog that guarded the entrance of hell to see that the living did not enter or the dead escape. Also called the **hound of hell.**

Charlotte Three, the, three black activists jailed in 1972 for the burning of a whites-only stable in Charlotte, North Carolina, namely: James Earl Grant, Jr.; Charles Parker; and Thomas J. Reddy. They received sentences ranging from 10 to 25 years. The case acquired notoriety for the length of these terms, the lapse of time between the alleged offense and the trial (when the fire took place in 1968 there was no suggestion of arson), and the benefits—sums of federal relocation aid and immunity from prosecution for other offenses—received by the chief witnesses for the prosecution. All of the prisoners were free by mid 1979, following reductions in their sentences that allowed for parole.

Christian Creeds, there are three forms of the statement of beliefs expressing the same tenets at differing length, namely: the Apostles', the Nicene, and the Athanasian Creeds.

Columbus, Christopher, the three ships that formed his fleet in the voyage to the New World were the *Niña,* the *Pinta*, and the *Santa Maria.*

3

consolidated threes, *or* **three percent consolidated annuities,** formerly, government securities of Great Britain, created under an act of Parliament in 1751. Compare **3: new threes.**

cubic equation, (in mathematics) an equation of the third degree, of the form, $x^3 + 3ax^2 + 6bx + 2c = 0$.

Cyclopes, the, (in Greek mythology,) three giants, namely, Arges, Brontes, and Steropes, each of whom had only one eye, in the middle of his forehead. Titans, the offspring of Uranus (Heaven) and Gaea (Earth), they forged the thunderbolts of Zeus, Pluto's helmet, and Poseidon's trident. They were regarded as the primeval patrons of all smiths. (These are not the same as the race of giants referred to in the *Odyssey*, one of which was Polyphemus.) They appear in the third voyage of Sindbad. Compare **1: Polyphemus.**

D, (in music) the third string of a violin, tuned to the tone D.

D, d, the third consonant of the English alphabet and in the Latin, Greek, and Phoenician, as well.

dactyl, (in prosody) **1.** (in modern, or accentual versification) a foot of three syllables, the first long, the second and third short.
2. (in ancient metrics) a tribrach, a cretic, or an anapest.

delta, **1.** a triangular island or alluvial tract included between the diverging branches of the mouth of a great river, as the delta of the Nile or of the Mississippi.
2. (in anatomy) a triangular space or surface.

deltoid, *or* **deltoideus,** the large triangular muscle of the shoulder.

diastyle, (in architecture) intercolumniation of 3 diameters, in the classification of Vitruvius. See also **2: intercolumniation.**

diet, a meeting for legislative, political, ecclesiastical, or municipal purposes which, in the old Roman-German empire, sat in three colleges: that of the electoral princes; that of the princes, in two benches, the temporal and the spiritual; and that of the imperial cities.

diminished triad, (in music) a triad consisting of a tone with its minor third and its diminished fifth.

dom, the three of trumps in the game dom pedro, which is the name given to the game sancho pedro when the trey and the joker are used and counted toward game.

dreibund, a triple alliance.

dreikanter, (in geology) a pebble polished by the wind, typically having three faces. [From German *drei* 'three' + *Kant* 'edge' + *-er* nominal suffix]

dummy, a game of whist in which three play, the fourth hand being exposed.

eaglet, (in heraldry) the name given to an eagle on an escutcheon when three or more are borne.

earth, the planet, being the third in order from the sun.

earths, the three essential principles assumed by Becher, in the seventeenth century, to be present in all substances, namely: terra mercurialis; terra vitrea; and terra pinguis. The last was regarded as the essence of combustibility—the "phlogiston" of the eighteenth-century chemists—and was said to escape or be lost when a substance burned.

excelsior, a size of type on a three-point body.

Fates, the, *or* **Moirae, the,** (in Greek mythology) three daughters of Zeus and Themis who controlled the length of life, namely: Clotho, who spun the thread of life; Lachesis, who measured it; and Atropos, who cut it. Also (in Roman mythology), **Parcae.**

feminine rime, (in prosody) a rime of three syllables in which the first syllable only is stressed, as *flattering/battering, spurious/furious.* Also called **triple rime.**

"fiddlers three," from the old nursery rime, "Old King Cole":
> Old King Cole
> Was a merry old soul
> And a merry old soul was he.
> He called for his pipe
> And he called for his bowl
> And he called for his fiddlers three.

French star, a cluster of three asterisks, arranged in this form ⁂, used as a mark of division between different articles in print.

Frimaire, the third month of the French revolutionary calendar, 1793, beginning on November 21st and ending on December 20th.

Furies, the, (in classical mythology) three goddesses, horrible to behold, with heads wreathed with serpents, who punished those who escaped or defied public justice, namely: Alecto; Megaera; and Tisiphone. Also called **Erinyes, Eumenides.**

gamma, the third letter of the Greek alphabet, corresponding to the English *g.*

gimel, the third letter of the Hebrew alphabet, corresponding to the Greek *gamma,* and the English *g* in *go;* it has a numerical value of 3.

gleek, three cards of the same kind in the old card game of gleek.

3

Golden Apples of the Hesperides, (in Greek mythology) golden apples given by Earth (Gaea) to Hera on her marriage to Zeus and guarded by women (the Hesperides), assisted by the hydra, a dragon named Ladon. Hippomenes was given three of them by Aphrodite to help him win a footrace with Atalanta (and, hence, her hand) by dropping them, delaying her when she stopped to pick them up.

"Goldilocks and the Three Bears," a nursery tale about a little girl who disobeys her mother and runs into the forest where, during their absence, she visits the house where live a "Mama Bear," a "Papa Bear," and a "Baby Bear," who return to find her asleep. She awakens and runs home, terrified.

Graces, the, three goddesses of everything that lends charm and beauty to nature and human life, namely: Aglaia (brilliance); Euphrosyne (joy); and Thalia (bloom), the last not to be confused with a Muse of the same name. Also called **Charites, Gratiae.**

Graeae, the, three old women, the daughters of Phorcys and Ceto, namely: Dino; Enyo; and Pephredo. They were born with gray hair and shared among them one eye and one tooth. They advised Perseus on his slaying of the Medusa. Also called **the Phorcydes.**

Great Triumvirate, the, *or* **Immortal Trio, the** John C(aldwell) Calhoun (1782–1850), Henry Clay (1777–1852) called the Great Compromiser, and Daniel Webster (1782–1852), so called because they were the great American law-makers of their time and their lives were so similar: they were about the same age; they became Representatives, then Senators at about the same time; each served as Secretary of State; and each had unfulfilled aspirations to the presidency. All were lawyers and inspiring speakers.

Hakatist, one of the three members of the league organized in Germany for the support of Prussian measures against the Poles in Posen. [From the initials, *H, K, T,* of the three founders of the league, Von Hansemann, Von Kennemann, and Von Tiedemann.]

Harpies, the, (in classical mythology) vultures with the head and breasts of a woman, fierce and loathsome, living in filth and stench, and contaminating everything they came near. Homer mentions one Harpy, Hesiod two, and Vergil three; they were named Aello ('storm'), Ocypete ('rapid'), and Celeno ('blackness').

hat trick, (in hockey and soccer) the scoring of three goals in a game by one player, originally from cricket in reference to the feat of a bowler who dislodges three wickets consecutively.

Hecate, (in Greek mythology) a triformed goddess of Thracian origin, akin to Artemis, combining the attributions of Demeter or Ceres, Rhea, Cybele, Artemis or Diana, and Persephone or Proserpine. She was associated with the teaching and practice of witchcraft and sorcery.

Hermes Trismegistus, the Egyptian god Thoth, said to have invented hieroglyphic writing, the art of harmony, the science of astrology, the lute and lyre, magic, etc., and to have written the first Egyptian code of laws. [*Trismegistus* 'thrice-greatest']

Horae, (in Greek mythology) the Hours, three females, Eunomia, Dike, and Eirene, who presided over meteorological phenomena which regulate vegetable and animal life and who served Zeus by opening and closing the doors of Heaven. Their number, somewhat vague in Homer, was later fixed at four when they were merged with the seasons.

I, i, the third vowel in the English alphabet.

Iambic Poets, the three iambic poets included in the Alexandrian Canon, namely: Archilochus; Hipponax; and Simonides.

Ionian, a member of one of the three (including the Dorian and Æolian) or four (adding the Achaean) great divisions of the ancient Greeks, or Hellenes.

"I saw three ships . . . ," from an early nineteenth-century nursery rime:

> I saw three ships come sailing by,
>   Come sailing by, come sailing by;
> I saw three ships come sailing by,
>   On New-Year's day in the morning.
> And what do you think was in them then,
>   Was in them then, was in them then?
> And what do you think was in them then,
>   On New Year's day in the morning?
> Three pretty girls were in them then,
>   Were in them then, were in them then.
> Three pretty girls were in them then,
>   On New-Year's day in the morning.
> One could whistle and one could sing,
>   And one could play on the violin;
> Such joy there was at my wedding,
>   On New-Year's day in the morning.

Judges of Hades, the, the three Judges of Hades were Minos (the chief baron), Æacus (the judge of Europeans), and Rhadamanthus (the judge of Asians and Africans).

June, the third planetoid, discovered in 1804.

kingdom, one of the three great divisions into which natural objects are classified, namely: the animal, vegetable, and mineral kingdoms.

3

Kings, the eleventh and twelfth books of the Bible, originally undivided, first separated in the Septuagint and retained in the Vulgate, in both of which they are named the third and fourth books of Kings, the two books of Samuel being the first and second.

kit, a miniature violin, about sixteen inches long, having three strings, once used by dancing-masters.

kowtow, *or* **kotow,** *or* **kotoo,** *or* **kootoo,** *or* **kotou,** an act of obeisance, reverence, worship, respect, subservience, etc., formerly practised in China. Before the Emperor and in worship, the person knelt three times and touched the head to the ground with the forehead three times after each kneeling.

League of the Three Emperors, the, an alliance among William I (1797–1888), Emperor of Germany and Prussia (1861–88), Francis Joseph I (1830–1916), Emperor of Austria (1848–1916), and Czar Alexander II (1818–81), Emperor of Russia (1855–81), aimed at reducing tensions in the Balkans, signed October 22, 1873. It was renewed on June 18, 1881, then again in 1884. It became ineffective in 1885, upon the invasion of Bulgaria by the Serbs. Also called **Dreikaiserbund, Three Emperors' Alliance.**

lever of the third class *or* **kind,** a straight lever in which the fulcrum is at one end and the weight at the other, with the power between.

Little Three, the, 1. the basketball teams of Canisius College, Niagara University, and St. Bonaventure College.
2. the athletic teams of Amherst College, Wesleyan University, and Williams College.

Magi, the, *or* **Three Kings of Cologne, the,** three kings, Gaspar 'the white one,' Melchior 'king of light,' and Balthazar 'king of treasures,' who, drawn by the star of Bethlehem, attended the infant Jesus. They are also known, according to the tradition, under the names of Apellius, Amerus, and Damascus; Magalath, Galgalath, and Sarasin; or Ator, Sator, and Peratoras. Compare **6: Magi.** Also called the **wise men of the east.**

magpies, to see three magpies denotes a successful journey. [Brewer, *Reader's Handbook*, 1889.] See also **4: magpies.**

March, the third month of the modern calendar.

Marx Brothers, the, a family of American comedians successful in vaudeville, on Broadway, and in motion pictures, originally five, namely: Chico (Leonard); Harpo (Adolph); Groucho (Julius); Gummo (Milton); and Zeppo (Herbert). The last two left fairly early on.

master mason, a freemason who has reached the third degree.

matador, (in ombre or quadrille) one of the three principal cards, namely, the ace of clubs, the ace of spades, and either the two of trumps, should clubs or spades be trumps, or the seven of trumps, should diamonds or hearts be trumps.

ménage à trois, a household where three persons live together, usually either two men and a woman, or two women and a man.

M-3, *U. S.* a measurement of the amount of money in circulation, defined by the Federal Reserve as consisting of M-2 plus large denomination deposits at all depository institutions and term repurchase agreements at commercial banks and savings and loan institutions. See also **1: M-1A; 1: M-1B; 2: M-2.**

Muses, three Muses who are from a pantheon older than that which produced the nine Muses (See **9: Muses**), namely: Melete (of meditation); Mneme (of remembrance); and Aoide (of song.)

Needles, the, three tall chalk pillars that form the western end of the Isle of Wight, off the southern coast of England.

neuter, a (third) grammatical gender in some languages, in addition to masculine and feminine. [From Latin *ne* 'not' + *uter* 'other,' meaning 'neither of the other (two genders).']

new threes, *or* **reduced threes,** securities similar to consolidated threes but at a lower rate of interest. Compare **3: consolidated threes.**

Newton's Third Law of Motion, the actions of two bodies on one another are always equal and opposite, i.e., reaction is equal and opposite to action. Compare **1: Newton's First Law of Motion, 2: Newton's Second Law of Motion.**

nictitating membrane, a filmy retractile covering of the eye, occurring in certain reptiles and birds, in addition to the normal eyelids, hence called the **third eyelid.**

Norn, (in Scandinavian mythology) one of three Fates who tend the root of Yggdrasil that extends into Asgard, the realm of the gods, namely: Urdur (the past); Verdandi (the present); and Skuld (the future). It is their responsibility, also, to engrave the runes of fate upon a metal shield. See also **3: Yggdrasil.**

Oculi Sunday, the third Sunday in Lent. [From the first word, *Oculi* 'eyes' in the Latin text of the introit, beginning with the 15th verse of the 25th Psalm, "Mine eyes are ever toward the Lord."]

oculomotor nerve, either of two nerves involved in the muscle control of the eyes. Also called **third cranial nerve.**

Olynthiac orations, a series of three orations delivered by Demosthenes to induce the Athenians to support Olynthus against Philip; they constitute a part of the Philippics.

3

pair, (in archery) a set of three arrows.

pair royal, three similar things; specifically, in certain card games, three cards of a kind, as three kings or three queens. Also, *Obsolete* **parial, pairial, prial.**

Parliament, (in the United Kingdom) the supreme legislative body, consisting of the three estates of the realm, the Lords Spiritual, the Lords Temporal, and the Commons. Compare **4: fourth estate.**

patriarch, (in the Bible) one of three, namely: Abraham; Isaac; or Jacob.

pawnbroker, the three balls symbolizing the trade of the pawnbroker are said to have been adopted from those appearing on the arms of the Medici family, who, in turn, are said to have adopted them from the three-balled mace of the giant Mugello who was slain by Averardo de Medici, a commander under Charlemagne.

phlebotomous fever, a disease characterized by fever, various pains, conjunctivitis, and, in some cases, a rash. It is caused by a sandfly infected by one of five arboviruses. Also called **pappataci, sandfly fever, three-day fever.**

Primalia, the third of the three kingdoms of organized beings, including those that are neither true plants (*Vegetabilia*) nor true animals (*Animalia*).

primary color, one of the three colors, red, yellow, or blue, which yield other colors when mixed in pairs. Also called **fundamental color.** See also **2: secondary color.**

Prince of Wales's feathers, the crest of the Prince of Wales, first borne by Edward the Black Prince, consisting of three ostrich-plumes with the motto *Ich dien,* 'I serve.'

Pythagorean letter, the letter Y, so called because its Greek original represented the sacred triad, formed by the duad, proceeding from the monad.

Pythagorean triangle, a triad of whole numbers proportional to the sides of a right triangle, the sum of the squares of the two smaller being equal to the square of the largest, as 3, 4, 5 ($3^2 + 4^2 = 5^2$; $9 + 16 = 25$); 12, 35, 37 ($12^2 + 35^2 = 37^2$; $144 + 1225 = 1369$).

queer as a three-dollar bill, very odd or peculiar (as there is no money of that denomination).

Ritz Brothers, the, three nightclub comedians also featured in musical motion pictures in the 1930s, namely: Al; Jim; and Harry (the leader). Their real surname was Joachim.

Rogation days, (in Christianity) the Monday, Tuesday, and Wednesday before Ascension day, set aside for special intercessions. See also **4: Ember days.**

rule of three, the method of finding the fourth term of a proportion when three are given. If the numbers are arranged so that the first is to the second as the third is to the fourth, the fourth is found by multiplying the second and third terms together and dividing the product by the first: $3/9 = 6/x$; $(9 \times 6) \div 3 = x$ and $x = 18$; or $3:9 = 6:x$, yielding the same result.

"Rumpelstiltskin," a fairy tale about a miller's daughter, wedded to a king, who is ordered by him to spin straw into gold; she accomplishes this with the aid of Rumpelstiltskin, a dwarf who demands that she give him, in return, her first-born. On the birth of the child, she refuses, and the dwarf promises to relent if she can guess his name in three days. She guesses vainly for two days but on the third her servants overhear a voice singing, "Little dreams my dainty dame Rumpelstiltskin is my name." They tell her, she saves her child, and the dwarf kills himself.

secondary color, one of three colors created by mixing together two primary colors, as green (a mixture of blue and yellow), violet (a mixture of red and blue), and orange (a mixture of red and yellow).

Shadrach, Meshach, and Abednego, by the command of Nebuchadnezzar, they were cast into a "fiery furnace," but were uninjured. [Dan. iii.22.]

Siren, one of the female temptresses said to lure seamen to a bad end by their sweet song. Homer refers to two, but later writers to three, namely: Leucosia; Ligea; and Parthenope.

Soledad Brothers, the, three black militants who, while serving a jail term for burglary, were charged with murdering a guard at the California State Prison at Soledad, in 1970, namely: John W. Cluchette; Fleeta Drumgo; and George Jackson. See also **6: San Quentin Six.**

terdiurnal, occurring three times a day.

tern, 1. something that consists of three things or numbers, as a lottery prize won by drawing three favorable numbers; the numbers so drawn. 2. a three-masted schooner.

ternary, proceeding by threes or consisting of three.

tertian, occurring every third day, including the first and last, as a fever.

Tertiary, (in geology) a series of geological formations that lies between the Mesozoic (Secondary) and the Quaternary, and including the Eocene, Miocene, and Pliocene to which the Oligocene was later added between the first two. It is characterized as a period of land increase and elevation, the disappearnce of the cephalopods and gigantic reptiles, the devel-

3

opment of mammals, the diminution of importance of ferns, lycopods, and cycads, the zonal distribution of life and climate, and, finally, the appearance of man.

tertium quid, something that mediates or serves as a point of compromise between two things that are essentially opposite. [From Latin *tertium* 'third' + *quid* 'what, who' in the sense of 'neither mind nor matter,' especially an idea regarded not as a mere modification of the mind nor a purely external thing in itself.]

terza rima, a form of iambic verse in which lines of ten or eleven syllables are arranged in sets of three that are closely connected so that the middle line of the first tiercet rimes with the first and third lines of the second, the middle line of the second with the first and third lines of the third, and so on; there is an extra line at the end of the poem or canto that rimes with the middle line of the preceding tiercet, the rime scheme being *aba/bcb/cdc/d*. It is the form of Dante's *Divina Commedia* and of Byron's *Prophecy of Dante.*

Third Avenue, a street in New York City formerly associated chiefly with poverty and seedy bars. Since the removal of the elevated subway tracks that dominated the street before the 1950s, it has become a high-rent office area (between 40th and 59th Streets) and a site of expensive apartment houses (chiefly from 60th Street northward).

third base, (in baseball) the base between second base and home plate.

third class, a classification of accommodation on board ship or at a hotel, below second class, offering few amenities, and usually the cheapest.

third class mail, (in the U. S.) mail consisting of advertising matter, carried at a preferential rate.

Third-day, Tuesday, the third day of the week, so called by the Friends.

third degree, 1. (in Freemasonry) the degree of master mason, demanding of the initiate a very elaborate and severe test of his craftsmanship. See also **32: thirty-second degree Mason, 33: thirty-third degree Mason.**
2. hence, a severe method of examination or treatment applied to prisoners by the police in order to extract information or a confession.

third-degree burn, a severe burn exhibiting necrosis through the full thickness of the skin.

Third Department, a governmental department, created by Emperor Nicholas I in 1826, for secret police operations. It was abolished in 1880 and its functions were transferred to the Ministry of the Interior. Also called **Third Section of His Imperial Majesty's Own Chancery in Russia.**

third estate, the common people. See also **4: fourth estate.**

third hour, the, the third of twelve hours reckoned from sunrise to sunset; the hour midway between sunrise and noon; the canonical hour of terce.

third house, the lobby connecting the two houses of a legislature.

Third Law of Thermodynamics, the relative disorder of ordered solids reaches zero at absolute zero. Compare **0: Zeroth Law of Thermodynamics, 1: First Law of Thermodynamics, 2: Second Law of Thermodynamics.**

third man, (in cricket) a fielding position, analogous to that of shortstop in baseball.

Third Order of St. Francis, an order of lay men and women, founded before 1230, whose members are known as Tertiaries. Compare **2: Second Order of St. Francis.**

third party, 1. a person involved in a legal proceeding only incidentally or by chance.
2. (in British insurance) any person injured or killed or whose property is damaged by the insured, the insurance company and the insured being the first two parties.

third person, (in grammar) a category of pronouns and of verb forms referring to one or more objects or persons other than the speaker or speakers or the person or persons addressed.

Third Philippic, Demosthenes' third, and generally described as his most successful speech against Philip of Macedon, 341 BC.

third position, (in ballet) a position in which the feet are parallel and together, with the toes of each foot against the heel of the other.

third rail, 1. (in electric railroads) a horizontal rail, additional to the two bearing rails, that carries the electric power, fed to the train by means of a sliding shoe.
2. a cocktail made with equal parts of rum, apple brandy, and brandy with a small amount of Pernod, shaken with ice and strained.

third-rate, inferior in quality, talent, luxury, rank, etc.

third reading, (in British Parliament) the discussion of a report on a bill submitted by the committee assigned to it.

Third Reich, the Nazi regime in Germany (1933–45) under Adolf Hitler. Compare **1: First Reich, 2: Second Reich.**

Third Republic, (in France) the government from 1870, at the fall of Napoleon III in the Franco-Prussian War, till 1940, the occupation by Germany.

3

third sex, *Slang.* homosexuals.

Third World, nations that are underdeveloped and not aligned politically with the western nations or the Soviet Union.

Thor, (in Scandinavian mythology) Odin's eldest son, who has three precious possessions, namely: a hammer, which, when thrown, returns to his hand; a belt, which doubles his strength; and iron gloves, which he wears when using the hammer.

three, 1. the whole number after two and before four, sometimes considered lucky, possibly because of its association with the Trinity.
2. the number three, or the triad, contains the mystery of mysteries, for everything is composed of three substances. It represents God, the soul of the world, and the spirit of man. [Pythagoras, *On the Pentad.*]

three-a-cat, *or* **three-o'-cat,** *or* **three old cat,** a simplified form of baseball, formerly played by children, in which there are three bases and three players. Compare **1: one-a-cat, 2: two-a-cat, 4: four-a-cat.**

"three bags full . . . ," a nursery rime goes:
Baa, baa, black sheep,
Have you any wool?
Yes, sir, yes, sir,
Three bags full.
One for my master,
One for my dame,
And one for the little boy
Who lives down the lane.

three-banded armadillo, a species of armadillo having three bands.

"Three blind mice," from an old nursery rime and round:
Three blind mice! Three blind mice!
See how they run! See how they run!
They all ran after the farmer's wife,
Who cut off their tails with a carving knife!
Did you ever see such a thing in your life—
As three blind mice.

three-color Ming ware, decorated porcelain, stoneware, and earthenware objects in which a combination of three colors (*san ts'ai*) is used, usually from dark violet blue, turquoise, eggplant purple, yellow, and white overlaid on a ground of dark blue or turquoise. The technique is typical of the later part of the Ming period, 1368–1644.

Three Days' Battle, a naval battle in the English Channel, 1666, in which the English were defeated by the Dutch.

three-dimensional, noting the magnitude of a solid, which has length, breadth, and thickness. Also called **tridimensional.**

three-dog night, (especially in arctic conditions) a very cold night [From the number of sled dogs one has to sleep with in order to keep warm.] Compare **2: two-dog night.**

Three Emperors, Battle of the, the Battle of Austerlitz, December 2, 1805, was so called because Napoleon of France, Alexander I of Russia, and Francis I of Austria were all present.

threefold ministry, in the Orthodox and Roman Catholic churches, and those of the Anglican Communion, the ordained ministry consists of three parts, namely: bishops, priests, and deacons.

Threefold Refuge, (in Buddhism) the ceremony of ordination into Buddhism by three affirmations, namely: *I go to the Buddha for refuge; I go to the Doctrine for refuge;* and *I go to the Order for refuge.* Also called the **Three Jewels,** the **Three Treasures.**

three F's, the, the three demands of the Irish Land League, namely: free sale; fixity of tenure; and fair rent.

three hours, *or* **three hours' service,** *or* **three hours' agony,** a service held on Good Friday from noon to 3 p.m. in Roman Catholic and many churches of the Anglican communion, in commemoration of Christ's sufferings on the cross, the time recorded in Mat. xxvii.45, Mark xv.33, Luke xxiii.44.

three-legged mare, *or* **three-legged stool,** the gallows, which formerly consisted of three posts, over which was laid a transverse beam. Also called **three trees.**

"Three Little Pigs, the," a nursery tale about three little pigs, each of whom built a house, one of straw, one of sticks, and one of bricks. The "Big Bad Wolf" blew down the first and the second, but he could not budge the third. In the earlier versions, only the third pig survives; in recent versions, all survive, sheltered in the brick house.

three L's, the, (in nautical use) lead, latitude, and lookout: a phrase used to signify that a careful use of the lead, in sounding, a knowledge of the latitude, and a vigilant performance of the third will prevent a vessel from running ashore.

"Three men in a tub . . . ," a nursery rime goes:
> Rub-a-dub-dub,
> Three men in a tub,
> And who do you think they be?
> The butcher, the baker,
> The candlestick-maker;
> Turn 'em out—knaves all three!

Three Mile Island, a nuclear power plant near Harrisburg, Pennsylvania, where, in 1979, human and mechanical failures resulted in a breakdown of the core cooling system of the reactor. The event occasioned

3

great concern for the safety of such installations and has provided much fodder for antinuclear activists ever since.

three-mile limit, for many years (after 1703) the limit of a state's jurisdiction over the waters of its coastline, originally based on the effective range of land-based arms.

Three Musketeers, The, a romantic novel (1844) by Alexandre Dumas, based on the *Mémoires de M. d'Artagnan* (1701), by Giles de Courtilz de Sandras. The three musketeers in the novel are named Athos, Porthos, and Aramis; but the main protagonist is a Gascon adventurer, D'Artagnan, who is a rogue at the beginning of the book but dies, eventually, as the Comte D'Artagnan, Commander of the Musketeers and Marshal of France.

three on a match, a superstition that it is unlucky to light three person's cigarettes or cigars on one match.

threepenny nail,
1. a nail that, before 1500, cost threepence a hundred.
2. (in modern use) a nail 1¼ inches long.

Threepenny Opera, The, (*Die Dreigroschenoper*) an opera by Kurt Weill, book by Bertolt Brecht, first staged in 1928.

three-point landing, (in aviation) a perfect landing, with the two main wheels and the nose or tail wheel all touching down simultaneously.

three-ring circus, 1. a circus with three arenas in which acts are simultaneously under way.
2. *Informal.* any scene of confusion and unorganized or uncoordinated activity; a Chinese fire-drill.

three R's, the, (in humorous use) reading, writing, and arithmetic, regarded as basic to education. [Attributed to Sir William Curtis (1752–1829), an eminent but illiterate alderman and lord mayor of London who, on being asked to give a toast, said, "I give you the three R's, Riting, Reading, and Rithmetic."] See also **4: fourth R, the.**

three sheets in the wind, be, be drunk. Also, **be four sheets in the wind.**

threesome, any set of three, especially a golf match in which one player, playing his own ball, competes against two others who together play one ball, each playing alternate strokes.

three-square file, an ordinary tapering handsaw file, triangular in cross-section. Also called **triangular file.**

Three Stooges, the, American comics, specializing in a violent form of slapstick, who became world famous in motion picture shorts during the thirties, forties, and fifties; they are still being shown on television. The original trio were Larry Fine, Moe Howard, and Jerry Howard; Jerry was subsequently replaced variously by Shemp Howard, Joe Besser, and Joe de Rita.

three-time loser, *U. S.* (formerly) a person automatically condemned to life imprisonment on conviction of three felony indictments.

three-toed sloth, a sloth that has three toes on each front and hind foot. See also **2: unau, 2: ai.**

three trees, a gallows, so called after the two upright beams and the one horizontal. Also called **three-legged mare.**

three-way switch, one of a pair of electrical switches, usually mounted at opposite ends of a stairway or hallway, by means of which a light or other electrical device can be turned on or off at either place.

"Three wise men of Gotham," from the old nursery rime:
Three wise men of Gotham,
They went to sea in a bowl,
And if the bowl had been stronger
My song had been longer.

thrice, three times, as, *Thrice is he armed that hath his quarrel just.* —Shak. *2 Henry VI,* III, ii, 233.

tierce, the third of eight parries or guards against thrusts in fencing. Compare **1:prime, 2:seconde, 4:quarte, 5:quinte, 6:sixte, 7:septime, 8:octave.**

tiercet, a triplet; three riming lines of poetry.

treble fitché, (in heraldry) terminating in three points, as the arms of a cross.

treble-tree, a combination of whiffle-trees for three horses.

trefoil, a three-leaved plant, ornament, or architectural accessory.

tri-, a prefix borrowed from Latin or Greek, meaning 'three.'

triactinal, *or* **triactine,** *or* **triact,** having three rays.

triad, **1.** a union or conjunction of three; a group or class of three closely associated persons or things.
2. (in Welsh literature) a form of composition in which the contents are in groups of three.

trialogue, a colloquy among three people.

triangle, a figure, plane or curved, with three angles.

Triangulum, an ancient constellation shaped like an equilateral triangle.

trianthous, (in botany) having three flowers.

triapsidal, having three apses, as most Greek churches.

3

triarchy, rule by three persons, or a government so organized.

Trias, (in German history) the old German empire, consisting of Austria, Prussia, and a group of smaller states.

Triassic, (in geology) the lowest of three major divisions of fossiliferous rocks which together make up the Mesozoic or Secondary series, namely: Triassic; Jurassic; and Cretaceous. It lies above the Permian and below the Jurassic.

triathlon, an intense endurance competition in which participants must complete an ocean swim, a bicycle ride, and a long-distance run.

triaxial, having three axes.

tribasic, (in chemistry) having three hydrogen atoms, as some acids.

tribrach, (in ancient prosody) a foot consisting of three short times or syllables.

tribrachial, a figure or utensil with three arms, especially a three-branched flint.

tribracteate, (in botany) having three bracts.

triceps, (in anatomy) a muscle having three heads.

tricerion, a candlestick with three lights, symbolizing the Triunity and used by Greek bishops blessing people. See also **2: dicerion.**

trichotomy, division into three parts, especially (in theology) division of human nature into body, (soma), soul (psyche), and spirit (pneuma).

trichroism, the property of some crystals of exhibiting different colors in three different directions when viewed by transmitted light. Compare **2: dichroism.**

trichromatic, characterized by three colors, especially having the three fundamental color sensations of red, green, and violet of the normal eye.

triclinium, (in ancient Rome) a diningroom with three couches, one at each of three sides of a rectangular table, the fourth side left open for access by servants.

tricolor, the red, white, and blue flag of France, representing the white of the Bourbons, and the blue and red of the city of Paris, adopted by Louis XVI at the Hôtel de Ville, July 17, 1789, as the national symbol.

triconsonantal, *or* **triconsonantic,** composed of or containing three consonants, as *habit.*

tricorn, **1.** having three horns or horn-like parts.
2. a hat with three points formed by turning the brim upward to the crown on three sides. Compare **2: bicorn.**

tricycle, a velocipede with three wheels, one in the front and two in the rear.

tridactyl, *or* **tridactyle,** *or* **tridactylous,** having three digits, either fingers or toes. Also, **tridigitate.**

tridimensional theory, (in psychology) Wundt's theory of the simple feelings (1896) according to which there are three dimensions or directions of the affective life, namely: pleasantness-unpleasantness; strain-relaxation; and excitement-depression.

triduum, a period of three days, especially (in Roman Catholicism) prayers for three days in preparation for keeping a saint's day.

triennial, 1. of or pertaining to a period of three years.
2. a commemoration or celebration of an event that occurred three years earlier; a third anniversary.
3. lasting for three years, as *a triennial plant*; occurring once every three years, as *a triennial celebration.*
4. a Mass performed daily for three years for the soul of a deceased person.

Triennial Act, (in England) an act of Parliament (1694) requiring that a new Parliament be summoned at least once every three years. It was superseded by the Septennial Act (1716).

triens, 1. (in ancient Rome) a coin worth a third of an as.
2. (in the Roman empire) a coin worth a third of a solidus.

trierarch, (in Greek antiquity) the commander of a trireme.

trifacial, pertaining to or involving the face in a threefold manner, as the trigeminal nerve.

trifid, divided into three parts; trichotomous. Compare **2: bifid.**

trifold, threefold; triple; triune.

trifoliate, *or* **trifoliated,** *or* **trifoliolate,** having three leaves or leaflets; trefoil.

triforium, (in medieval church architecture) a gallery, usually in the form of an arcade with three openings, above the arches of the nave and choir.

triformed, *or* **triform,** *or* **triformous,** 1. formed of three parts or of three divisions or lobes, as *the triformed wreath of laurel symbolizing England, Scotland, and Ireland.*
2. having three shapes or bodies or assuming three forms, as the triple Hecate. See also **3: Hecate.**

trifurcate, *or* **trifurcated,** divided into three forked parts. Compare **2: bifurcate.**

3

trigamist, a person who has three spouses simultaneously. Compare **2: bigamy, 2: deuterogamist.**

triglot, pertaining to, involving, or containing three languages, as *a triglot dictionary.*

triglyph, (in classical architecture) a structural member in the Doric frieze, usually repeated over each column and over the middle of each columniation, having two incised grooves framed with three fillets and with a semigroove on each side.

trigoneutic, (in entomology) having three broods a year.

trigonometry, the branch of mathematics dealing with plane and spherical triangles and other related figures.

trigrammatic, *or* **trigrammic,** consisting of three letters or of three sets of letters.

trigraph, a combination of letters representing a single sound, as *eau* in *plateau.*

trihedral, having three faces.

trilateral, having three sides.

trilemma, a dilemma involving three possibilities. Compare **2: horns of the dilemma, 4: tetralemma.**

trilingual, expressed in or involving languages, as *a trilingual person, a trilingual book.*

triliteral, 1. consisting of three letters, as a word or a syllable, as *set.*
2. consisting of three letters, as the morphology of the words of a language, as *KTB* 'writing' in Hamito-Semitic.

trilith, *or* **trilithon,** a massive prehistoric stone structure, resembling a doorway, consisting of two vertical stones with a third supported horizontally by the other two.

trilogy, 1. (originally) a written work consisting of three parts.
2. any collection of three works by the same author or by different authors dealing with the same theme.

trimaran, a light-displacement vessel having three parallel hulls, usually one main hull flanked by two lighter hulls which may be merely outriggers, used as stabilizers. [From *tri-* 'three' + *-maran* from Tamil *catamaran,* by false analysis.] Compare **2: catamaran.**

trimensual, occurring every three months.

trimester, a period of three months.

trimeter, (in prosody) a verse consisting of three measures, especially of three iambs, the usual verse form of the dialogue in ancient Greek drama.

trimorphism, the state or condition of having three distinct forms, as a plant or an insect.

Trimurti, the Hindu trinity consisting of Brahma, Siva, and Vishnu, viewed as an inseparable unity and symbolically represented as a body with three heads, Vishnu at the right, Siva at the left, and Brahma in the middle.

trinary, threefold; triple; ternary.

trine, (in heraldry) a group of three, especially three animals, in a bearing.

trine immersion, *or* **trine aspersion,** (in baptism) the immersion or sprinkling of a person three times.

Trinity, the, the union of three persons (the Father, the Son, and the Holy Ghost, or Holy Spirit) in one Godhead.

trinomial, *or* **trinominal,** consisting of three terms, as a scientific name.

trio, 1. a musical composition for three singers or musicians.
2. three singers or musicians who perform trios.

trionym, a name, especially a scientific name, consisting of three terms. Compare **1: mononym, 2: binomial, 2: dionym.**

triphthong, a combination of three vowels forming a single or compound sound, as *ieu* in *lieu, eau* in *beauty,* etc. Compare **2: diphthong.**

Tripitaka, (in Tibetan Buddhism) the complete collection of the Rules, Discourses, and Treatises that form a part of the Kagyur, namely: Abhidharma; Sutra; and Vinaya. Also called **the Threefold Collection.**

triple, threefold; consisting of three; treble.

Triple Alliance, 1. an alliance (1668) between England, Sweden, and the Netherlands designed to curb the aggressions of France.
2. an alliance (1717) between France, Great Britain, and the Netherlands directed chiefly against Spain. See also **4: Quadruple Alliance.**
3. an alliance (1882) between Germany, Austria-Hungary, and Italy designed to check Russia and France. Compare **3: League of the Three Emperors.**

triple bogey, (in golf) the holing of a ball in three strokes more than par for the hole. See also **1: bogey, 2: double bogey.**

triple-decker, 1. a ship with three decks above the waterline.
2. a sandwich made with three slices of bread and two layers of filling. Also called **double-decker sandwich.**
3. a bus or train carriage with two sets of seats, arranged in compartments one above the other.

3

triple double, (in basketball) the achievement of a player who, in a single game, compiles statistics in double figures in three areas: scoring, rebounds, and assists.

Triple Entente, an alliance (1907) among France, Russia, and Great Britain, considered an element in bringing about World War I.

triple play, (in baseball) a play in which three men are put out before the ball is returned to the pitcher for delivery, as by catching a batted fly and then successively touching two bases before runners can return to them, or by tagging, or both.

triplet, 1. one of three born at the same birth.
2. a set or combination of three of a kind.

tripod, a stand, table, support, or other object with three legs or feet. See also **2: bipod, 3: tripos.**

tripos, 1. a tripod.
2. a three-legged stool, especially (in Cambridge University) in reference to the seat assumed originally by the student, called "Mr. Tripos," who disputed with the "Father" in the Philosophy School on Ash Wednesday. See also **10: tripos.**

trirectangular, (of certain spherical triangles) having three right angles.

trireme, (in ancient Greece) a vessel with three banks of oars on each side.

trisacramentarian, a person who maintains that the three sacraments necessary to salvation are Baptism, the Eucharist, and Absolution.

trisect, to divide or cut into three parts, especially (in geometry) into three equal parts.

triskelion, *or* **triskele,** a figure composed of three branches radiating from a center, as the symbol of the Isle of Man, in which the branches are running legs. Compare **4: tetraskele.**

trispast, *or* **trispaston,** a machine using a tackle with three pulleys, used for lifting heavy weights.

trisplanchnic, of or pertaining to the sympathetic nervous system or the viscera of the three great cavities of the body, namely: the cranial; the thoracic; and the abdominal.

trisula, *or* **trisul,** (in Hindu mythology) the trident symbol of Siva.

trisyllabic, *or* **trisyllabical,** containing three syllables, as the word *syllable*.

tritagonist, (in ancient Greek drama) the third actor, whose part is usually that of the evil genius or as promoter of the sufferings of the protagonist. According to Aristotle, he was first introduced by Sophocles. See also **1: protagonist, 2: deuteragonist.**

tritheism, the doctrine that there are three Gods, especially that the Father, Son, and Holy Spirit are three distinct Gods.

tritium, an isotope of hydrogen, having a nucleus (called a triton) triple the mass of the nucleus of hydrogen. Also called **heavy hydrogen.** Compare **2: deuterium.**

tritocere, the third tine of a deer's antler in order of development, or the one developed after the third year.

tritubercular, having three cusps, as a molar tooth.

triumvirate, an office held by three men (*triumvirs* or *triumviri*) especially (in ancient Rome) one of several groups of joint magistrates chosen for various purposes, as for establishing colonies, revising the lists of knights, guarding against fires at night, etc. See also **1: First Triumvirate, 2: Second Triumvirate, 3: triumviri capitales.**

Triumvirate, the, **1.** (in 17th-century England) the Duke of Marlborough (controlling foreign affairs), the Duchess of Marlborough (controlling the court and queen), and Lord Godolphin (controlling council and parliament).
2. (of English poets) Gower, Chaucer, and Lydgate.
3. (of Italian poets) Dante, Boccaccio, and Petrarch.

triumviri capitales, (in ancient Rome) an important triumvirate, elected by the people, whose duty it was to enquire into capital crimes, arrest offenders, superintend prisons, and see to the execution of condemned persons.

trivalent, (in chemistry) having a valence of three.

triverbial, designating those days in the Roman calendar appointed to the praetor for deciding causes; dies fasti. [From the three characteristic words of the praetor's office, *do, dico, addico* 'I give, I speak, I award.']

trivet, a three-footed stand or tripod, especially such a device, made of iron, for resting a hot cooking vessel to avoid marring a surface.

Trivia, the Greek goddess Diana, so called because she had three faces: Luna in heaven; Diana on earth; and Hecate in hell.

trivium, (in schools of the Middle Ages) the first three liberal arts, namely: grammar; rhetoric; and logic. See also **4: quadrivium, 7: seven arts.**

troika, a team of three horses abreast, the carriage drawn by such a team, or the team and carriage together.

3

Tuesday, the third day of the week.

ugly man, *Thieves' slang.* of three persons involved in garroting, the one who actually commits the crime and whose escape is covered by the **forestall** and **backstall.** Also called **nastyman.**

Vegetabilia, the second of the three kingdoms of organized beings, including those that are true plants and not true animals (*Animalia*) or neither (*Primalia*).

vice-chancellor, (in England) one of three judges in the chancery division of the High Court of Justice holding a separate court, whose decisions are subject to appeal to the lords-justices of appeal and to the House of Lords, of which the lord chancellor is head.

War of the Third Coalition, a war (1805–07) between France and an alliance of European monarchies, in which Napoleon Bonaparte was the victor, at the decisive battle of Austerlitz. The result was the fall of the Holy Roman Empire.

Yggdrasil, (in Scandinavian mythology) the tree of life, knowledge, time, and space that supports the entire universe and that has three huge roots. One extends into Asgard, the dwelling of the gods; the second into Jotunheim, the dwelling of the giants; and the third, the deepest, into Niflheim, the realm of death, where the dragon Nidhug dwells in the spring Hvergelmir and where Uller, god of winter, lives during the summer. See also **3: Norn.**

3.1415927 . . .

pi, the ratio of the circumference of a circle to its diameter, the irrational number 3.1415927 . . . .

3½

brilliant, a size of type on a three-and-one-half-point body.

4

April, the fourth month of the modern calendar.

areostyle, (in architecture) intercolumniation of 4 or 5 diameters, in the classification of Vitruvius. See also **2: intercolumniation.**

be on all fours with, to be squarely in agreement with; to be in full conformity with.

best-ball foursome, (in golf) **1.** a match in which the best individual score is used as the team's score.
2. a match in which one player plays against the best score of two or three players making up the other side.

4

Big Four, the, 1. the men considered to be responsible for construction of the first transcontinental railroad in America, the Central Pacific and the Union Pacific railroads, namely: Charles Crocker, Collis P. Huntington, Mark Hopkins, and Leland Stanford.
2. Great Britain, France, the Soviet Union, and the United States, so designated after World War II because of their influence throughout the world.

bob-tail flush, (in poker) four cards of one suit and one of another, so called because there is a chance of filling the flush by drawing one card.

cardinal point, one of the four points of the horizon, being north, east, south, and west. See also **32: point of the compass.**

carré, (in roulette) a wager placed on a group of four numbers that form a square on the layout.

cater, (in cards or dice) the four.

coach and four, a coach and the four horses by which it is drawn.

D, d, 1. the fourth letter of the English alphabet and in the Latin, Greek, and Phoenician, as well.
2. the fourth item in a system where A = 1, B = 2, C = 3, etc.

daleth, the fourth letter of the Hebrew alphabet, corresponding to the Greek *delta* and the English *d.*

dearborn, a light four-wheeled country horse-drawn vehicle formerly used in the U. S.

decuman, one of the four gates of a Roman camp, the furthest from the enemy. The others were, clockwise, in plan, the porta principalis sinistra, the pretorian, and the porta principalis dextra. Compare **10: decuman.**

delta, the fourth letter of the Greek alphabet, corresponding to the English *d.*

Devil's four-poster, the (in cards) the four of clubs, thought to be unlucky.

diamond, a geometrical figure bounded by four equal straight lines or by four lines each opposing pair of which is of equal length, forming two acute and two obtuse angles.

diatessaron, 1. a harmony of the four Gospels.
2. (in old pharmacology) an electuary composed of four medicines: gentian; birthwort; bayberries; and myrrh.

dimension, magnitude, usually considered as described by length, breadth, and thickness, to which time has been added.

Dorian tetrachord, (in ancient Greek music) a descending tetrachord, in which two whole steps are followed by a half step.

4

double doubloon, a former gold coin of Spain and the Spanish-American states, equivalent in value to four pistoles.

double pair royal, (in cribbage) four cards of the same denomination.

droshky, *pl.* **droshkies. 1.** a kind of light, four-wheeled carriage, usually without a top, formerly used in eastern Europe and Russia.
2. a very low four-wheeled carriage of the cabriolet type.

Elegiac Poets, the four elegiac poets included in the Alexandrian Canon, namely: Callimachus; Callinus; Mimnermus; and Philetas.

elemental quality, *or* **first quality,** one of the four qualities, namely, hot and cold, moist and dry, which, according to Aristotle, distinguish the four elements (earth, water, air, and fire.) See also **5: quintessence.**

Eloquence, the Four Monarchs of, Demosthenes, the Greek; Cicero, the Roman; and Sadi and Zoroaster, the Persians. [Brewer, *Reader's Handbook,* 1889.]

Ember days, (in the Roman Catholic Church and those of the Anglican communion) the Wednesdays, Fridays, and Saturdays during four weeks of the year, coinciding with the seasons, that are set aside for prayer, fasting, and the ordination of the clergy, namely, the weeks following the Feast of the Exaltation of the Cross (September 14); the Feast of St. Lucy (December 13); the first Sunday in Lent; and Pentecost (Whitsunday). See also **3: Rogation days.**

erminé, (in heraldry) composed of four ermine spots, as a cross.

Evangelists, the, the ancient symbols described in Revelation as being associated with the four evangelists, namely: Matthew, a man's face (said to symbolize Christ's human genealogy); Mark, a lion (said to symbolize Christ's royalty); Luke, an ox or calf (said to symbolize Christ's sacrifice or sacerdotal office); and John, a flying eagle (said to symbolize the grace of the Holy Ghost).

evenly even, divisible by four with no remainder.

exon, (in England) one of the four officers of yeomen of the royal bodyguard. Also called **exempt.**

F, f, the fourth consonant of the English alphabet.

fa, (in solmization) the syllable used for the fourth tone of the scale, the subdominant.

faction, (in Roman antiquity) one of the four regular classes into which the charioteers of the circensian games were divided, distinguished by the colors of their dress, green, red, blue, and white, as representing spring, summer, autumn, and winter. In a dispute in Constantinople, in AD 532, between the green and blue factions and their partisans, the emperor Justinian favored the latter, leading to a civil war of five days, which cost 30,000 lives and nearly overthrew the government.

Fast of the Fourth Month, (in Judaism) a fast observed on the seventeenth day of Tammuz (the fourth month of the Jewish ecclesiastical calendar) for on that day Moses destroyed the tablets, the daily sacrifices ceased, and Apostemus burned the law and placed an idol in the sanctuary.

Fearsome Foursome, the, members of the defensive line of the Los Angeles Rams football team in the mid 1960s, namely: Roosevelt Grier; Deacon Jones; Lamar Lundy; and Merlin Olsen.

figure-of-four trap, a trap for catching wild animals the trigger of which is in the shape of the number 4.

flag, (in old thieves' slang) a groat; fourpence.

four, **1.** a four-oared boat, or its crew.
2. a playing-card with four pips or spots on it.
3. a domino with four spots on it.
4. a team of four horses harnessed together to draw a coach or other vehicle.
5. (in cricket) **(a)** a boundary. **(b)** Also called **fourer.** a ball, hit by a batsman, which reaches this boundary, resulting in a score of four points.
6. See **4: fourings.**

four-a-cat, *or* **four-o'cat,** *or* **four old cat,** a simplified form of baseball, formerly played by children, in which there are four batters and with four bases. Compare **1: one-a-cat, 2: two-a-cat, 3: three-a-cat.**

four-bagger, (in baseball) *Slang.* a home run.

four bells, (on a motorship telegraph) a signal calling for flank, or full speed from the engines.

four-boater, a whaling vessel capable of carrying four boats. Compare **5: five-boater.**

Four Branches of the Mabinogi, the, four related stories, written in the late eleventh century by an unknown author, that are considered the best in the Mabinogion, a collection of tales of Welsh bards.

four-color map problem, a problem of topology, originally set forth in the mid nineteenth century, requiring that four different colors be used to color a political map so that no two regions sharing a boundary at any point have the same color.

four corners, the limits of the contents of a document.

four corners of the earth, everywhere.

4

four corners rule, (in law) a finding that the intentions of the parties to a document must be determined from the entire instrument, not from separate parts of it.

Four Ends, terms set forth by President Woodrow Wilson in a speech, July 4, 1918, that further define objectives of a peace settlement with Germany originally promulgated in his Fourteen Points. See also **4: Four Principles, 5: Five Particulars, 14: Fourteen Points.**

foureye butterfly fish, a Caribbean fish, about 6 inches long, with a large, eye-like patch on each side near its tail, making it appear to be much larger and formidable to predators.

four-eyed fish, a South American fish, about 12 inches long, with eyes that are divided into an upper part for vision in the air and a lower part for vision under water.

four-eyes, *Derogatory.* a person who wears eyeglasses.

Four-F, *U. S.* **1.** (in the Selective Service System) unfit for active duty in the armed services.
2. a person so classified.

fourflusher, a cheat, braggart, pretender.

Four Freedoms, freedom of speech and expression, freedom of every individual to worship God in his own way, freedom from want, and freedom from fear, as promulgated by President Franklin D. Roosevelt in his State of the Union address to Congress on January 6, 1941.

four-handed, 1. done, played by, or suitable for four persons, as a piano piece, card game, etc.
2. See **4: quadrumanous.**

Four-H Club, *or* **4-H Club,** *U. S.* an association organized and sponsored by the Department of Agriculture for the instruction of young people in rural areas in modern techniques of farming, rearing livestock, etc.

four horsemen, (in tenpin bowling) after one ball has been played, the pins remaining upright, either the 1, 2, 4, and 7 pins, or the 1, 3, 6, and 10 pins.

Four Horsemen, the, (in football) the 1924 back field of Notre Dame, coached by Knute Rockne, so named by Grantland Rice, of the *New York Herald Tribune*, namely: Jim Crowley, halfback; Elmer Layden, fullback; Don Miller, halfback; and Harry Stuldreher, quarterback.

Four Horsemen of the Apocalypse, four horsemen riding white, red, black, and pale horses symbolizing pestilence, war, famine, and death, respectively. [From the Revelation of St. John, in the New Testament.]

fourings, *or* **four,** *Dialectal.* (in England) a meal eaten at 4 o'clock in the afternoon during harvest-time. Compare **11: elevenses.**

four-in-hand, 1. a vehicle drawn by four horses.
2. a team of four horses harnessed to a single vehicle or matched for the purpose.
3. a method of tying a necktie in which the longer end is wound twice over the shorter, passed up between the neck and the tie, and then brought down through the loop thus formed.
4. a necktie or scarf for tying in this way.

four-leaf clover, a leaf of clover with four instead of the usual three leaves, symbolizing good luck to its possessor.

four-letter word, an English word that is usually considered taboo for use in polite company and is often subject to censorship because it refers to sex, excrement, or to certain parts of the body.

Four Lings, the, the four convents at Lhasa, capital of Tibet, from which the regent was chosen who ruled during the Dalai Lama's minority, namely: Tsamo Ling (or Chomo Ling); Kundä Ling (or Kundeling); Tsemchong Ling (or Tsecholing); and Tengyeling (dissolved).

Four Marys, the, Mary, Queen of Scots, and her three attendants, namely: Mary Seton; Mary Beaton; and Mary Carmichael.

Four Masters of Anhwei, Chinese artists of the Ch'ing dynasty (17th century) known for their original styles of painting landscapes of Anhwei Province, namely: Sun I; Wang Chih-jui; Hung-jen; and Ch'a Shih-piao; Hsiao Yün-ts'ung and Mei Ch'ing are sometimes substituted for the first two.

Four Masters of the Yüan Dynasty, Chinese artists of differing styles who painted during the Yüan dynasty (1279–1368), emphasizing individual expression rather than visual appeal, and were highly respected during the Ming dynasty, namely: Huang Kung-wan; Wu Chen; Ni Tsan; and Wang Meng.

four-minute mile, a milestone of athletic performance long believed unattainable. The first to achieve a time of less than four minutes for a mile in competition was Roger Bannister, who ran the distance in 3 minutes, 59.4 seconds on May 6, 1954. The achievement is now quite common.

Four Noble Truths, (in Buddhism) the doctrines that all life is suffering, that the cause of suffering is ignorant desire, that this desire can be destroyed, and that the means to do it can be found by following the Eightfold Path. Compare **8: Noble Eightfold Path.**

four-o'clock, 1. an Australian bird, the friar-bird or leatherhead, so called from the sound of its cry.
2. a garden flower that grows on a vine, so called because its flowers

4

open in the afternoon.
3. See **4: fourings.**

four of a kind,   (in poker) a strong hand, consisting of four cards of the same denomination.

four-on,   (in printing) noting an arrangement of pages for presswork by which four copies in quadruplicate can be printed together on the same sheet in the same operation.

four on the floor,   an automobile with four forward gears and with a manual stick shift mounted on the floor.

four orders,   the four orders of the mendicant friars, namely: the Dominicans, or Black Friars; the Franciscans, or Gray Friars; the Carmelites, or White Friars; and the Augustinian or Austin Friars.

four-part,   (in music) composed or arranged for four voices or parts in harmony.

fourpence,   (in Great Britain) formerly, a small silver coin worth four pence. Also called **fourpenny bit, fourpenny piece, groat.**

fourpenny nail,   **1.** a nail that, before 1500, cost fourpence a hundred. **2.** (in modern use) a common nail $1\frac{1}{2}$ inches long, or certain fine nails $1\frac{3}{8}$ inches long.

four-point average,   the highest average in an academic grading system where the top grade given in any course is counted as 4 points.

four-poster,   a bed with a tall post at each of its four corners, usually supporting a tester and sometimes curtains.

four-pounder,   a cannon capable of firing a ball weighing four pounds.

Four-Power Pacific Treaty,   a ten-year agreement involving the United States, Great Britain, Japan, and France, in which each party was to respect the others' property rights in the Pacific Ocean and to confer in the event of conflict. It was signed on December 13, 1921, and ratified in 1923.

Four Principles, the,   terms set forth by President Woodrow Wilson in a speech, February 11, 1918, that further define objectives of a peace settlement with Germany, originally promulgated in his Fourteen Points. See also **4: Four Ends, 5: Five Particulars, 14: Fourteen Points.**

four questions,   the four questions traditionally asked by the youngest boy at the Jewish Passover Seder concerning the meaning of Seder customs, which are answered by readings from the *Haggadah,* the liturgy of the Seder service.

four seas,   the seas surrounding England, namely: the Western (including the Scotch and Irish seas); the Northern (North Sea); the Eastern (the German Ocean); and the Southern (the English Channel).

foursome, a group of four people acting together, as a team of two players on each side in golf.

four species, the, **1.** the four fundamental operations of arithmetic, namely: addition; subtraction; multiplication; and division.
2. (in logic) the four principal forms of argumentation, namely: syllogism; induction; enthymeme; and example. [Blundeville, 1599]

foursquare, **1.** noting a figure with four equal sides; square.
2. *Informal.* straightforward; honest; on the up-and-up.

four-striper, a captain in the navy.

four-tailed bandage, a cloth bandage having two strips at each end for tying over a prominent part, as the knee, elbow, chin, etc.

fourth, (in music) **1.** a tone four diatonic degrees above or below any given tone.
2. the interval between any tone and a tone four degrees distant from it.
3. the harmonic combination of two such tones.
4. (in a scale) the fourth tone from the bottom; the subdominant; solmizated *fa.*

fourth best, (in whist and bridge) the fourth card of any suit counting from the top.

fourth class mail, (in the U. S.) mail consisting of parcels or merchandise.

Fourth-day, Wednesday, the fourth day of the week, so called by the Society of Friends.

Fourth Estate, the, the three estates of the realm are the nobility, the clergy, and the common people, to which the press has been added as the fourth.

fourth position, (in ballet) **(a) open fourth position.** a position in which the feet are parallel and slightly apart, the heels of the feet being in line from front to back. **(b) crossed fourth position.** a position in which the feet are parallel and slightly apart, the toes of each foot parallel to the heel of the other.

fourth R, the, reasoning. See **3: three R's, the.**

Fourth Republic, the republic proclaimed in France in 1945, after liberation from the rule of Germany 1940–45, succeeded in 1958 by the Fifth Republic.

four-wheel drive, (in a motor vehicle) an arrangement by which the power can be transmitted to both axles, often at the option of the operator.

4

Four Wheeled Hussars, the, *Brit.* a nickname for the Royal Horse Artillery, namely, two wheels for the gun, and two on the detachable limber.

Franciade, (in the French revolutionary calendar) a four-year period between leap years, which were scheduled to occur one year earlier than provided for in the Gregorian calendar.

Glyconic, (in ancient prosody) a meter similar to a trochaic tetrapody catalectic, but differing from it by the substitution of a dactyl for the second trochee.

hand, a unit of four inches, used in measuring the height of a horse at the withers.

History of the Four Kings, the, *Slang.* a deck of cards.

Inauguration day, March 4th, formerly (until 1933) the day on which the President elect of the United States took the oath of office. See **20: Inauguration day.**

Independence day, the day on which the Congress of the North American colonies of Great Britain (afterward the United States) passed the Declaration of Independence, July 4, 1776, since celebrated as a national holiday in the U. S. and its possessions.

Ionic, (in ancient prosody) a foot consisting of four syllables, two long followed by two short, or vice versa.

Kepler-Poinsot solid, *or* **Kepler solid,** one of four regular geometric solids which enwrap their centers more than once, namely: great icosahedron (twenty faces); great dodecahedron (twelve faces); small stellated dodecahedron (twelve faces); and great stellated dodecahedron (twelve faces).

last things, the, (in some Christian denominations) the four last things to be remembered, namely: death; judgment; heaven; and hell.

Lesbian Poets, the, four poets of Lesbos, namely: Alcaeus; Arion; Sappho; and Terpander.

Limehouse Four, the, an appellation for the founding members of the Social Democratic Party in England, namely: Roy Jenkins; David Owen; William Rodgers; and Shirley Williams; so called because they met in the Limehouse district of London.

Lydian tetrachord, (in ancient Greek music) a descending tetrachord in which a half step is followed by two whole steps.

magpies, to see four magpies denotes good news or, according to another superstition, death. [Brewer, *Reader's Handbook*, 1889.] See also **5: magpies.**

4

Mid-Lent, the middle, or fourth Sunday in Lent.

Mills Brothers, the, a close-harmony singing group popular during the 1940s and 1950s, namely: Donald; Harry; Herbert; and John, Jr.

nibble, (in computer technology) a sequence of adjacent bits that is half the length of a byte, that is, usually four bits. It is treated as a unit for data manipulation in some computer systems. See also **2: bit, 8: byte.**

Noble Truths, (in Buddhism) four basic truths that constitute the enlightenment of the middle path, namely: the Noble Truth of Pain (or Suffering), which is made up of five groups of pain involved with ''clinging to existence''; the Noble Truth of the Cause of Pain, which involves various cravings; the Noble Truth of the Cessation of Pain, involving the cessation of those cravings; and the Noble Truth of the Path that Leads to the Cessation of Pain, which consists of the Noble Eightfold Path. See also **8: Noble Eightfold Path.**

O, o, the fourth vowel of the English alphabet.

Olympiad, a period of four years reckoned from one celebration of the Olympic games to the next, by which the Greeks computed time from 776 BC, the reputed year of the first Olympiad.

on all fours, on one's hands and knees.

parents, house of, (in astrology) the fourth house. See **12: house.**

perfect women, according to Muhammad, the four perfect women were: Asia, wife of Pharaoh, who reared Moses; Mary, daughter of Imran; Khadijah, Muhammad's first wife; and Fatima, Muhammad's daughter.

plus fours, wide knickerbockers popular as sports apparel during the early 1920s and particularly associated with golfers. They were cut four inches longer than ordinary knickerbockers to give more freedom of movement at the knees.

proceleusmatic, (in prosody) consisting of four short syllables, as a metrical foot.

pythiad, the period of four years between successive celebrations of the Pythian games.

Pythian games, one of the four great national festivals of Greece, celebrated once in four years in honor of Apollo at Delphi.

quadral, divided into four parts.

quadrangle, 1. a plane figure having four angles, especially a square or rectangle (with four right angles).
2. a square or rectangular courtyard surrounded or nearly surrounded by buildings.

quadrate, a square; a plane figure having four equal and parallel sides.

4

quadratoquadratic, (in algebra) noting an equation or expression of the fourth degree.

quadrennial, 1. of or pertaining to a period of four years.
2. a commemoration or celebration of an event that occurred four years earlier; a fourth anniversary.
3. lasting for four years; occurring once every four years.

quadrennium, *or* **quadrenniate,** a period of four years.

quadri-, *or* **quadra-,** a prefix borrowed from Latin, meaning 'four.'

quadriceps, a muscle having four heads, or origins, as the quadriceps extensor cruris of the thigh. Compare **2: biceps, 3: triceps.**

quadricorn, 1. having four horns; quadricornous.
2. an animal with four horns. Compare **1: monocerous, 2: bicorn, 3: tricorn, 4: quadricorn.**

quadricostate, having four ribs.

quadricrescentic, *or* **quadricrescentoid,** having four crescents.

quadricuspidate, having four cusps, as a tooth.

quadricycle, a four-wheeled vehicle similar to a bicycle or tricycle.

quadridigitate, having four fingers or finger-like parts; tetradactyl; quadrisulcate.

quadrifarious, set or arranged in four rows or series. Compare **1: unifarious, 2: bifarious, 3: trifarious.**

quadrifid, divided into four parts by deep clefts. Compare **2: bifid, 3: trifid.**

quadrifoliate, having four leaves or leaflets.

quadrifrons, having four faces. Compare **2: bifrons.**

quadriga, (in classical antiquity) a two-wheeled chariot drawn by four horses harnessed abreast, used in racing by the Greeks in the Olympian games and by the Romans in the circensian games.

quadrigemina, the fourfold bodies of the brains of mammals.

quadrigeminous, having four parts; fourfold; quadrigeminate; quadrigeminal.

quadrilaminate, *or* **quadrilaminar,** having four layers or plates.

quadrilateral, 1. having four sides; composed of four lines or verses.
2. a figure formed of four straight lines.
3. an enclosed space between and defended by four fortresses, as that in northern Italy defended by the fortresses of Peschiera, Mantua, Verona, and Legnago.

quadrilingual, involving four languages, as *a quadrilingual inscription, a quadrilingual translator.*

quadriliteral, 1. consisting of four letters.
2. a word or root of four letters, as *good*, or containing four consonants, as *habitat.*

quadrille, 1. a card game played by four people with forty cards, the tens, nines, and eights having been removed.
2. a square dance for four couples consisting of five parts, namely, le pantalon, l'été, la poule, la trénise (or la pastourelle), and la finale.
3. the music accompanying such a dance.

quadrimembral, having four limbs, as most vertebrates.

quadrinomial, (in algebra) 1. an expression consisting of four terms.
2. consisting of four terms. Also, **quadrinomical, quadrinominal.**

quadripara, a woman who is bearing a fourth child.

quadriparous, (in ornithology) laying exactly four eggs.

quadripartite, 1. divided into four parts.
2. referring to a unit consisting of or divided into four parts, especially a book or other written work divided into four parts, as the *Quadripartite of Ptolemy,* the *quadripartite (four Gospels)* of the New Testament; tetrabiblion.
3. having four participants.

quadripennate, (in entomology) having four wings, or a four-winged insect.

quadriphyllous, (in botany) having four leaves; quadrifoliate.

quadrireme, a galley with four banks of oars, as used by the ancient Greeks and Romans.

quadrisacramentarian, *or* **quadrisacramentalist,** one of a small body of German Protestants in the middle of the sixteenth century who held that the four sacraments of Baptism, the Eucharist, Holy Orders, and Absolution are required for salvation.

quadriseptate, having four septa, or partitions.

quadriserial, set or arranged in four rows or series; quadrifarious; tetrastichous.

quadrisulcate, having four grooves or furrows, especially (of mammals) having a four-parted hoof; four-toed; quadridigitate.

quadrisyllabic, 1. (of a word) consisting of four syllables.
2. a word of four syllables, as *interpreting.*

4

quadrivium, 1. the four branches of mathematics according to the Pythagoreans, namely, arithmetic (number in itself), music (applied number), geometry (stationary number), and astronomy (number in motion). See also **3: trivium, 7: seven liberal arts.**
2. a crossroads.

quadrumane, a quadrumanous animal.

quadrumanous, quadrimanous, having four hands; having the four feet equipped for use as hands, as opossums, primates, etc. Compare **4: quadruped.**

quadruped, 1. having four feet or legs.
2. a quadruped animal. Compare **2: biped, 4: quadrumanous.**

quadruple, 1. fourfold.
2. a number, sum, quantity, etc., that is four times an amount taken as standard, regular, or normal.
3. to increase fourfold.

quadruplet, one of four born at a single birth.

quadruplex, (of a system of telegraphy) designed so that four messages can be transmitted simultaneously over one wire.

quadruplicate, 1. to increase fourfold.
2. increased fourfold.

quarrel, 1. one of several small panes of glass, square or of diamond shape, used in glazing a window; a paving stone of similar shape.
2. a bolt or arrow having a square or four-edged head.

quart, (in card-playing) a sequence of four cards.

quartan, having to do with the fourth, especially occurring every fourth day, as quartan fever.

quartan fever, intermittent fever in which the paroxysm recurs every fourth day (both paroxysmal days being counted).

quarte, the fourth of eight parries or guards against thrusts in fencing. Compare **1: prime, 2: seconde, 3: tierce, 5: quinte, 6: sixte, 7: septime, 8: octave.**

quarterly, 1. occurring or appearing four times a year, as a periodical publication.
2. such a periodical.

quartet, *or* **quartette,** 1. a musical composition for four vocal or instrumental parts.
2. four singers or players who perform quartets.

quartic, (in mathmatics) of the fourth degree, especially of the fourth order. See also **2: quantic.**

quartic equation, (in mathematics) an equation of the fourth degree.

quarto, a book consisting of signatures, or formes, each of which is a sheet of paper folded into quarters to make four leaves, or eight pages.

quaternion, 1. a set or group of four.
2. a word of four syllables; a quadrisyllabic, as *quaternion.*

Quaternity, the union of four persons in one godhead. Compare **3: Trinity.**

quatrain, a group of four lines of verse, usually with alternate lines riming.

quatre, the number four, especially in dice, cards, or dominoes.

quatrefoil, 1. a leaf with four leaflets, as a four-leaf clover.
2. (in architecture) an opening or panel divided by cusps or foliations into four foils.

rectangle, a plane figure having four sides and four right angles, the opposite sides being parallel and of equal length. When the adjacent sides are equal, it is a square.

rivers of Eden, a river (unnamed) flowed out of Eden and divided into four rivers, namely: Pishon; Gihon; Tigris; and Euphrates. [Gen. 2.10-14.]

Scotch foursome, a golf match in which two pairs of players compete, with each pair playing one ball and the partners striking alternately. Compare **3: threesome.**

secondary, one of the qualities considered by Aristotle as derived from the four primary qualities, namely: hot; cold; wet; and dry.

semi-brevier, a size of type on a four-point body.

Serrano, the Spanish name for four groups of Uto-Aztecan-speaking Indians, originally inhabiting a mountainous region of southern California, who constituted a branch of the Shoshoni tribe, namely: the Alliklik; the Kitanemuk; the Serrano proper; and the Vanyume. By the early 1970s only a few Vanyume remained, and fewer than 400 Serrano proper.

Speech day, (in England) June 4th, the anniversary of the birthday of George III, celebrated at Eton College by various events.

Tebeth, the fourth month of the Jewish secular year. See also **10: Tebeth.**

tetra-, a prefix borrowed from Greek, meaning 'four.'

tetrachloride, a chloride which contains four atoms of chlorine.

tetrachord, 1. a diatonic series of four notes or tones, with the first and last separated by an interval of a perfect fourth.
2. a four-stringed musical instrument.

4

tetractomy, a division into four parts.

tetracycline, $C_{22}H_{24}N_2O_8$, an antibiotic derived from chlortetracycline and used medically to treat a broad spectrum of infections. Tetracycline can be produced by a soil actinomycete or synthetically.

tetrad, 1. the number four.
2. a collection of four things. Also called **quadrad.**

tetradactyl, *or* **tetradactyle,** *or* **tetradactylous.** having four fingers or toes, as the forefeet and hindfeet of some quadrupeds. Also called **quadridigitate.**

tetradrachm, a silver coin of ancient Greece having the value of four drachmas.

tetragammadion, *or* **tetragammation,** an ornament consisting of four Greek gammas arranged to form a cross, formerly frequent on certain vestments of Greek prelates and of the western church.

tetragon, 1. a plane geometric figure having four sides and four angles. Also called **quadrangle, quadrilateral.**
2. (in astrology) an aspect of two planets with regard to earth when they are one quarter of the zodiac, three signs, or 90° distant from each other. Also called **quartile aspect.**

tetragram, a word or root composed of four letters, as *four,* or Hebrew *YHVH.*

Tetragrammaton, the four Hebrew letters that make up the name of God. The four letters, *yod, he, vav,* and *he,* are usually transliterated *YHVH* and pronounced as *Adonai* or *Elohim* since the pronunciation of the original name is forbidden.

tetrahedron, a geometric solid with four plane faces; a triangular pyramid.

tetralemma, a dilemma involving four possibilities. Compare **2: horns of the dilemma, 3: trilemma.**

tetralogy, 1. a group of four dramatic compositions, three tragic and one satiric, which were staged in Athens for the prize at the festival of Bacchus, or Dionysus.
2. a group of four operatic works, treating of related themes, intended to be staged in connection.

tetrameter, a verse consisting of four measures, as, "Once upon a midnight dreary,/ as I pondered weak and weary." [Poe, *The Raven.*]

tetramorph, (in Christian art) the union of the four attributes of the evangelists in one figure, winged, and standing on winged fiery wheels, the wings covered with eyes.

4

tetrapharmacon, *or* **tetrapharmacum,** an ointment composed of wax, resin, lard, and pitch.

tetraphyllous, having four leaves or leaflets.

tetrapolis, a group or association of four towns.

Tetrapolitan, belonging to the tetrapolis composed of Constance, Lindau, Memmingen, and Strasbourg. The Tetrapolitan Confession was a confession of faith presented at the Diet of Augsburg in 1530 by representatives of these four cities, resembling the Augsburg Confession but somewhat inclined to Zwinglian views.

tetrapteran, having four wings; an insect so equipped; tetrapterous.

tetraptych, an altar-piece or other arrangement having four compartments or panels of pictures. Compare **2: diptych, 3: triptych.**

tetrarch, 1. (in the Roman empire) the ruler of the fourth part of a country or province in the east, or one of the rulers who share power jointly; a subordinate ruler; a viceroy.
2. the commander of a subdivision of a Greek phalanx.

tetrarchate, *or* **tetrarchy,** the district governed by a tetrarch.

tetraskele, *or* **tetrascele,** *or* **tetraskelion,** a figure composed of four branches radiating from a center; the swastika or fylfot. Compare **3: triskelion.**

tetrastigm, a plane figure formed by four points.

tetrastyle, (in ancient architecture) 1. having or consisting of four columns, as a portico.
2. a portico with four columns.

tetrasyllable, a word which contains four syllables, as *intelligent.*

tetratheism, (in theology) the doctrine that the Godhead consists of four elements, namely: the Divine Essence; the Father; the Son; and the Holy Spirit.

tetravalent, (in chemistry) having a valence of four. Also **quadrivalent.**

trochlear nerve, either of two small nerves involved with the muscles of the eye. Also called **fourth cranial nerve.**

Vesta, the fourth and brightest planetoid, discovered in 1807.

watch, (in nautical use) a period of time during which a part of a ship's crew is on duty, nominally four hours, traditionally marked by the ringing of the ship's bell at half-hour intervals, the number of rings indicating the half hour, from one to eight, just ended.

Wednesday, the fourth day of the week.

4½

diamond, a size of type on a four-and-one-half-point body.

fourpence-halfpenny, a small Spanish coin, formerly used in New England, worth 4½d. in old New England currency.

5

Adoptive Emperors, the collective name for five Roman emperors, namely: Nerva (AD 96–98); Trajan (98–117); Hadrian (117–138); Antonius Pius (138–161); and Marcus Aurelius (161–180), so called because after Nerva, who was elected on the death of Domitian (AD 96) each was the adopted son of his predecessor. According to Gibbon, the 85 years of their reigns were the happiest in the history of the world.

alauda, [lit. 'lark'] the Fifth Roman Legion from the time of Julius Caesar till the third century AD, so called because its soldiers wore helmets bearing the device of a lark.

Alkoremmi, a palace, built by Motassem, to which his son, Vathek, added five wings, one for the gratification of each of the senses: the eternal banquet, in which were tables covered with tempting foods; the nectar of the soul, filled with the best poets and musicians; the delight of the eyes, filled with the most enchanting objects; the palace of perfumes, always filled with sweet odors; and the retreat of joy, filled with seductive, beautiful houris. [*Vathek*, by William Beckford.]

basketball team, called "five" because of the number of players on a side.

bat-fives, a variation of fives, played with a wooden bat.

Big Five, the, 1. the nickname for the five major powers in World War II, namely: China; France; Great Britain; the Union of Soviet Socialist Republics; and the United States.
2. the basketball teams of five institutions in the area of Philadelphia, Pennsylvania, namely: La Salle College; the University of Pennsylvania; St. Joseph's College; Temple University; and Villanova University.

Big Five Packers, the, five companies, originally meat packers, found in 1919 to control butter and eggs, oleomargarine, cottonseed oil, fertilizer, perfumes, and leather novelties, namely: Armour; Cudahy; Morris; Swift; and Wilson. In 1920 the Supreme Court ordered them to discontinue all operations other than meat packing.

Buddha, (in Buddhism) 1. the five earthly forms of a Buddha, namely: Dharmakaya (abstract body); Dhyanibuddha (meditative body); Manusibuddha (mortal body); Nirmanakaya (ascetic); and Sambhogakaya (heavenly body).
2. the five celestial Buddhas, namely: Bhaisajyaguru (master of healing); Dipamkara (enlightener); Kasyapa (luminous protector); Maitreya (loving

5

one); and Sakyamuni (Gautama, the enlightened one). [From Jobes, *Dictionary of Mythology, Folklore and Symbols*.]

Charter Oath, a statement of principle, promulgated by the Meiji emperor of Japan, Mutsuhito, in 1868, and aimed at moving the country out of its feudal past and into a modern form of government. It contained five articles, namely: deliberative assemblies shall be established on an extensive scale, and all governmental matters shall be determined by public discussion; all classes, high and low, shall unite to carry out vigorously the plan of government; all classes shall be permitted to fulfill their just aspirations so that there will be no discontent; evil customs of the past shall be discontinued and new customs shall be based on just laws of nature; and knowledge shall be sought throughout the world in order to promote the welfare of the empire. Also called **Five Articles Oath.**

children, house of, (in astrology) the fifth house, a succedent house. See **12: house.**

cinquain, an order of battle for drawing up five battalions into three lines, namely, a van, main body, and reserve.

cinque, a set of five objects or five separate items treated as one.

cinquefoil, a design resembling a five-leafed clover, used in the Pointed style of architecture.

cinque-pace, an old French dance involving five steps.

Cinque Ports, originally, five ports or havens on the southern shore of England, namely, Dover, Hastings, Hythe, Romney, and Sandwich; Rye and Winchelsea were added later, together with some lesser places. They were considered so important in the defense of England against an invasion from France that they received royal grants of special privileges on provision of their providing a certain number of ships at their own expense in the event of war.

cinques, the changes that can be run on a chime of eleven bells, so called because five pairs of bells change places in the order of ringing every time a change is rung.

cinque-spotted, having five spots.

circular number, the number 5, so called because the last digit is always the same, regardless of the power considered: $5^2 = 25$; $5^3 = 125$; $5^4 = 625$, etc. Compare **6: circular number.**

Darnel's Case, a celebrated case, in 1627–28, that tested clause 39 of the Magna Carta, namely, that no man should lose his liberty without due process of law. It involved the refusal of an application for a writ of habeas corpus in behalf of Sir Thomas Darnel and four other knights, namely: Sir John Corbet; Sir Walter Earl; Sir Edmund Hampden; and Sir John Heveningham. Also called **the Five Knights' Case.**

5

defective fifth, (in music) an interval containing a semitone less than the perfect fifth.

dekass, a unit of mass; ten asses: in the grand duchy of Baden equal to 5 decigrams, or 7.7 grains troy.

Deuteronomy, the fifth (and last) book of the Pentateuch, consisting of three addresses purported to have beem made by Moses to Israel shortly before his death. See also **2: Deuteronomy.**

Dhyanibuddha, (in Buddhism) one of the five meditative Buddhas, namely: Aksobhya (unagitated); Amitabha (infinite light) or Amitayus (infinite life); Amoghasiddhi (infallible power); Ratnasambhava (precious birth); and Vairocana (brilliant light). See also **5: Buddha, 1.** [From Jobes, *Dictionary of Mythology, Folklore and Symbols.*]

diapente, (in pharmacology) a composition of five ingredients.

E, e, 1. the fifth letter of the English alphabet.
2. the fifth item in a series where A = 1, B = 2, C = 3, etc.

elegiac, (in prosody) a pentameter, or line of five syllables, consisting of two dactylic penthemims or written in elegiac meter.

Epic Poets, the five epic poets included in the Alexandrian Canon, namely: Antimachus; Hesiod; Homer; Panyasis; and Pisander.

epsilon, the fifth letter of the Greek alphabet, corresponding to the English *e.*

Fast of the Fifth Month, (in Judaism) a fast observed on the ninth day of Ab (the fifth month of the Jewish ecclesiastical calendar) for on that day, according to the Talmud, it was decreed that the children of Israel should not enter the Promised Land, and the destruction of the first and second temples occurred.

Fifth, the, *U. S. Informal.* the Fifth Amendment to the Constitution, which provides, essentially, that no one shall be required to testify against himself in a criminal proceeding and that no one shall be tried for a second time on a charge for which he was tried before.

Fifth-Amendment Communist, *U. S.* a defendant in an action brought against a person suspected of being a communist, especially by Senator Joseph E. McCarthy in the late 1950s, so called because such defendants often invoked the Fifth Amendment of the Constitution. See also **5: Fifth.**

Fifth Avenue, a street in New York City associated, for much of its length, with wealth. Expensive shops and department stores are situated between 49th and 59th streets; luxurious apartment houses can be found between Washington Square (its origin at the southern end) and 12th Street, then again between 60th and 96th Streets. Formerly, much of its length was occupied by private mansions.

fifth column, a subversive group, operating within a country to destroy its unity by means of sabotage and espionage, especially a group loyal to an enemy. The term was coined by General Gonzalo Queipo de Llano y Sierro, a Fascist revolutionary: as four of his army's columns closed on Madrid during the Spanish Civil War (1936–39), he referred to his sympathizers within Madrid as his "fifth column."

Fifth-day, Thursday, the fifth day of the week, so called by the Society of Friends.

fifth disease, erythema infectiosum, a children's disease of unknown origin that begins with a facial rash which spreads to other parts as earlier parts clear, taking about ten days to run its course. It disappears, leaves no ill effects, and requires no treatment.

Fifth Monarchy men, members of an extreme Puritan sect in England during the Commonwealth and Protectorate. who believed that the Second Coming of Christ was at hand and that it was the duty of Christians to be prepared to establish his reign by force and, in the meantime, to repudiate all allegiance to any other government. [After the identification of Christ as the leader of the fifth monarchy, the first four having been Assyrian, Persian, Greek, and Roman.]

fifth position, (in ballet) a position in which the legs are crossed with the feet together and pointing in opposite directions, left and right, with the heel of each foot touching the joint of the big toe of the other.

fifth quarter, the hide, fat, and other less valuable parts of a slaughtered sheep, ox, etc.

Fifth Republic, the republic proclaimed in France in 1958, succeeding the Fourth Republic.

fifth wheel, 1. *Informal.* a person who is considered as unwanted or unneeded; excess baggage.
2. a spare wheel carried on a four-wheeled vehicle.
3. a horizontal wheel-like bearing supporting the front wheels of a carriage while allowing the entire axle to pivot freely.

fin, *U. S. Slang.* a five-dollar bill.

fine as fivepence, *Informal.* very smartly or gaily dressed.

fingers, (in chiromancy) the thumb is associated with Venus, the index finger with Jove, the middle finger with Saturn, the ring finger with Sol, and the little finger with Mercury. [Ben Jonson, *The Alchemist* I,i.]

fipenny bit, *Informal.* **1.** fivepence.
2. (formerly, in Pennsylvania and several southern states) the Spanish half-real.

5

five, 1. a playing card with five pips on it. Also called, *Slang.* **five-spot.**
2. a throw of five on a pair of dice.
3. the number five stops the power of poisons and is dreaded by evil spirits. As the sum of the first of the equals and the first of the unequals (2 + 3), it is also the combination of good and evil in nature. [Pythagoras, *On the Pentad.*]

Five Agents, *or* **Five Elements,** (in Chinese philosophy) the primary elements, earth, fire, water, wood, and metal.

five-and-ten-cent store, a retail shop where, formerly, many items were for sale for between five and ten cents. Also called **five-and-dime store.** Compare **10: dime store.**

Five Articles and the Five Points, the, statements of the distinctive doctrines of the Arminians and Calvinists, respectively, the former promulgated in 1610 in opposition to the restrictive principles of the latter, which were sustained in 1619 and which are the following: particular predestination; limited atonement; natural inability; irresistible grace; and the perseverance of saints. Also called the **Quinquarticular controversy.**

Five Blessings, (in Chinese art) serenity, virtue, wealth, longevity, and easy death, symbolized as bats.

Five Bloods, the, the five principal families of Ireland: 1, O'Neils of Ulster; 2, O'Connors of Connaught; 3, O'Briens of Thomond; 4, O'Lachlans of Meath; and 5, McMurroughs of Leinster.

Five Blossoms, a pattern in which four stylized blossoms appear as a square with a fifth forming the third in a straight row with another two, a motif typical of carpets woven in Chinese Turkestan.

five-boater, a relatively large whaling vessel capable of carrying five boats. Compare **4: four-boater.**

five by five, *U. S. slang.* very obese, ostensibly five feet tall and five feet in width or circumference.

five-cant file, a file having one angle of 108° and two of 36° each, used to file M-toothed saws.

Five Civilized Nations, *or* **Five Civilized Tribes,** the Cherokee, Chickasaw, Choctaw, Creek, and Seminole Indians, collectively, who were settled in the Indian Territory, 1830–40.

five-dollar word, *Jocular.* a multisyllabic or unfamiliar word. Compare **10: ten-dollar word.**

Five Dynasties, (in Chinese history) a fifty-year period between the end of the T'ang dynasty (AD 907) and the beginning of the Sung dynasty (960), when five families vied for the rule of North China. Also called **Ten Kingdoms.**

"five farthings, . . ." an old English song sung when playing a game:
>   Oranges and lemons say the bells of St. Clement's,
>   You owe me five farthings say the bells of St. Martin's.
>   When will you pay me? Say the bells of Old Bailey,
>   When I grow rich say the bells of Shoreditch.
>   When will that be? Say the bells of Stepney.
>   I'm sure I don't know says the great bell at Bow.
>   Here comes a candle to light you to bed.
>   Here comes a chopper to chop off your head.

five-finger, any of several species of plants having quinate leaves.

fivefold, five times the number or quantity.

Five Freedoms of the Air, the rights that form the basic considerations in drawing up agreements between countries regarding air transport, namely: to fly across a foreign country without landing; to land for refueling or repair; to discharge in a foreign country passengers, mail, and cargo from the home country; to pick up in a foreign country passengers, mail, and cargo for transportation to the home country; and to transport passengers, mail, and cargo from one foreign country to another.

five-gaited, denoting a horse, used mainly for showing, that has been trained to perform, in addition to the walk, trot, and canter, the rack and slow-gait.

Five Great Kings, (in Tibetan Buddhism) five deified heroes worshiped as protectors, generally taken to be brothers from Mongolia who traveled to Tibet: Pe-har (king of the deeds); Brgya-byin (king of the mind); Mon-bu-pu-tra (king of the body); Shing-bya-can (king of virtue); and Drga-lha skyes-gcig-bu (king of speech).

Five Mile Act, an English law (1665) that forbade any clergyman who was not of the Anglican Church to remain less than five miles away from where he used to preach unless he had sworn an oath of nonresistance to the crown.

Five Nations, the, the five confederated American Indian tribes: Cayuga; Mohawk; Oneida; Onondaga; Seneca. Also called the **Iroquois Confederation.** See also **6: Six Nations.**

five o'clock shadow, the short stubble that grows on the face of a man who has shaved in the morning, and that is sometimes evident at about 5 o'clock in the evening. It was so named in an advertising campaign for Gem Blades in the 1930s and was adopted into the language.

Five Particulars, terms set forth by President Woodrow Wilson in a speech, September 27, 1918, that further define objectives of a peace settlement with Germany originally promulgated in his Fourteen Points. See also **4: Four Principles, 4: Four Ends, 14: Fourteen Points.**

5

Five Pecks of Rice, a popular rebellion led by Chang Lu towards the end of the Han dynasty (206 BC–AD 220) against the rulers in what is today Szechuan Province. It is named for Chang Ling, Chang Lu's grandfather, first patriarch of the Taoist church in China, who, as a faith healer in the early part of the second century AD, was paid five pecks of rice a year by his clients.

fivepence, 1. an English coin worth five pence.
2. formerly, an American five-cent piece, or nickel.

fivepenny morris, nine-men's morris played with five counters or pegs on each side.

fivepenny nail, 1. a nail that, before 1500, cost fivepence a hundred.
2. (in modern use) a nail 1¾ inches long.

five percenter, *U. S.* a person who accepts a commission of five percent for helping to obtain contracts with the government.

Five Points Gang, a gang of hoodlums in New York City in the late 1890s who attacked and terrified passers-by. They were named for the Five Points, a slum area on the Lower East Side where they lived and congregated.

Five-pound Act, a statute of the colony of New York (1759) that gave jurisdiction over civil cases involving sums not exceeding five pounds to justices of the peace and to other local magistrates. Compare **10: Ten-pound Act.**

"five-pound note," Edward Lear's poem, "The Owl and the Pussycat." (1871), begins:
> The Owl and the pussycat went to sea
> In a beautiful pea-green boat.
> They took some honey and plenty of money,
> Wrapped up in a five-pound note.

Five-Power Constitution, a system of government proposed for China in 1905 by Sun Yat-sen, providing for five branches of government, namely: legislative; executive; judicial; examination (to select civil servants); and control (to check on honesty and efficiency). It was put into effect in 1928 by the Kuomintang under Chiang Kai-shek, but the executive branch, or yüan, dominated the others and allowed them little independence. The system is still in effect in Taiwan.

fiver, *Slang.* a five-dollar bill or five-pound note.

fives, a game resembling tennis, played with the hands instead of racquets.

fives-court, the court where fives is played.

five social relations, (in Chinese Confucianism) the five basic relationships that ensure the well-being of a country, namely: between sovereign and subject; father and son; husband and wife; brother and brother; and friend and friend.

fivespot, *U. S. slang.* **1.** a five-dollar bill.
2. the five of any suit at cards.

five-star admiral, *U. S.* the highest level of the rank of admiral, called Fleet Admiral. It was created by Congress in 1944 and awarded to four heroes of World War II, namely: William F. Halsey; Ernest J. King; William D. Leahy; and Chester W. Nimitz.

five-star general, *U. S.* the highest level of the rank of general, called General of the Army or Air Force. It was created in 1866 by Congress and awarded to Ulysses S. Grant for his service in the Civil War, then again to five heroes of World War II, namely: Henry H. Arnold; Omar N. Bradley; Dwight D. Eisenhower; Douglas MacArthur; and George C. Marshall.

five-twenty bond, *or* **five-twenty,** a bond issued by the United States government in 1862, 1864, and 1865, redeemable after five years and payable in full in twenty years.

five wits, the, **1.** the five senses.
2. sometimes, as fancifully enumerated, common wit, estimation, fantasy, imagination, and memory.

five W's, (in journalism) the five basic questions, implicit in the mind of a reader or listener, that must be answered in a properly prepared news article, namely, Who, What, Where, When, and Why.

Five-Year Plan, any of a number of national economic and social governmental plans scheduled to effect changes in a country over a five-year period, especially those promulgated in the Soviet Union (under Josef Stalin, beginning in 1928), China, Hungary, or Malaysia.

Fort Dix Five, the, five of the military prisoners who rioted in June 1969 at the stockade at Fort Dix, New Jersey, and were brought before general courts-martial for destruction of government property, arson, and conspiracy to riot, namely: William Brakefield; Thomas Catlow; Carlos Rodrigues-Torres; Jeffrey Russell; and Terry Klug, who was acquitted.

Frankfurt Five, the, five composers of modern English music who studied with Iwan Knorr (1853–1916) at the Hoch Conservatorium in Frankfurt-am-Main, namely: Henry Balfour Gardiner (1877–1950); Percy A. Grainger (1882–1961); Norman O'Neill (1875–1934); Roger Quilter (1877–1953); and Cyril Scott (1879–1970). Also called **the Frankfurt Group**.

5

Fundamentalism, five points of, five doctrines set forth as the traditional affirmations of the Fundamentalist creed, namely: preservation of the Bible as free from error in the original manuscripts; the Virgin Birth; Atonement; the Resurrection; and the power of Christ to work miracles.

G, 1. the fifth consonant of the English alphabet.
2. in solmization, the fifth tone of the scale, called *sol.*

geilfine, one of the four groups of five persons into which the ancient Irish clans or families were organized. The others were: **deirbhfine, iarfine,** and **indfine.**

gobang, *or* **go-moku-narabe,** a Japanese board game using counters or beads in which the object is to get five counters in a row.

grandsire, (in change-ringing) changes on five bells. Also called **double,** because two pairs of bells change places.

hand, five things sold together, as oranges or herrings.

Harvard Classics, fifty books containing selections from the world's literature chosen by Charles W. Eliot, president emeritus of Harvard University and published by P. F. Collier & Son. It was called the "five-foot shelf" from an off-hand comment by Eliot, "All the books needed for a real education could be set on a shelf five feet long!"

he, the fifth letter of the Hebrew alphabet, corresponding to the English *e.*

heaven, (in the Ptolemaic system) one of the five spheres, namely: that containing the planets; that containing the fixed stars; the vibrating crystalline sphere; the prime mover, which makes the lower spheres move; and the empyrean sphere, where dwell the supreme deity and the angels.

Herukabuddha, (in Buddhism) one of the five manifestations of the Dhyanibuddhas, namely: Kamaheruka (of Amoghasiddhi); Padmaheruka (of Amitabha); Ratnaheruka (of Ratnasambhava); Vairocana; and Vajraheruka (of Aksobhya). See also **5: Dhyanibuddha.**

Intolerable Acts, the, five laws passed by the British Parliament in 1774, four of which were intended as punishment for the Boston Tea Party and to strengthen British authority in Massachusetts, which helped unite the thirteen colonies against Britain, namely: the Boston Port Bill, closing the port until such time as the colonists demonstrated proper respect for British authority; a law providing for any British officer or soldier arrested for murder to be tried in England; a law to change the Massachusetts charter, providing for the appointment of a Crown council, and prohibiting town meetings without the governor's permission; a law requiring that British soldiers be fed and quartered by the colonists; and the Quebec Act, extending the province of Quebec south to the Ohio River and granting Roman Catholics freedom of worship there. Also called **Coercive Acts.**

kissar, a five-stringed lyre of northern Africa and Abyssinia, similar in form to an instrument represented in the hands of captives on Assyrian bas-reliefs.

know how many (blue) beans make five, be reasonably intelligent; not be a complete idiot.

litany, a series of petitions offered to God, originally composed to be sung in Solemn Procession, and consisting of five parts, namely: invocations, addressed to the individual parts of the Trinity; deprecations, for deliverance from various evils; obsecrations, for mercy; intercessions, for others; versicles and prayers.

lustrum, a period of five years. [From the ceremonial purification of the Roman people, performed at the end of five years in classical times.]

magpies, to see five magpies denotes that company is coming. [Brewer, *Reader's Handbook*, 1889.] See also **1: magpie.**

Manusibuddha, (in Buddhism) one of the five Buddhas who are creators of the universe, namely: Avalokitesvara; Ratnapani; Samantabhadra; Vajrapani; and Visvapani. See also **5: Buddha, 1.** [From Jobes, *Dictionary of Mythology, Folklore and Symbols.*]

May, the fifth month of the year.

M. I. 5, (in Britain) a former name of the Security Service, its director codenamed K, that is responsible to the Home Office and deals with counterintelligence. Compare **6: M. I. 6.** See also **1: M. O. 1.**

nickel-and-dime (someone) to death, persist in doing business with (someone) for amounts of money that yield no profit; to pursue (someone) with trifling matters.

Passion Sunday, the fifth Sunday of Lent. See also **2: Passion Sunday.**

Passion Week, the fifth week in Lent, from Passion Sunday to Palm Sunday.

pearl, a size of type on a five-point body.

penta-, a prefix borrowed from Greek, meaning 'five.'

pentachord, a musical instrument with five strings.

pentacle, *or* **pentangle,** *or* **pentagram,** a five-pointed star, used in magical ceremonies and considered a defense against demons. Its construction depends on an abstruse proposition discovered in the Pythagorean school, and it seems to have been adopted as their seal.

pentact, *or* **pentactinal,** having five rays, arms, or branches.

pentad, **1.** the number five, in the abstract; a set of five of anything. **2.** a period of five consecutive years.

5

pentadactyl, *or* **pendactyl,** *or* **pentadactylous,** having five fingers or toes; quinquedigitate.

pentagon, a plane closed figure with five sides and five angles. Also called **pentalpha.**

Pentagon, the, an enormous pentagonal office building in Arlington, Virginia, near Washington, D. C., housing most of the U. S. Department of Defense.

pentahedron, a solid figure with five plane surfaces.

pentamerous, five-parted; five-jointed.

pentameter, (in modern prosody) a line of poetry containing five measures.

Pentapolis, a confederation of five cities, specifically the ancient Pentapolis of Cyrenaica, in northern Africa, which consisted of Appolonia; Arsinë; Berenice; Cyrene; and Ptolemais.

pentarchy, government by five persons.

pentasyllabic, consisting of five syllables, as the word *syllabication.*

Pentateuch, the first five books of the Old Testament, namely: Genesis; Exodus; Leviticus; Numbers; and Deuteronomy, considered as a connected group.

pentathlon, 1. **(a)** an athletic contest in which competitors are awarded points for their performances in five track and field events, namely (for women): 800-meter run; 100-meter hurdles; high jump; long jump; and shot put; (for men): 200-meter run; 1500-meter run; long jump; javelin throw; and discus throw. **(b)** (in ancient Greece) an athletic contest consisting of five events, namely: a footrace; long jump; javelin throw; discus throw; and wrestling.
2. (in modern Olympic games) Also called **modern pentathlon.** an athletic contest in which competitors are awarded points for their performances in five events, namely: fencing; horseback riding; pistol shooting; cross-country running; and swimming.

penteteric, 1. occurring once in five years or at five-year intervals.
2. occurring every fifth year, the years of two consecutive occurrences being both reckoned.

Philistines, the five chief cities of the Philistines, namely: Ashdod; Askalon; Ekron; Gath; and Gaza.

Philosophers, the five philosophers included in the Alexandrian Canon, namely: Æschines; Aristotle; Plato; Theophrastus; and Xenophon.

Platonic solid, one of the five regular solids which enwrap the center only once, namely: tetrahedron (four faces); cube (six faces); octahedron (eight faces); twenty-vertexed dodecahedron (twelve faces); and icosahedron (twenty faces).

plead the Fifth, *U. S. Informal.* to invoke the Fifth Amendment in response to a charge or a question in court. See **5: Fifth.**

Pluviôse, the fifth month of the French revolutionary calendar, from about January 20th to February 18th, inclusive.

prayer, (in some Christian denominations) the five principal kinds of prayer, namely: praise and adoration; thanksgiving; penitence; intercession; and petition.

Punjab, a region in India, now divided between India and Pakistan. [After Hindi *Panjāb*, Persian *panj* 'five' + *āb* 'water,' because the area has five rivers.]

quinarius, an ancient Roman coin, first issued in 177 BC, equivalent to five asses.

quinary, 1. divided into five parts or sets of five.
2. based on five, as a number system. Compare **2: binary system, 10: decimal system.**

quincunx, an arrangement of five objects in a square, with one at each corner and the fifth in the center.

quinquangular, having five angles; quinque-angled.

quinque-, a prefix borrowed from Latin, meaning 'five.'

quinquefarious, arranged in five rows, sets, or series; quinqueserial; pentastichous.

quinquefid, cleft into five segments.

quinquefoliate, *or* **quinquefoliated,** having five leaves or leaflets.

quinquelateral, five-sided.

quinqueliteral, consisting of five letters; a five-letter word, as *often*.

quinquennalis, (in Roman antiquity) public games celebrated every fifth year. In antiquity, both the first and last years of a cycle were reckoned; thus, the Olympian, Pythian, and Isthmian games, celebrated once in four years, were regarded as quinquennalia.

quinquenniad, a period of five years; a quinquennium.

quinquennial, 1. of or pertaining to a period of five years.
2. a commemoration or celebration of an event that occurred five years earlier; a fifth anniversary.

5

3. lasting for five years; occurring once every five years. See **5: quinquennalis.**

quinquepartite, divided into or consisting of five parts.

quinqueradiate, *or* **quinqueradial,** having five rays, as a starfish; pentameral.

quinquereme, an ancient galley having five banks of oars.

quinquesect, to cut into five parts.

quinqueseptate, having five septa, or partitions.

quinqueserial, arranged in five series, or rows.

quinquesyllabic, 1. having exactly five syllables, as a word.
2. a five-syllable word, as *interpretation.*

quinquevalent, (in chemistry) (of an element) having a valence of five.

quinquevir, (in Roman antiquity) one of five commissioners appointed from time to time under the republic to carry any measure into effect, as to provide relief in time of public distress, to direct the establishment of a colony, or to provide for the repair of fortifications.

quint, 1. a set or sequence of five.
2. (in music) a fifth.
3. the E string, or chanterelle, of a violin.

quintain, an object to be tilted at, consisting (in England) of an upright post on top of which was a horizontal bar turning on a pivot; to one end of this a sandbag was attached, to the other a board. It was a trial of skill to strike the board with a lance and pass on before the sandbag could swing round to strike the tilter on the back. [After Latin *quintana*, a street in a camp, between the fifth and sixth maniples, where it is supposed that martial exercises took place.]

quintan, occurring or recurring every fifth day, as a quintan fever.

quintan fever, intermittent fever in which the paroxysm recurs on the fifth day (both paroxysmal days being counted).

quinte, the fifth of eight parries or guards against thrusts in fencing. Compare **1: prime, 2: seconde, 3: tierce, 4: quarte, 6: sixte, 7: septime, 8:octave.**

quintessence, 1. the fifth essence or substance of which the heavenly bodies are composed, according to Aristotle (following Pythagorean doctrine), specifically, not composed of earth, water, air, or fire.
2. the essential part or essence of anything. See also **4: elemental quality.**

106

quintet, *or* **quintette,** (in music) **1.** a composition for five vocal or instrumental parts.
2. the singers or players of such a composition.

quintic, of the fifth degree.

quintic equation, (in mathematics) an equation of the fifth degree.

quintuple, 1. fivefold.
2. to increase fivefold.

quintuplet, 1. a set of five.
2. one of five children born at one birth.

quintuplicate, 1. to increase fivefold.
2. consisting of or relating to a set of five.

rivers of Hell, the five rivers of Hell were Cocytus (of weeping), Styx (of hate), Acheron (of grief), Phlegethon (of liquid fire), and Lethe (of oblivion). See also **8: Hell.**

Sans-culottides, (in the French revolutionary calendar) the five days (from September 17–21, inclusive) called the Festivals of Virtue, Genius, Labor, Opinion, and Rewards. See also **10: décade.**

slip me five, *Slang.* 'Shake hands.'

sol, (in solmization) the fifth tone of the scale.

St. Petersburg Five, the, *or* **Russian Five, the,** *or* **Five, the,** five Russian composers who, about 1875, united their efforts and created a national school of Russian music, namely: Mily A. Balakirev (1837–1910); Aleksandr P. Borodin (1833–87); César A. Cui (1835–1918); Modest P. Mussorgsky (1839–81); and Nikolay A. Rimsky-Korsakov (1844–1908). A newspaper article in 1867 nicknamed them *Moguchaya Kuchka* "The Mighty Handful." Compare **6: Six.**

Thursday, the fifth day of the week.

trench fever, a disease characterized by a rash, fever, weakness, and pains in the legs. Also called **quintana fever,** *Informal.* **five-day fever.**

trigeminal nerve, one of a pair of cranial nerves, so called because they connect to three areas of the brain. Also called **fifth cranial nerve.**

U, u, the fifth vowel of the English alphabet.

V, the Roman numeral for the number five.

Vancouver Five, the, five young political activists who were convicted of a series of bombings across Canada in 1982, namely: Julie Belmas; Gerry Hannah; Ann Hansen; Doug Stewart; and Brent Taylor. They bombed the Litton Industries plant and the Cheekye-Dunsmuir power plant because of environmental issues, and, because of the exploitation of

5

women, the Red Hot Video warehouse. Also called **Direct Action** and **Wimmin's Fire Brigade.**

"What this country needs is a good five-cent cigar," a remark attributed to Thomas Riley Marshall, vice president (1913–21) under Woodrow Wilson.

5½

agate, a size of type on a five-and-one-half-point body.

6

abducens nerve, either of a pair of nerves involved in lateral eye movements. Also called **sixth cranial nerve.**

apostle, (in Muhammadanism) one of the six apostles, namely: 1, Adam; 2, Noah; 3, Abraham; 4, Moses; 5, Jesus; and 6, Muhammad.

apostolic fathers, the six early teachers and expounders of Christianity who, next to the apostles, were the founders, leaders, and defenders of the Christian church and who were, at some part of their lives, contemporary with the apostles, namely: 1, Barnabas (AD *c.*70–100); 2, Clement of Rome (d. *c.* 100); 3, Hermas (*fl. c.* 100); 4, Ignatius (d. 107); 5, Papias (*fl. c.* 130); 6, Polycarp (d. 155).

at sixes and sevens, 1. at loose ends; in a confused and upset state; disordered.
2. at odds; in dispute or disagreement; inharmonious.

Augustae Historiae Scriptores, six Roman biographers who wrote the lives of the emperors from Hadrian to Numerianus (AD 117–284), namely: 1, Julius Capitolinus; 2, Vulcatius Gallicanus; 3, Ælius Lampridius; 4, Trebellius Pollio; 5, Ælius Spartianus; and 6, Flavius Vopiscus.

benzene ring, (in chemistry) the six-sided representation of the bonding structure of carbon atoms present in the molecules of aromatic organic compounds; a fundamental concept in organic chemistry.

circular number, the number 6, so called because the last digit is always the same, regardless of the power considered: $6^2 = 36$; $6^3 = 216$; $6^4 = 1296$, etc. Compare **5: circular number.**

Cities of Refuge, (in the Bible) the six cities to which those persons who had inadvertently slain a human being might flee for refuge. Those east of the river Jordan, established by Moses, were: 1, Bezer; 2, Gotan; and 3, Ramoth; those west of the Jordan, established by Joshua, were: 4, Hebron; 5, Kadesh; and 6, Shechem.

Creation, God took six days to create the world, namely: 1, (first day) light; 2, (second day) heaven and the waters; 3, (third day) dry land and vegetation; 4, (fourth day) the sun, the moon, and the stars; 5, (fifth day) living creatures; and 6, (sixth day) man. On the seventh day God rested. [Gen. 1,1-31; 2,1-3.]

deep six,   *U. S. Slang.* to get rid of (something); discard (something) (as by throwing it overboard). Also, **give (something) the deep six.**

F, f,   1. the sixth letter of the English alphabet.
2. the sixth item in a system where A = 1, B = 2, C = 3, etc.

fathom,   a measure of length equal to six feet, used chiefly in nautical contexts.

Friday,   the sixth day of the week.

H, h,   the sixth consonant of the English alphabet.

Harrisburg Six, the,   six political activists, accused in January 1971 of conspiring to kidnap Henry A. Kissinger and of plotting to blow up the heating systems of federal buildings in Washington, namely: **1**, Eqbal Ahmed; **2**, Philip F. Berrigan; **3**, Elizabeth McAlister; **4**, Neil McLaughlin; **5**, Anthony Scoblick; and **6**, Joseph Wenderoth.

Hebe,   the sixth planetoid, discovered in 1847.

hexa-,   a prefix occurring in words from Greek and in modern coinages, meaning 'six.'

hexace,   a summit of a polyhedron formed by the concurrence of six faces.

hexachord,   1. (in Greek music) a diatonic series of six tones.
2. an instrument having six strings.

hexactinal,   having six rays.

hexad,   1. the sum of six units; the number six; a series of six numbers; a set of six of anything.
2. a hexagon.

hexadactylous,   having six fingers on each hand or six toes on each foot.

hexadrachm,   an ancient Greek coin having the value of six drachmas.

hexaëmeron, *or* **hexahemeron,   1.** a term of six days.
2. a history of the six days of creation, as contained in the first chapter of Genesis.

hexafoil,   having six foils or lobes.

hexagon,   a plane figure having six sides and six angles.

hexagram,   1. a figure with six lines.
2. a six-pointed star formed by placing two equilateral triangles with their sides parallel and on opposite sides of the center.
3. this figure as a seal or symbol of the Pythagorean school.
4. this figure as a symbol of the Jewish religion (Magen David, or Star of

6

David) and of the State of Israel.

5. one of the 64 figures which form the basis of the *I Ching,* or *Book of Changes,* each of which is made up of a pattern of six parallel lines, some whole and some divided.

hexahedron, a solid having six faces, especially the regular hexahedron, or cube.

hexamerous, *or* **hexameral,** *or* **hexapartite,** divided into six parts; containing six parts.

hexameter, (in prosody) **1.** containing or consisting of six measures; having a length of six feet, or dipodies.
2. composed of six feet of which the first four are dactyls or spondees, the fifth ordinarily a dactyl, sometimes a spondee, and the last a spondee or trochee:
In the hexameter rises the fountain's silvery column
[Coleridge, tr. of Schiller's *Ovidian Elegiac Meter.*]

hexandrous, *or* **hexandrian,** (in botany) having six stamens.

hexane, a paraffin containing six atoms of carbon.

hexangular, having six angles.

hexapartite, having or consisting of six parts; hexamerous.

hexapetalous, (in botany) having six petals.

hexaphyllous, (in botany) having six leaves.

Hexapla, an edition of the Bible in six versions, especially a collection of texts of the Old Testament collated by Origen, which contained in six parallel columns the Hebrew text in Hebrew characters and in Greek characters, the Septuagint with critical emendations, and versions by Symmachus, Aquila, and Theodotion, along with fragments of other versions.

hexaplar, *or* **hexaplaric,** **1.** sextuple; containing six columns.
2. (*cap.*) of or pertaining to the Hexapla.

hexapod, *or* **hexapodous,** *or* **hexapodan,** **1.** having six feet, as any adult insect.
2. a six-footed insect.

hexapody, (in prosody) a group, series, or verse consisting of six feet.

hexapsalmus, *or* **hexapsalmos,** (in the Greek Church) a group of six invariable psalms (Ps. iii, xxxviii, lxiii, lxxxviii, ciii, cxliii, according to the numbering in English Bibles) said daily at lauds (orthron), in the earlier part of that office.

hexapterous, having six wings or winglike parts.

hexastemonous, (in botany) having six stamens.

hexaster, (of a sponge) a star or stellate spicule having six generally equal rays.

hexastich, *or* **hexastichon,** (in prosody) a strophe, stanza, or poem consisting of six lines.

hexastichous, (in botany) having the parts arranged in six vertical rows or ranks.

hexastyle, *or* **hexastylar,** (in architecture)
1. (of a portico) having six columns in front.
2. such a portico.

hexasyllabic, *or* **sexisyllabic,** containing or consisting of six syllables, as the word *irreconcilable.*

Hexateuch, the first six books of the Old Testament, namely: 1, Genesis; 2, Exodus; 3, Leviticus; 4, Numbers; 5, Deuteronomy; and 6, Joshua.

hexatomic, (in chemistry) consisting of six atoms.

hexavalent, (in chemistry) having a valence of six.

Holy Week, the sixth week of Lent, from Palm Sunday through Holy Saturday; it includes Maundy Thursday and Good Friday.

Hyksos, the, the six Shepherd-Kings [literal translation of *hyksos*], driven from Assyria by Aralius and the Shemites, namely: 1, Saites, or Salates; 2, Beon; 3, Apachnas; 4, Apophis; 5, Janias; and 6, Asseth.

"I've got sixpence, . . . ," a traditional English song, adapted in modern times as an Australian Air Force song (1940s):
> I've got sixpence, jolly, jolly sixpence.
> I've got sixpence to last me all my life:
> I've got twopence to spend, and twopence to lend,
> And twopence to send home to my wife. (Poor wife!)
> No cares have I to grieve me,
> No pretty little girls to deceive me
> I'm happy as a lark—Believe me,
> As I go rolling, rolling home.
> Rolling home! (Rolling home!)
> Rolling home! (Rolling home!)
> By the light of the silvery moon . . .
> Happy is the day that an airman gets his pay,
> As he goes rolling, rolling home!

la, (in solmization) the syllable used for the sixth note of the scale.

lexiarchus, one of a board of six officers at Athens who attended to the registration of citizens, assigned the young men to their place on the list of the deme, and were stationed at the entrance to the Pnyx to prevent the intrusion of those who were not citizens.

6

long six, candles, about eight inches long, that weigh six to the pound. Compare **6: short six.**

Lucifera, daughter of Pluto and Proserpine whose chariot was drawn by six beasts, each ridden by one of the queen's councilors, namely: 1, an ass, ridden by Idleness; 2, a swine, by Gluttony; 3, a goat, by Lechery; 4, a camel, by Avarice; 5, a wolf, by Envy; and 6, a lion, by Wrath. Satan was the charioteer, and the footman was Vanity. [Spenser, *Faerie Queen* (1590).]

Magi, six kings, according to Klopstock, in *The Messiah*, who brought gifts to the infant Jesus, guided by the star of Bethlehem, namely: 1, Beled; 2, Hadad; 3, Mirja; 4, Selima; 5, Sunith, and 6, Zimri. Compare **3: Magi.**

M. I. 6, (in Britain) a former name of the Secret Intelligence Service (SIS), its director codenamed C; it was the successor to M. I. 1, established at the end of World War I, and it deals with secret service and security. Compare **5: M. I. 5.** See also **1: M. O. 1.**

nonpareil, a size of type on a six-point body.

octagon loop, the mesh of pillow lace, as the ground of Brussels lace: the term is a misnomer, the mesh being hexagonal.

palle, six balls (five red and one white, with a bearing on it) forming the cognizance of the family of the Medici.

Palm Sunday, the sixth Sunday in Lent.

Poseideon, the sixth month of the ancient Athenian year, from about December 15th to about January 15th.

poyou, the six-banded armadillo.

rhombohedron, a regular solid having six plane faces each of which is a rhomb.

roseola infantum, a disease of infants and young children, characterized by sustained or intermittent high fever, sore throat, enlargement of the lymph nodes, and, in some cases, convulsions, with a later development of a rash. Also called **exanthem subitum, sixth disease, Zahorsky's disease.**

San Quentin Six, the, six men killed during an attempted escape from the California State Prison at San Quentin in 1971, namely: 1, George Jackson, one of the Soledad brothers, who led the escape and was killed by a guard. After order was restored, five other men were found with their throats slashed: two white prisoners, 2, John Lynne and 3, Ronald Kane; and three guards, 4, Frank DeLeon; 5, Gere Graham; and 6, Paul Krasnes. See also **3: Soledad Brothers.**

semidiurnal, accomplished in or lasting half a day (six hours). Compare **12: semidiurnal.**

senocular, having six eyes, as some spiders.

sestet, 1. the last six lines of a sonnet.
2. See **6: sextet.**

sestiad, one of six parts or divisions, especially as applied by Chapman to his divisions of Marlowe's (unfinished) narrative poem, "Hero and Leander," including four parts added by himself.

sestina, *or* **sestine,** a poem of fixed form, consisting originally of six stanzas of six unrimed lines with a final half-stanza (triplet) also unrimed, all the lines being of the same length. It was borrowed from the French and was said to have been invented by the Provençal troubadour Arnaut Daniel, in the thirteenth century. Sestinas were written by Dante, Petrarch, Cervantes, Camoëns, Swinburne, and others.

sex-, *or* **sexi-,** a prefix borrowed from Latin, meaning 'six.'

sexangle, a hexagon.

sexdigitate, having six fingers on one or each hand or six toes on one or each foot.

sexennial, *or* **sextennial,** 1. of or pertaining to a period of six years.
2. a commemoration or celebration of an event that occurred six years earlier; a sixth anniversary.
3. lasting for six years.
4. occurring once every six years.

sexivalent, (in chemistry) having a valence of six.

sext, *or* **sexte,** (in the Roman Catholic and Greek churches, in religious houses, and as a devotional office in the Anglican communion) the office of the sixth hour, originally and properly said at midday.

sextain, a six-line stanza.

sextan, recurring every sixth day, as a fever.

sextan fever, intermittent fever in which the paroxysm recurs on the sixth day (both paroxysmal days being counted).

sextet, *or* **sextette,** *or* **sestetto,** 1. a musical composition for six instruments or voices.
2. a group of six musicians or singers who perform sextets.

sexto, a book in which the signatures are formed by folding each sheet into six leaves; half a duodecimo. See also **12: duodecimo.**

sextumvirate, a group of six men holding the same office; the office held by such a group.

sextuple, 1. sixfold.
2. to multiply by six.

6

sextuplet, one of six offspring born at the same birth.

sextuplicate, in six copies.

short six, candles, four to five inches long, that weigh six to the pound. Compare **6: long six.**

sickness, house of, (in astrology) the sixth house, a cadent house. See **12: house.**

"Sing a song of sixpence," an old nursery rime. See **24: "Four and twenty blackbirds."**

six, the whole number coming after five and before seven.

Six, les a group of musicians and composers who gave new stimulus to French music in the 1920s by breaking away from the Romanticism of Wagner and Strauss and the Impressionism of DeBussy, namely: **1,** Darius Milhaud (1892–1974), **2,** Georges Auric (1899–1983), **3,** Arthur Honegger (1892–1955), **4,** Francis Poulenc (1899–1963), **5,** Louis Durey (1888–1979), and **6,** Germaine Tailleferre (1892–1983). The group started under the name *Les nouveaux jeunes,* about 1918, and thrived under the promotion of Jean Cocteau (1889–1963), its mentor, who wrote and choreographed *Les mariés de la tour Eiffel,* in 1921. [They were so named by Henri Collet, the French critic, in an article, "The Russian Five, The French Six, and Erik Satie" (1920).] Compare **5: St. Petersburg Five.**

Six Acts, the, English statutes of 1819 that restricted the rights of public assembly and military organization and of the freedom of the press.

sixain a stanza consisting of six lines, or verses.

Six Articles, the, the six articles imposed in England by a statute of 1539, during the reign of Henry VIII, that decreed: **1,** the acknowledgment of transubstantiation; **2,** the sufficiency of communion in one kind; **3,** the obligation of vows of chastity; **4,** the propriety of private masses; **5,** celibacy of the clergy; and **6,** auricular confession. The act was relaxed in 1544 and repealed by Parliament in 1549. Also called **whip with six strings.**

Six Companies, the, six great organizations of Chinese merchants in San Francisco who at one time controlled the immigration of Chinese into the United States and the immigrants, as well.

Six Counties, (in Ireland) the six counties which make up Northern Ireland, namely: **1,** Armagh; **2,** Antrim; **3,** Down; **4,** Derry; **5,** Fermanagh; and **6,** Tyrone.

six-day bicycle race, a sporting event popular in the U. S. from 1891, when the first race was held at Madison Square Garden, in New York City, till the late 1930s.

6

Six-Day War, a war that broke out on June 5th, 1967, between Israel and the Arab states of Iraq, Jordan, Syria, and the United Arab Republic. It ended with Israel victorious on June 10th. Also called **the Arab-Israeli War.**

Six Dynasties, the Chinese dynasties between the Han and the Sui, namely: 1, Wu (AD 222–280); 2, Eastern Chin (317–419); 3, Liu-Sung (420–79); 4, Southern Ch'i (479–502); 5, Liang (502–57); and 6, Ch'en (557–89). It was a period that saw the introduction of tea, gunpowder, the wheelbarrow, the kite, and the use of coal for fuel.

sixfold, six times as much or as many.

Six Idlers of the Bamboo Streams, the, six Chinese poets of the eighth century, namely: 1, Chang Shuming; 2, Han Chun; 3, King Chao; 4, Pei Chêng; 5, Tao Mien; and 6, Li Po, the last of whom is considered by many to be the greatest poet of China. See also **8: Eight Immortals of the Wine Cup.**

Six Islands, the, before the Saxon period, "Great Brittany" was said to consist of six islands: 1, Dacia (Denmark); 2, Gothland; 3, Iceland; 4, Ireland; 5, Norway; and 6, the Orkneys.

Six Months' War, the war in which Prussia defeated France (July 28, 1870–January 28, 1871).

Six Nations, the, the five confederated American Indian tribes with the addition of the Tuscarora. See **5: Five Nations.**

six of one, half a dozen of the other, (used in reference to a choice between two alternatives) equivalent; virtually equal; of no important or discernible difference.

six-pack, a package containing six cans or bottles of beer, a soft drink, or, occasionally, some other product.

sixpence, an obsolete English silver coin worth six pence, or half a shilling. After decimalization (1971), it was equivalent to five new pence.

sixpenny, worth or costing sixpence; cheap; worthless.

sixpenny nail, 1. a nail that, before 1500, cost sixpence a hundred. 2. (in modern use) a nail 2 inches long.

Six Schools of Hindu Philosophy, namely: 1, Mimamsa; 2, Nyaya; 3, Samkhya; 4, Vaisesika; 5, Vedanta; and 6, Yoga.

six-shooter, a pistol for firing six bullets in succession, without reloading; usually a revolver with six chambers; a six-gun.

sixte, the sixth of eight parries or guards against thrusts in fencing. Compare **1: prime, 2: seconde, 3: tierce, 4: quarte, 5: quinte, 7: septime, 8: octave.**

6

Sixth-day, Friday, the sixth day of the week, so called among the Society of Friends.

sixth hour, the sixth of twelve hours reckoned from sunrise to sunset; noon; the canonical hour of sext.

six-year molar, the first permanent molar tooth, which appears normally at about the age of six.

St. Nicholas day, December 6th, the day observed in honor of St. Nicholas, patron saint of sailors, merchants, travelers, and captives, and revered especially by the Dutch (under the name of Santa Claus, made familiar in America by the Dutch settlers) as the guardian of children.

Troy, the ancient city of Troy had six gates, namely: 1, Antenorides; 2, Dardan; 3, Ilia; 4, Scaea; 5, Thymbria; and 6, Trojan.

vav, the sixth letter of the Hebrew alphabet, corresponding to the English *v., w.*

Ventôse, the sixth month of the French revolutionary calendar, beginning about February 19th and ending about March 20th.

Vestal Virgins, (in classical mythology) six priestesses who tend the sacred fire in the temple of Vesta (Greek Hestia), which was linked to the safety of the city: if it went out, the priestesses were punished severely, and the fire could be relit only by the rays of the sun.

zeta, the sixth letter of the Greek alphabet, corresponding to English *z.*

7

alchemy, the seven bodies in alchemy, namely: 1, the sun (gold); 2, the moon (silver); 3, Jupiter (tin); 4, Mars (iron); 5, Mercury (quicksilver); 6, Saturn (lead); and 7, Venus (copper).

Associated Counties, the seven English counties that combined in 1642–46 to join Parliament's side in the English Civil War and to save their territory from invasion, namely: 1, Cambridge; 2, Essex; 3, Hertford; 4, Huntingdon; 5, Lincoln; 6, Norfolk; and 7, Sussex.

B, the seventh musical tone of the scale of C in the diatonic scale.

Barbarossa, said to change his position in sleep every seven years. [Brewer, *Reader's Hand-Book,* 1889.]

candelabrum, 1. (in Christian tradition) a seven-branched candelabrum symbolizing the seven gifts of the Holy Ghost. See **7: seven gifts of the Holy Ghost.**
2. See **7: menorah.**

Canonical Hours, (in Christianity) there are seven hours of prayer offered daily in fulfillment of the Scriptures, "Seven times a day will I praise thee," [Ps. 119:164], namely: 1, matins and lauds, which together form the Night Office; 2, prime, for the first hour of the day, or early morning; 3, terce, for the third hour of the day, or midmorning; 4, sext, for the sixth hour, or midday; 5, none, the ninth hour, or midafternoon; 6, vespers, for sunset; and 7, compline, at the end of the day to complete the round of prayer and praise. While these Offices are still said or sung in many religious houses, the Prayer Book has combined matins and lauds with elements of prime to form the Office of Matins, or Morning Prayer, and vespers and compline to form that of Evensong, or Evening Prayer. See also **3: Apostolic Hours of Prayer.**

Champions of Christendom, the, seven Champions, namely: 1, St. George (England); 2, St. Andrew (Scotland); 3, St. Patrick (Ireland); 4, St. David (Wales); 5, St. Denys (France); 6, St. James (Spain); and 7, St. Anthony (Italy).

Chapter 7, a provision of the U. S. federal Bankruptcy Act for the relief of insolvent debtors and of their creditors in which, essentially, the debtor is relieved of the obligation to pay creditors, who receive the proceeds realized from the sale of assets. Compare **11: Chapter 11, 13: Chapter 13.**

Charlemagne, said to "start in his chair from sleep" every seven years. [Brewer, *Reader's Hand-Book,* 1889.]

Chicago Seven, the, seven defendants who were convicted (1970) of conspiracy to incite a riot and, subsequently, of contempt of court and on various other counts in connection with disturbances at the Democratic National Convention (Chicago, 1968), namely: 1, Rennie Davis; 2, David T. Dellinger; 3, John Froines; 4, Thomas Hayden; 5, Abbie Hoffman; 6, Jerry C. Rubin; and 7, Lee Weiner. An eighth defendant, Bobby Seale, was tried separately. Originally called **the Chicago Eight.**

Christ on the cross, Christ spoke seven times while on the cross: 1, "Father, forgive them; for they know not what they do"; 2, "Today shalt thou be with Me in paradise"; 3, "Woman, behold thy son!"; 4,"My God, My God, why hast Thou forsaken Me?"; 5, "I thirst"; 6, "It is finished"; and 7, "Father, into Thy hands I commend My spirit."

church year, (in the Anglican Communion) there are seven seasons in the church year, namely: 1, Advent; 2, Christmas; 3, Epiphany; 4, Lent; 5, Easter; 6, Ascension; and 7, Pentecost. Formerly, the last season was Trinity, but now Trinity Sunday is part of the season of Pentecost.

Cibola, Seven Golden Cities of, a collection of seven towns, a pueblo of the Zuñi Indians in what is now Zuñi, New Mexico, fabled by Fray Marcos de Niza to be a source of great riches, which prompted Francisco

7

Vásquez de Coronado, in 1540, to head an expedition of 1300 men to conquer them. No riches were found.

commit the seventh, break the seventh commandment; commit adultery.

corporal works of mercy, (in some Christian denominations) acts of charity that render physical aid, namely: 1, to feed the hungry; 2, to give drink to the thirsty; 3, to clothe the naked; 4, to harbor the homeless; 5, to visit the sick, widows, and orphans; 6, to visit the imprisoned; and 7, to bury the dead. See also **7: spiritual works of mercy.**

Dance of the Seven Veils, a sensuous dance, performed by Salome in Oscar Wilde's play of that name (refused a performance license in 1893) and in Richard Strauss's opera *Salome* (1905), to entertain Herod before the beheading of John the Baptist. There is no evidence of a "dance of the seven veils" in the Bible, which does not mention Salome by name.

day, any one of the seven days of the week.

day of rest, the seventh day, when God rested, after having spent six days creating the world. See **6: Creation.**

death-watch beetle, a species of beetle that abounds in old houses, where they bore into the wood and make a clicking sound by standing up on their hind legs and knocking their heads against the wood quickly and forcibly, the number of distinct strokes being in general from seven to eleven, as a mating call. They are so called because they are supposed by superstitious persons to be ominous of death.

die, *pl.* **dice.** the sum of the values of the marks on opposite sides of a die is seven, as 1 and 6, 2 and 5, 3 and 4.

doctor, *Brit.* a nickname given to the seventh child of a seventh child, regarded as lucky.

dysis, (in astrology) the seventh house of the heavens, which relates to love, litigation, etc.

eta, the seventh letter of the Greek alphabet, corresponding to English *e*.

facial nerve, either of a pair of nerves involved in the muscles of the face, scalp, etc. Also called **seventh cranial nerve.**

Fatima, the seventh (and last) wife of Bluebeard, spared from the death meted out to her predecessors by the arrival of her brothers, who kill her husband.

Fundamentalism, seven points of, seven doctrines set forth as affirmations of the Fundamentalist Christian faith. Those promulgated at the Niagara Conference, in 1878, were: 1, the preservation of the Bible as free from error in the original manuscripts; 2, the Trinity; 3, the total depravity of man; 4, the necessity of a new birth for salvation; 5, substitutionary atonement; 6, assurance of salvation to the believer; and 7, the premillennial Second Coming of Christ. See also **5: Fundamentalism, five points of.**

G, g, 1. seventh letter of the English alphabet.
2. the seventh item in a system where A = 1, B = 2, C = 3, etc.

Gamelion, the seventh month of the Attic year, corresponding to the latter part of January and the first part of February.

Germinal, the seventh month of the French revolutionary calendar, commencing March 21st and ending April 19th.

Group of Seven, 1. a committee of seven finance ministers, representing Great Britain, West Germany, the United States, Japan, France, Italy, and Canada, that meets to discuss and plan international monetary and economic strategy, foreign exchange rates, etc., in an attempt to foster stability and equitable change.
2. a Toronto-based group of Canadian landscape artists, namely: 1, Tom Thompson; 2, J. E. H. MacDonald; 3, Lawrence Harris; 4, Arthur Lismer; 5, Frederick Varley; 6, Franklin Carmichael; and 7, Frank H. Johnston, with Alexander Young. In 1913, they joined to create a national style of painting. The group, which became influential in the 1920s and '30s, changed its name (1933) to the Canadian Group of Painters.

hebdomad, 1. the number seven; the idea of seven, or the quality of being seven in number.
2. the sum of seven things; a collection of seven persons or things, especially a group of seven days; a week.
3. (in some Gnostic systems) a group of seven superhuman beings, angels, or divine emanations, apparently developed from the idea of seven planets or planetary heavens, or that of gods, spirits, or angels personifying, indwelling, ruling, or creating them; collectively, the whole sublunary sphere or its ruler.

hebdomadal, consisting of seven days or occurring or appearing every seven days; weekly, as a periodical magazine or newspaper. Also, **heptal.**

hebdomadal council, (at Oxford University) a board of twenty-one members, ordinarily meeting weekly, elected by the senate to regulate the business of the university.

Hell, according to Muslim belief, Hell is divided into seven compartments: 1, for Muslims; 2, for Jews; 3, for Christians; 4, for Sabians; 5, for Magians; 6, for idolaters; and 7, for hypocrites.

hepta-, a prefix borrowed from Greek, meaning 'seven.'

heptace, a summit of a polyhedron formed at the concurrence of seven faces.

7

heptachord, 1. (in Greek music) **(a)** a diatonic series of seven tones. **(b)** the interval of the major seventh.
2. an instrument with seven strings.

heptad, the sum of seven units; the number seven; a set of seven of anything.

heptaglot, 1. a book in seven languages.
2. using, knowing, or written in seven languages.

heptagon, a plane figure having seven sides and seven angles.

heptahedron, a solid having seven plane faces.

heptameride, anything consisting of seven parts.

heptameron, a book describing the events of seven days, as the *Heptameron* of Margaret of Angoulême, Queen of Navarre (1492–1549). Compare **10: decameron.**

heptameter, (in prosody) a verse consisting of seven measures. Also, *in modern usage,* **heptapody.**

heptandrous, *or* **heptandrious,** *or* **heptandrian,** (in botany) having seven stamens.

heptane, a paraffin containing seven atoms of carbon.

Heptanesus, the Ionian islands, seven in number.

heptangular, having seven angles.

heptapetalous, (in botany) having seven petals in the corolla.

heptaphyllous, (in botany) having seven leaves.

heptapody, (in prosody) a meter, period, or verse consisting of seven feet. Compare **7: heptameter.**

heptarchy, 1. government by seven persons.
2. a group of seven kingdoms or governments, especially the seven principal Anglo-Saxon kingdoms of 1, Kent; 2, Sussex; 3, Wessex; 4, Essex; 5, Northumbria; 6, East Anglia; and 7, Mercia, which ended in 829 when Egbert, king of Wessex, became overlord of the other kingdoms.

heptasepalous, (in botany) having seven sepals.

heptastich, (in prosody) a strophe, stanza, or poem consisting of seven lines, or verses.

Heptateuch, the first seven books of the Old Testament, namely: 1, Genesis; 2, Exodus; 3, Leviticus; 4, Numbers; 5, Deuteronomy; 6, Joshua; and 7, Judges.

heptathlon, an athletic contest, usually for women, in which competitors are awarded points for their performances in seven track and field events, namely: 1, 100-meter run; 2, 800-meter run; 3, 100-meter hurdles; 4, high jump; 5, long jump; 6, javelin throw; and 7, shot put.

heptavalent, (in chemistry) having a valence of seven.

Holy Sacraments, the, (in some Christian denominations) religious acts deemed to lead to spiritual betterment, namely: 1, Baptism; 2, Confirmation; 3, Holy Communion; 4, Penance; 5, Holy Orders; 6, Matrimony; and 7, Extreme Unction. See also 3: **trisacramentarian.**

Homer, seven cities boasted of being the birthplace of Homer:
Seven cities warred for Homer being dead,
Who living had no roof to shroud his head.
[Thomas Heywood, "Hierarchie of the Blessed Angels," 1635.]
Seven wealthy towns content for Homer dead,
Through which the living Homer begged his bread.
[Thomas Seward, (1708–90) "On Homer,"; found only in *Bartlett's Familiar Quotations,* 1980.]
Seven cities strive for the learnèd root of Homer:
Smyrna, Chios, Colophon, Ithaca, Pylos, Argos, Athens.
[anon., from *The Greek Anthology.*]
The foregoing are listed in *Bartlett;* the *Encyclopaedia Britannica* substitutes Salamis and Rhodos for Ithaca and Pylos, and comments that Kyme, Ios, Sparta, Egypt, and Babylon "also compete," but gives no source.

Hotei, one of the seven beneficent beings of the Japanese pantheon, represented as a fat, smooth-faced man, with a protruding naked abdomen, and usually carrying a large hempen bag.

Hyades, *or* **Hyads,** 1. (in astronomy) a constellation of about seven stars, of which the principal is Aldebaran, in the head of the Bull, supposed by the ancients to indicate the approach of rainy weather when it rose with the sun.
2. (in Greek mythology) nymphs who nursed the infant Bacchus and were transformed into stars in compasssion for their incessant weeping for the fate of their brother, who was torn to pieces by a wild beast.

Ionian Islands, seven islands off the west coast of Greece, namely: 1, Cephalonia; 2, Corfu; 3, Cythera; 4, Ithaca; 5, Leucas; 6, Paxos; and 7, Zacynthus. Also called **Seven Islands, Septinsular Republic, Heptanesus,** and in French, **Sept-îles.**

Island of the Seven Cities, a fabled isle where seven bishops, who quit Spain during the Moorish occupation, were said to have founded seven cities. The legend has it that though many traveled there, none has ever returned. [Brewer, *Dictionary of Phrase & Fable,* 1894.]

J, j, the seventh consonant of the English alphabet.

7

Jarvik-7, a two-chambered mechanical heart developed by Dr. Robert Jarvik and first used in December 1982 in Salt Lake City on Barney Clark, who suffered from a form of heart disease not treatable except by a heart transplant. He survived 112 days. It may be used only with U. S. government approval on an experimental basis and, since its tubes must always connect with the electronically controlled air compressor that drives it, it is usually implanted to keep a patient alive only until a human heart becomes available.

Joys of Mary, the seven joys of Mary, namely: 1, the Annunciation; 2, the Visitation; 3, the Nativity; 4, the adoration of the Magi; 5, the presentation in the temple; 6, the finding of the lost Child; and 7, the Assumption. See also **7: Sorrows of Mary.**

July, the seventh month of the modern calendar.

lease, (in Britain) many leases run for seven years or multiples thereof, said to go back to the ancient notion of "climacteric years," or those in which life was supposed to be in special peril. [Brewer, *Dictionary of Phrase & Fable,* 1894.]

Libra, the seventh sign of the zodiac, a pair of scales.

manille, (in omber or quadrille) the seven of diamonds or hearts, the second highest card.

marriage, house of, (in astrology) the seventh house. See **12: house.**

menorah, (in Judaism) the seven-branched candelabrum associated with Hanukkah, or the Feast of Dedication, which celebrates the defeat of Antiochus IV by the Maccabees (165 BC). On the first evening, one candle is lighted, on the second, two, and so on. After the desecration of the temple, only a small amount of oil remained, enough for one evening; miraculously, it lasted for eight days, and the lighting of the candles commemorates that event. According to some sources, the seven candles symbolize the archangels, namely: 1, Cassiel; 2, Gabriel; 3, Haniel; 4, Madimial; 5, Michael; 6, Raphael; and 7, Zadkiel. [Jobes, *Dictionary of Mythology, Folklore and Symbols*.] See also **8: Hanukkah.**

minion, a size of type on a seven-point body.

muni, one of the seven stars (the seven sages, or rishis) of the Big Dipper, in the constellation Ursa Major.

names of God, (in Judaism) the seven divine names, namely: 1, El; 2, Elohim; 3, Adonai; 4, YHWH; 5, Ehyeh-Asher-Ehyeh; 6, Shaddai; and 7, Zeba'ot. See also **10: names of God.**

natural, (in craps) a winning throw of dice showing a score of seven on the first roll. See also **11: natural.**

number 7 steak, a cut of beefsteak from which half of the small crossbone of the "T" has been removed, leaving a bone shaped like a "7."

odd trick, the, (in whist) the seventh trick won by either side out of a possible thirteen.

Ogier, the Dane, said to "stamp his iron mace on the floor" every seven years. [Brewer, *Reader's Hand-Book,* 1889.]

Olaf Redbeard, (of Sweden) said to "unclose his eyes" every seven years. [Brewer, *Reader's Hand-Book,* 1889.]

Pearl Harbor day, the 7th of December, 1941, when the Japanese launched a surprise air attack against the U. S. naval base at Pearl Harbor, Hawaii, killing thousands and wreaking great destruction on ships, planes, and other equipment. By this action, the U. S. active participation in the war in Europe, against Germany and Italy, was precipitated, and the U. S. immediately declared war on Japan. Compare **14: V-J Day.**

perfect color, one of the seven colors used in the manufacture of oriental carpets, namely: 1, blue; 2, green; 3, indigo; 4, orange; 5, red; 6, rose; and 7, yellow.

Pleiad, the, 1. **the Pleiad of Alexandria,** consisted of: 1, Apollonios Rhodios; 2, Aratos; 3, Callimachos; 4, Homer the Younger; 5, Lycophron; 6, Nicander; and 7, Theocritus, all of Alexandria in the time of Ptolemy Philadelphos. Also called **Poetic Pleiades.**
2. **the Pleiad of Charlemagne,** consisted, in addition to 1, Charlemagne (called "David"), of: 2, Alcuin ("Albinus"); 3, Adilard ("Augustine"); 4, Angilbert ("Homer"); 5, Eginhard; 6, Riculfe ("Damaetas"); and 7, Varnefrid.
3. **the First French Pleiad,** consisted of: 1, Antoine de Baïf; 2, Joachim du Bellay; 3, Remi-Belleau; 4, Jodelle; 5, Ronsard; 6, Ponthus de Thiard; and 7, either Dorat or Amadis de Jamyn, all under Henri III, in the sixteenth century.
4. **the Second French Pleiad,** consisted of: 1, Commire; 2, Dupérier; 3, Larue; 4, Ménage; 5, Petit; 6, Rapin; and 7, Santeuil.

Pleiades, the, *or* **Pleiads, the,** the seven daughters of Atlas, who appear as a cluster of seven stars in the constellation Taurus, namely: 1, Alcyone; 2, Asterope; 3, Celeno; 4, Electra; 5, Maia; 6, Merope; and 7, Taygete. Also called **Seven Sisters.**

relapsing fever, any of a number of acute louse- or tick-borne infectious diseases characterized by recurrent bouts of fever recurring between periods, lasting typically for about seven days, of normal temperature and blood pressure. Also called **seven-day fever, seventeen-day fever.**

rime-royal, a seven-line stanza introduced by Chaucer with the rimescheme *ababbcc.*

Rome, ancient Rome was built on seven hills, originally surrounded by fortifications, namely: 1, Aventine; 2, Caeline; 3, Capitoline; 4, Esquiline; 5, Palatine; 6, Quirinal; and 7, Viminal. Rome was called *Urbs Septicollis* ('Seven-hilled City') in Latin.

7

S, the medieval Roman numeral 7. See also **70: S.**

Sabbath, (in the Jewish calendar) the seventh day of the week, Saturday.

sacred books, *or* **bibles,** the seven major works of religion, namely: 1, the Christian Bible; 2, the Scandinavian Eddas; 3, the Chinese Five Kings; 4, the Muhammadan Koran; 5, the Hindu Three Vedas; 6, the Buddhist Tri Pitikes; and 7, the Persian Zendavesta.

Samson, (in the Bible) the strength of Samson was taken from him when Delilah cut seven locks of hair from his head. [Judges, xvi.19.]

Saturday, the seventh, or last, day of the week.

sennight, the period of seven days and nights; a week. [A contraction of *seven* and *night*.]

senses, according to *Ecclesiasticus,* there are seven senses, namely: 1, seeing; 2, hearing; 3. tasting; 4, feeling; 5, smelling; 6, understanding; and 7, speech.

septa-, a prefix borrowed from Latin, meaning 'seven.'

septan fever, intermittent fever in which the paroxysm recurs on the seventh day (both paroxysmal days being counted). Also called **seven-day fever.**

septangle, a plane figure with seven sides and seven angles; a heptagon.

September, the seventh month of the Roman calendar, which began March first.

septempartite, divided or separated into seven parts.

septemvir, *pl.* **septemviri.** one of seven men joined in any office or commission, as the septemviri epulones, one of the four chief religious corporations of ancient Rome.

septenary, 1. referring to seven, especially seven years.
2. a commemoration or celebration of an event that occurred seven years earlier; a seventh anniversary.

septennial, 1. of or pertaining to a period of seven years.
2. a commemoration or celebration of an event that occurred seven years earlier; a seventh anniversary.
3. lasting for seven years; occurring once every seven years.

Septennial Act, (in England) an act of Parliament, superseding the Triennial Act (1694), requiring that a new Parliament be summoned at least once every seven years. This period was reduced to five years in 1911.

septennium, a period of seven years.

septentrional, northern; boreal; hyperborean. [So called in reference to the *septentriones,* the seven stars that make up Ursa Major, or Charles's Wain, a constellation near the North Pole.]

septet, *or* **septette,** 1. a musical composition for seven instruments or voices.
2. a group of seven musicians or singers who sing or play septets.

septfoil, a seven-lobed figure, used especially in architecture. Compare **3: trefoil, 4: quatrefoil, 5: cinquefoil, 6: sexfoil.**

septiform, sevenfold.

septimal, relating to the number seven.

septime, the seventh of eight parries or guards against thrusts in fencing. Compare **1:prime, 2:seconde, 3:tierce, 4:quarte, 5:quinte, 6:sixte, 8:octave.**

septisyllable, 6or heptasyllable, a word consisting of seven syllables, as *intercontinentally.*

septivalent, (in chemistry) having a valence of seven.

septuplet, one of seven offspring born at the same birth.

seven, 1. the whole number following six and preceding eight; a prime number.
2. (in the Bible) a number frequently used, for example: the seventh day, on which God rested [Gen. ii.2]; the seven-day plague of bloody waters in Egypt [Ex. vii.25]; the seven weeks between Passover and the Pentecost [Lev. vii.15]; the symbol of "many sons" [Ruth iv.15, 1 Sam. ii.5]; and the seven loaves with which Jesus fed the multitudes [Matt. xv.34-38]. See also **7: seven years of plenty.**

Seven Against Thebes, (in Greek mythology) upon the fall of Oedipus, king of Thebes, his twin sons, Polyneices and Eteocles, agreed to share rule of the kingdom by alternating annually as king, with Eteocles to begin. At the end of the first year, when Eteocles refused to give up the throne, Adrastus, king of Argos and father-in-law of Polyneices, mobilized an army under the leadership of **1**, Amphiaraus; **2**, Capaneus; **3**, Eteoclus; **4**, Hippomedon; **5**, Parthenopaeus; **6**, Polyneices; and **7**, Tydeus (son of Oeneus, exiled king of Calydon). All were slain, and Eteocles and Polyneices killed each other, fulfilling their father's curse. When the sons of the Seven, called the Epigoni, had grown to manhood, Adrastus again attacked Thebes and occupied it. In some legends, Adrastus is substituted for one of the Seven, either Hippomedon or Polyneices.

seven arts *or* **sciences,** the trivium and quadrivium taken together, namely: **1**, grammar; **2**, rhetoric; and **3**, logic, together with **4**, music; **5**, arithmetic; **6**, geometry; and **7**, astronomy.

7

seven bishops, case of the (in England) seven bishops who protested against the Declaration of Indulgence of James II (1688), namely: 1, Archbishop Sancroft; 2, Bishop Ken; 3, Bishop Lake; 4, Bishop Lloyd; 5, Bishop Trelawney; 6, Bishop Turner; and 7, Bishop White.

seven blocks of granite, the, members of the 1936 Fordham University football team, namely: 1, Al Babartsky; 2, Johnny Druze; 3, Ed Franco; 4, Vince Lombardi; 5, Leo Paquin; 6, Nat Pierce; and 7, Alex Wojciechowicz.

seven-branched candlestick, a candelabrum having a central shaft and three branches on each side, common in the churches of the Middle Ages in allusion to the candlestick of the tabernacle (Ex. xxv.31) and the seven lamps of the Apocalypse.

Seven Cities, the, the great cities of all time, namely: 1, Egypt; 2, Jerusalem; 3, Babylon; 4, Athens; 5, Rome; 6, Constantinople; and 7, either London (for commerce) or Paris (for beauty). [Brewer *Reader's Handbook,* 1889.]

Seven Days' Battles, a series of battles in the War Between the States, (June 25–July 1, 1862), at Mechanicsville, Gaines' Mill, Savage's Station, Frayser's Farm (now Glendale), and Malvern Hill, near Richmond, Virginia, in which the Confederate Army under General Robert E. Lee prevented the Army of the Potomac, under General George B. McClellan from capturing Richmond, the Confederate capital, thus ending the Peninsular Campaign.

Seven Dials, a formerly run-down section of London, named for a column with seven dials that stood there, facing the seven streets which radiated from it.

seven gifts of the Holy Ghost, namely: 1, wisdom; 2, understanding; 3, counsel; 4, ghostly strength or fortitude; 5, knowledge; 6, godliness; and 7, fear of the Lord. Also called the **Seven Spirits of God.**

sevengills, a cow-shark. [So called from the seven gill-slits on each side.]

seven-handed euchre, a variety of euchre in which seven players play with a full pack of fifty-two cards.

seven holes, *or* **seven eyes,** a river-lamprey. [So called from the seven gill-slits on each side.]

Seven Holy Founders, the founders, in the thirteenth century, of the Order of Friar Servants of St. Mary, called Servites, a Roman Catholic order of mendicant friars dedicated to apostolic work, namely: Saints 1, Bonfilius; 2, Alexis Falconieri; 3, John Bonagiunta; 4, Benedict dell'Antella; 5, Bartholomew Amidei; 6, Gerard Sostegni; and 7, Ricoverus Uguccione. They were canonized in 1888 by Pope Leo XIII.

Seven Lamps of Architecture, a critical work (1847), by John Ruskin, in which architecture is regarded as an enlightening influence that reflects man's culture, which has seven manifestations, namely: 1, Sacrifice; 2, Truth; 3, Power; 4, Beauty; 5, Life; 6, Memory; and 7, Obedience.

seven-league boots, in many fairy tales, boots donned by the hero that enable him to traverse seven leagues at a stride. The length of the league varies considerably, depending on the country and the period; in England it was often three miles.

seven men of good repute, the disciples chosen to preach, after Pentecost, while the remaining worked to feed the populace, namely: 1, Nicolaus; 2, Nicanor; 3, Parmenas; 4, Philip; 5, Prochorus; 6, Stephen; and 7, Timon.

Seven Oaks Massacre, the destruction (June 1816) of a settlement of the Hudson's Bay Company, near Fort Douglas, by a band of halfbreeds under the leadership of an employee of the (competing) North West Company, resulting in the death of twenty soldiers and settlers.

Seven Odes, the the seven most famous pre-Islamic poems, dating from the mid 6th century and considered to be of perfect beauty. In Arabic they are called *al-Mu'allaqāt* 'the Suspended Ones,' because, according to legend, they had been hung in the most important religious sanctuary in Mecca in honor of their eloquence and beauty.

Seven Sages of Ancient Greece, seven men famous for their wisdom, usually listed as 1, Bias of Priene, 2, Chilo of Sparta, 3, Cleobulus of Lindos, 4, Periander of Corinth, 5, Pittacus of Mytilene, 6, Solon of Athens, and 7, Thales of Miletus. Compare 7: **Seven Wise Men.**

Seven Sages of the Bamboo Grove, *or* **Seven Worthies of the Bamboo Grove,** a group of Chinese poets and scholars who sought freedom of expression and escape from the perils of oppression in Shanyang (modern southern Shantung Province), in the mid second century AD, namely: 1, Hsi K'ang; 2, Hsiang Hsiu; 3, Liu Ling; 4, Kuo Hsiang; 5, Shan T'ao; 6, Wang Jung; and 7, Yüan Chi.

Seven Seas, essentially, all of the navigable waters of the world; traditionally, the Atlantic, Pacific, Indian, and Arctic oceans, together with the Mediterranean, Black, Ægean, and North, Arabian, or Baltic seas.

Seven Sisters, 1. seven women's colleges of the northeastern United States having a reputation for high scholastic achievement and social prestige, namely: 1, Barnard; 2, Bryn Mawr; 3, Mount Holyoke; 4, Radcliffe; 5, Smith; 6, Vassar; and 7, Wellesley.
2. the seven largest oil companies, namely: 1, British Petroleum; 2, Gulf; 3, Mobil Oil; 4, Shell Oil; 5, Standard Oil of California; 6, Standard Oil of New Jersey; and 7, Texaco.
3. seven chalk cliffs on the south coast of England, east of Brighton, the

7

tallest of which is the 500-foot Beachy Head.
4. See **7: Pleiades.**

Seven Sleepers, seven noble youths of Ephesus, namely, in western legend: 1, Constantine; 2, Dionysius; 3, John; 4, Malchus; 5, Maximian; 6, Martinian; and 7, Serapion, or, in eastern legend, 1, Charnouch; 2, Debermouch; 3, Jemlikha; 4, Keschetiouch; 5, Mechlima; 6, Mekchilinia; and 7, Merlima. They martyred themselves (AD 250) under the emperor Decius by fleeing to a cave in Mount Celion, where, having fallen asleep, they were found by Decius, who had them sealed up. The sleepers awoke after 230 years (309 years according to the *Oriental Tales,* 1743), lived for a few days, and, when they died, were transported in a stone coffin to Marseilles (according to Gregory of Tours in *De Gloria Martyrum,* AD 6th century) or (according to the *Koran*) were buried in the cave. They were accompanied by a dog (the Dog of the Seven Sleepers), named Kratim, or Kratimer, or Katmir, that spoke Greek.

Seventh Amendment to the Constitution of the United States, one of the ten amendments (also called articles) that make up the Bill of Rights, ratified on December 15, 1791: In Suits at common law, where the value in controversy shall exceed twenty dollars, the right of trial by jury shall be preserved, and no fact tried by a jury, shall be otherwise re-examined by any Court of the United States, than according to the rules of common law.

Seventh Avenue, a street in New York City, part of which, nicknamed "Fashion Avenue," with side streets between 35th and 40th Streets, is where most of the manufacturers of ladies' ready-to-wear clothing have showrooms and factories. Also called **the garment center.**

Seventh-day, the name for Saturday, the seventh day of the week, used by the Society of Friends.

Seventh-Day Adventist, a member of a Fundamentalist Christian sect, founded after 1844 but not officially so named till 1860, among the doctrines of which is the requirement that the seventh day of the week (Saturday) be observed as the Sabbath and kept holy.

Seventh Heaven, (in Muhammadanism) the highest heaven, formed of divine light beyond description, above the other six heavens, namely: 1, the heaven of the purest silver, where the stars are like lamps on golden chains, each attended by an angel and where Muhammad met Adam and Eve; 2, the heaven of splendorous polished steel, where Muhammad found Noah; 3, the third heaven, studded with precious stones "too brilliant for the eye of man," is inhabited by Azrael, the Angel of Death, who writes the names of those just born in a large book and deletes the names of those who have died; 4, a heaven of the finest silver where the Angel of Tears weeps endlessly for the sins of man; 5, the heaven of pure gold where the Avenging Angel abides and where Muhammad met Aaron; and

6, the Guardian Angel of heaven and earth dwells here, and it is where the Prophet met Moses.

seventh heaven, to be in,   to be supremely happy. See **7: Seventh Heaven.**

seventh-inning stretch,   a traditional pause during a baseball game, after the first half of the seventh inning, when fans stand, stretch, and often sing "Take Me Out to the Ballgame."

Seven Weeks' War,   a war between Prussia and Austria, June 8 to July 26, 1866, ending with the victory of Prussia. Terms were settled in the Treaty of Prague (August 23, 1866) and the Treaty of Vienna (October 3, 1866).

Seven Wise Men,   Plato listed seven sages of ancient Greece (late seventh, early sixth century BC) who were said to possess great wisdom: 1, Bias; 2, Chilon; 3, Cliobulus; 4, Myson of Chen; 5, Pittacus; 6, Solon; and 7, Thales. Compare **7: Seven Sages of Ancient Greece.**

Seven Wonders of the Middle Ages,   the seven wonders of the world during the Middle Ages, namely: 1, the Colosseum at Rome; 2, the catacombs at Alexandria; 3, the Great Wall of China; 4, Stonehenge; 5, the Leaning Tower of Pisa; 6, the Porcelain Tower of Nanking; and 7, the Mosque of St. Sophia at Constantinople. See also **7: Seven Wonders of Wales, 7: Wonders of the Ancient World.**

Seven Wonders of Wales, the,   they are listed as: 1, the mountains of Snowdon; 2, Overton churchyard; 3, the bells of Gresford church; 4, Llangollen bridge; 5, Wrexham steeple; 6, Pystyl Rhaiadr waterfall; 7, St. Winifred's well. See also **7: Seven Wonders of the Middle Ages, 7: Wonders of the Ancient World.**

seven-year itch,   1. (in medicine) scabies.
2. *Humorous.* an urge to engage in extramarital sexual activity after seven years of marriage, usually in reference to a man.

seven years of plenty, seven years of famine,   the meaning of Pharaoh's dreams, as interpreted by Joseph. According to an incident related in Genesis 41, Pharaoh prophetically dreamed of seven fat cows being devoured by seven lean ones, then of seven full ears of corn swallowed up by seven thin and shriveled ears. In all of Egypt only Joseph was able to interpret Pharaoh's dreams and, in addition, offer a plan to avoid the hardship of the famine by prudently storing grain during the years of plenty.

Seven Years' War, the,   war in which England and Prussia defeated Austria, Russia, Sweden, Saxony and France (1756–63).

Seven Years' War of the North, the   a war between Denmark and Sweden, 1563–1570, which ended inconclusively.

7

seven youths, seven maidens, (in classical mythology) the Minotaur, a monster with either a man's body and a bull's head, or vice versa, depending on the relater of the legend, was kept in a labyrinth designed by Daedalus for Minos, king of Crete. Each year, seven youths and seven maidens were sent to Crete as tribute from Athens for the Minotaur to devour. Theseus, with the aid of Ariadne, daughter of Minos, slew the Minotaur and escaped from the labyrinth and Crete.

Shichi Fukujin, the seven Japanese gods of luck, namely: 1, Benten, the only female (fine arts, female beauty, wealth); 2, Bishamonten (wealth); 3, Daikoku (wealth); 4, Ebisu (fishing, honest dealing); 5, Fukurokuju (fortune, longevity); 6, Hotei (treasure); and 7, Jorojin (health, longevity). [Leach, *Standard Dictionary of Folklore, Mythology, and Legend,* 1972.]

shiva, (in orthodox Judaism) the period of seven days of mourning for a parent, brother or sister, or husband or wife, marked by sitting on the floor, shoeless, in the home and devoting oneself to prayer and liturgical exercises.

Siberian Seven, the, a group of Soviet Pentecostalists who, in 1978, took refuge in the American Embassy, Moscow, namely: 1, Timofey, and 2, Maria Chmykhalova; and 3, Augustina, 4, Lidiya, 5, Lilya, 6, Lyuba, and 7, Pyotr Vashchenko. They sought help in obtaining permission to emigrate, and campaigns on their behalf were waged by Christians in the West. After returning in smaller groups to their home town, Chernogorsk, they were allowed to leave the USSR in 1983.

Siebengebirge, seven hills on the Rhine, above Bonn, namely: 1, Drachenfels; 2, Lohr-Berg; 3, Lowenburg; 4, Nonnen-Stromberg; 5, Ölberg; 6, Petersberg; and 7, Wolkenburg.

"Sindbad the Sailor, Seven Voyages of," (in the *Arabian Nights' Entertainments*) a merchant of Bagdad who gained fame and wealth and underwent many adventures in the course of seven voyages, namely: 1, in which, becalmed, he and his crew visit an island, which turns out to be a sleeping whale; 2, in which he acquires diamonds and escapes from a valley by deceiving a roc; 3, in which he encounters the Cyclops; 4, in which he narrowly escapes being buried alive; 5, in which he encounters and slays the Old Man of the Sea; 6, in which he visits the mountain whither Adam was exiled from Eden; and 7, in which he is enslaved by pirates but is finally freed on discovery of an "elephants' graveyard." See also **1001: Arabian Nights.**

sins, seven deadly, the sins of: pride, covetousness or avarice, lust, anger or wrath, gluttony, envy, and sloth. Also called **seven mortal sins.** See also **7: virtues.**

"Snow White and the Seven Dwarfs," a children's fairytale about a young girl who is cared for (and who looks after) seven dwarfs, till her Prince Charming comes along to marry her. The tale was made into a perennially popular animated motion picture by Walt Disney, in which the dwarfs were given distinctive characters and names to match them: 1, Doc; 2, Grumpy; 3, Sneezy; 4, Dopey; 5, Bashful; 6, Sleepy; and 7, Happy.

Sorrows of Mary,   the seven sorrows of Mary, namely: 1, Simeon's prophecy; 2, the flight into Egypt; 3, Jesus missed; 4, the betrayal; 5, the Crucifixion; 6, the taking down from the cross; and 7, the Ascension. See also **7: Joys of Mary.**

spiritual works of mercy,   acts of charity that render spiritual aid, namely: 1, to admonish sinners; 2, to instruct the ignorant; 3, to counsel the doubtful; 4, to pray for the salvation of thy neighbour; 5, to comfort the sorrowful; 6, to bear wrongs patiently; and 7, to forgive all injuries. See also **7: corporal works of mercy.**

stars,   **1.** the number of heavenly bodies known to the ancient astronomers was seven, namely: 1, Mercury; 2, Venus; 3, Mars; 4, Jupiter; 5, Saturn; 6, the sun; and 7, the moon.
2. Charles's Wain. See **7: septentrional.**
3. See **7: Pleiades.**

United Arab Emirates,   a federation of seven independent Arab states along the eastern coast of the Arabian Peninsula, namely: 1, Abu Dhabi; 2, Ajman; 3, Al Fujairah; 4, Dubai; 5, Ras al-Khaimah; 6, Sharjah; and 7, Umm al Qaiwain. Formerly called **Trucial States.**

virtues,   the seven virtues are: 1, faith; 2, hope; 3, charity; 4, prudence; 5, justice; 6, fortitude; and 7, temperance. See also **7: sins.**

Watergate Seven, the,   seven men convicted of breaking into and "bugging" Democratic National Committee headquarters in the Watergate office building in Washington, D. C., in 1972, and of engaging in other political espionage, namely: 1, Bernard L. Baker; 2, Virgilio R. Gonzalez; 3, E. Howard Hunt, Jr.; 4, G. Gordon Liddy; 5, James W. McCord; 6, Eugenio Martinez; and 7, Frank A. Sturgis.

weekly,   occurring or appearing every seven days, as a periodical; hebdomadal.

Whitsunday,   the seventh Sunday after Easter. See also **50: Pentecost, 2.**

Wonders of the Ancient World,   seven ancient buildings or works of art, distinguished for size, beauty, splendor, or as outstanding achievements, namely: 1, the Egyptian pyramids; 2, the hanging gardens of Babylon; 3, the temple of Artemis, at Ephesus; 4, the statue of Zeus, by Phidias, at Olympia; 5, the Mausoleum at Halicanarsus; 6, the Colossus of Rhodes; and 7, the lighthouse on the island of Pharos, at Alexandria, Egypt. Of all, only the pyramids remain. See also **7: Seven Wonders of the Middle Ages, 7: Seven Wonders of Wales.**

7

zayin, the seventh letter of the Hebrew alphabet, corresponding to the Greek *zeta* and the English *z.*

8

Agni, (in Hinduism) the eight forms of Agni, the ancient Aryan god, are: 1, Asani (lightning); 2, Bhava (existence); 3, Isana (the ruler); 4, Mahadeva (the great god); 5, Pasupati (the lord of the cattle); 6, Rudra (the roarer); 7, Sarva (all); and 8, Ugra (dread). See also **3: Agni.** [From Jobes, *Dictionary of Mythology, Folklore and Symbols.*]

Anthesterion, the eighth month of the ancient Greek year, corresponding to the last part of February and the beginning of March.

August, the eighth month of the modern calendar.

behind the eightball, in serious trouble or in a predicament. [From the game of Kelly pool in which striking the eightball occasions a penalty, putting a player who finds it blocking a pocket into a precarious position.]

Big Eight, the, 1. a college athletic conference composed of 1, the University of Colorado; 2, Iowa State University; 3, the University of Kansas; 4, Kansas State University; 5, the University of Missouri; 6, the University of Nebraska; 7, the University of Oklahoma; and 8, Oklahoma State University.
2. *U. S.* the major firms of Certified Public Accountants, namely: 1, Arthur Anderson & Co.; 2, Arthur Young & Co.; 3, Coopers & Lybrand; 4, Deloit, Haskins, & Sells; 5, Ernst & Whinney; 6, Peat, Marwick, Mitchell & Co.; 7, Price, Waterhouse; and 8, Touche, Ross & Co.

brevier, a size of type on an eight-point body.

byte, (in computer technology) a sequence of adjacent bits that is treated as a unit of data in a computer system. The most common are the 6- and 8-bit byte, although 16-bit bytes are also used. See also **2: bit.**

cheth, *or* **heth,** the eighth letter of the Hebrew alphabet, corresponding to the English *h,* and having a numerical value of 8.

classic orders, (in architecture) the Roman Composite, Corinthian, Doric, Ionic, and Tuscan orders and the Greek Corinthian, Doric, and Ionic orders.

death, house of, (in astrology) the eighth house, a succedent house. See **12: house.**

Dhyanibodhisattva, (in Buddhism) one of the eight Buddhas who are secondary in importance to the Manusibuddhas, namely: 1, Akasagarbha; 2, Avalokitesvara; 3, Ksitigarbha; 4, Mahasthamaprapta; 5, Maitreya; 6, Manjusri; 7, Sarvanivaranaveskambhin; and 8, Trailokyavyago. One of these is sometimes replaced by Samantabhadra or Vajrapani. See also **5: Bud-**

dha, 5: Manusibuddha. [From Jobes, *Dictionary of Mythology, Folklore and Symbols.*]

dollar, the (former) Spanish dollar, containing eight reals.

double quatrefoil, (in heraldry) an eight-leaved foil, used as the cadency mark of a ninth son. Also called **eightfoil, octofoil.**

Eight, the, a group of eight twentieth-century American painters, namely: 1, Arthur B. Davies; 2, William J. Glackens; 3, Robert Henri; 4, Ernest Lawson; 5, George Luks; 6, Maurice Prendergast; 7, Everett Shinn; and 8, John Sloan, who exhibited together, in 1908, in New York City. Although their styles of painting differed, because of the themes they selected—largely realistic street and other scenes, in rebellion against the academic aesthetics imposed by traditional European schools of art—they were later grouped, with George Bellows, Glenn Coleman, Eugene Higgins, Edward Hopper, and Jerome Myers, into what became known as the Ashcan School, because many critics found their subjects distasteful.

eight bells, (in nautical use) a striking of the ship's bell eight times to indicate the end of a four-hour watch.

eightfoil, (in heraldry) a plant or grass having eight rounded leaves. Also called **double quatrefoil.**

eight-foot law, (formerly) a regulation prohibiting cars from coming closer than eight feet to the door of a streetcar taking on or discharging passengers in the middle of a street.

eighth, (in music) 1. the interval between any tone and a tone on the eighth diatonic degree above or below it.
2. a tone distant by an eighth or octave from a given tone; an octave, or replicate.

eight-hour law, a law limiting to eight hours the day's work of certain classes of workers, passed by the U. S. Congress in 1868.

eighth wonder of the world, something sufficiently marvelous to warrant its being classed among the traditional seven wonders of the world, usually used sarcastically. See also **7: seven wonders of the world.**

Eight Immortals of the Wine Cup, the, eight Chinese poets of the eighth century, namely: 1, Chang Hsu; 2, Chio Sui; 3, Chu Chin; 4, Ho Chihchang; 5, Li Chin; 6, Li Shihchih; 7, Tsui Tsungchih; and 8, Li Po, the last of whom is considered by many to be the greatest poet of China. The group was so named because Li Po had been given an imperial edict enabling him to obtain free wine wherever he went. See also **6: Six Idlers of the Bamboo Streams.**

eightpenny nail, 1. a nail that, before 1500, cost eightpence a hundred.
2. (in modern use) a nail 2½ inches long.

8

Eights, boat races between the different colleges of Oxford and Cambridge Universities, held during the summer term. [After the number of oars in each boat.]

Eights Week, the week during which the Eights are held. See **Eights.**

figure of eight, a shape resembling the number 8, that is, consisting of two conjoined loops; lemniscate.

Furry-day, the 8th of May, so called in parts of Cornwall, where it is celebrated with ceremonies resembling those of ancient May-day feasts.

Grand Jury Eight, the, eight witnesses who were jailed for refusing to testify before the grand jury inquiring into the charges against the New York Eight, namely: 1, Olive Armstrong; 2, Lionel Jean Baptiste; 3, Jacqueline Bernard; 4, Dorie Clay; 5, John Ford; 6, Milton Parish; 7, Michelle Thomas; and 8, Wanda Wareham. All were freed after serving a few months of their sentences. See also **8: New York Eight, 8: New York Eight Plus.**

H, h, 1. the eighth letter of the English alphabet.
2. the eighth item in a system where A = 1, B = 2, C = 3, etc.

Hanukkah, *or* **Chanukah,** the important Jewish Festival of Lights, or dedication, commemorating the recovery of the temple after the Maccabees' defeat (164 BC) of the Syrians under Antiochus Epiphanes and the miraculous burning, for eight days and nights, of the golden candlestick from only a small quantity of oil. The ceremony requires that Hanukkah lights be lit by every adult male at home immediately after the evening prayer; one candle is lit on the first night, two on the second, and so on till the eighth night. A candelabra with nine candles is customarily used, the extra light, called *shamash* ('servant') serving to supply the light to the others. See also **7: menorah.**

Hell, according to Dante, Hell (Inferno) is in the form of a vast funnel in which there are eight circles, or ledges, each narrower than that above: in 1, bounded by the river Acheron, dwell the spirits of the heathen philosophers; in 2, presided over by Minos, are the spirits of those guilty of carnal and sinful love; gluttons dwell in 3, which is guarded by Cerberus; 4, guarded by Plutus, is the realm of the avaricious; in 5 is the Stygian Lake, where flounder those who in life restrained not their anger; 6 is the abode of those who perpetrated violence or fraud on their fellow man; in circle 7 dwell the spirits of suicides; and in 8 abide blasphemers and heretics. The eighth circle, Malebolge, is divided into ten chasms, the last of which is at the center of the earth where is found the frozen river Cocytus. See also **5: rivers of Hell.**

Ivy League, the, a college athletic conference composed of eight old institutions of the northeastern United States having a reputation for high scholastic achievement and social prestige, namely: 1, Brown University; 2, Columbia University; 3, Cornell University; 4, Dartmouth College; 5,

Harvard University; 6, the University of Pennsylvania; 7, Princeton University; and 8, Yale University.

K, k, the eighth consonant of the English alphabet.

Murrow's Boys, eight CBS newsmen who worked with newsman and commentator Edward R. Murrow and became well known in their own right, namely: 1, Cecil Brown; 2, Winston M. Burdett; 3, Charles Collingwood; 4, Richard C. Hottelet; 5, Larry Le Sueur; 6, Eric Sevareid; 7, William L. Shirer; and 8, Howard K. Smith.

New York Eight, the, eight political activists in Brooklyn, New York, charged in 1984 with conspiracy to rob armored cars, racketeering, and other acts of violence, namely: 1, Ruth Carter; 2, Coltrain Chimurenga; 3, Omowale Clay; 4, Yvette Kelly; 5, Viola Plummer; 6, Roger Taylor; 7, Roger Wareham; and 8, Howard Bonds, who became a government witness. All were acquitted of conspiracy and crimes of violence, but convicted of weapons possession and the use of false identity. See also **8: New York Eight Plus.**

New York Eight Plus, a ninth political activist, Pepe Ríos, was later indicted on the same charges as the New York Eight, and stood trial with them. Since posters and T-shirts had already been printed by their supporters to help finance their defense, it was decided to add a "+" to the "8," rather than incur the expense of reprinting as "New York Nine." Like the others, Ríos was convicted of weapons possession and the use of false identity but acquitted of the other charges. See also **8: New York Eight.**

Noble Eightfold Path, the, (in Buddhism) the eight elements involved in achieving the Noble Truth of the Path that Leads to the Cessation of Pain, namely: 1, right view; 2, right thought; 3, right speech; 4, right action; 5, right livelihood; 6, right effort; 7, right mindfulness; and 8, right concentration. See also **4: Noble Truths.**

octa-, *or* **octo-,** *or* **oct-,** a prefix borrowed from Greek and Latin, meaning 'eight.'

octachord, 1. a musical instrument having eight strings.
2. a diatonic series of eight tones.

octad, a system, set, or series of eight of anything.

octadrachm, *or* **octodrachm,** (in the coinage of certain ancient Greek systems, as those of the Ptolemies and Seleucids) a coin having the value of eight drachmae.

octaëteris, (in the ancient Greek calendar) a period or cycle of eight years, during which three intercalary months of 30 days were inserted after the sixth month in the third, fifth, and eighth years to bring the year of twelve lunar months, alternately of 29 and 30 days, into accord with the solar year. The average number of days in the year was thus made up

8

to 365^1/$_2$. In most states, the intercalary month took the name of the sixth month, which it followed, being distinguished from it by the epithet *second*. The system was devised by Cleostratus of Tenedos, about 500 BC.

octagon, 1. a plane figure having eight angles and eight sides.
2. (in fortification) a work with eight bastions.

octahedron, a solid figure having eight faces.

octal, 1. pertaining to eight.
2. a number system to the base eight. See also **8: octal number.**

octal number, a quantity represented in the octal system, in which 8 is the base, or radix. Figures used are 0 through 7. The quantity represented by 140 [(1 × 10^2) + (4 × 10^1) + (0 × 10^0)] in the decimal system (base 10) would be shown as 214 in the octal system: (2 × 8^2) + (1 × 8^1) + (4 × 8^0)). Compare **2: binary number, 16: hexadecimal number.**

octamerous, *or* **octomerous,** *or* **octameral,** *or* **octomeral,** (in zoology and botany) having parts in sets or series of eight.

octameter, a verse consisting of eight measures, a term little used except by some writers on modern versification who confound *measure* with *foot*.

octan, occurring every eighth day, as octan fever.

octan fever, intermittent fever in which the paroxysm recurs on the eighth day (both paroxysmal days being counted).

octangle, an octagon.

octarchy, government by eight persons, or a region occupied by eight affiliated communities each with its own chief or government.

octastich, *or* **octastichon,** a strophe, stanza, or poem consisting of eight verses.

octastyle, (in architecture) 1. having eight columns, as a portico.
2. such a portico.

Octateuch, *or* **Octoteuch,** the first eight books of the Old Testament, namely: 1, Genesis; 2, Exodus; 3, Leviticus; 4, Numbers; 5, Deuteronomy; 6, Joshua; 7, Judges; and 8, Ruth.

octave, 1. the eighth day of a festival, counting the feast-day as the first, as Low-Sunday, the octave of Easter.
2. a division, section, or group of tones consisting of eight tones (inclusive of the first and last).
3. the eighth of eight parries or guards against thrusts in fencing. Compare **1: prime, 2: seconde, 3: tierce, 4: quarte, 5: quinte, 6: sixte, 7: septime.**

8

Octavian, one of the eight members from the Secret Council appointed to a committee of finance by James VI, in 1595, to control the Royal Exchequer.

octavo, a book consisting of signatures, or formes, of eight leaves, or sixteen pages. When not otherwise specified, it is usually 6" × 9½".

octennial, 1. of or pertaining to a period of eight years.
2. a commemoration or celebration of an event that occurred eight years earlier; an eighth anniversary.
3. lasting for eight years; occurring once every eight years.

octic, (in mathematics) of the eighth degree or order.

octireme, an ancient vessel with eight banks of oars.

October, the eighth month of the year in the Roman calendar, reckoning from March.

octocerous, having eight arms or rays, as a cephalopod or octopus; octopod.

octofoil, 1. (in heraldry) a figure having eight lobes or subdivisions, like leaflets, used as a mark of cadency for the ninth son.
2. See **8: double quatrefoil.**

octoglot, 1. written in eight versions, each in a different language, as a lexicon.
2. a book so written.

octolateral, having eight sides.

octonary, *or* **octonal, 1.** consisting of eight.
2. computed by eights, as a numerical system based on eight. Compare **2: binary number, 8: octal number, 10: decimal system.**

octonocular, having eight eyes, as a spider; octophthalmous.

octopod, *or* **octopodous,** having eight feet or arms, as an octopus; octocerous.

octopus, a cephalopod with eight arms.

octosyllable, a word having eight syllables, as *antimaterialistic.*

octuple, eightfold.

octuplex, 1. of, pertaining to, or noting a system for sending eight different messages simultaneously over the same wire.
2. such a system.

ogdoad, 1. anything made up of eight parts, as an eight-line poem.
2. (in Gnosticism) **(a)** (in Basilidianism) a group of eight divine beings, the supreme god and seven direct emanations from him. **(b)** (in Valentinism) a group of eight divine beings called eons.

8

piece of eight, (formerly) the Spanish *peso duro,* worth eight reals.

Santa's reindeer, according to "A Visit from St. Nicholas," by Clement Moore, Santa Claus's sleigh is drawn by "eight tiny reindeer," namely: 1, Blitzen; 2, Comet; 3, Cupid; 4, Dancer; 5, Dasher; 6, Donder; 7, Prancer; and 8, Vixen.

section eight, *U. S.* a former Army regulation that provided for the discharge of a person found physically or mentally unfit to serve. Compare **8: ward eight, 1.**

"Tell me this riddle while I count eight," an old seventeenth-century riddling nursery rime:
> Purple, yellow, red, and green,
> The king cannot reach it nor yet the queen;
> Nor can Old Noll, whose power's so great:
> Tell me this riddle while I count eight.

"Old Noll" was the Royalists' nickname for Oliver Cromwell. The answer is a rainbow.

theta, the eighth letter of the Greek alphabet, corresponding to the English *th* of *thin.* It was sometimes considered unlucky because it was used by judges in condemning prisoners and it is the first letter of the Greek word for 'death,' *thanatos.*

truncated tetrahedron, a regular solid with eight plane faces. See also **13: Archimedean solid.**

vestibulocochlear nerve, either of a pair of nerves involved with the sense of hearing. Also called **acoustic nerve, eighth cranial nerve.**

ward eight, 1. *Military slang.* a hospital ward where the insane are treated. Compare **8: section eight.**
2. a cocktail made with lemon juice, sugar, grenadine, and whiskey, shaken with ice, strained into a glass with cracked ice, and decorated with slices of orange and lemon and with a cherry.

9

agretae, (in ancient Greece) the name of nine maidens chosen every year as priestesses of Athena in the island of Cos.

Amalthea, *or* **Amalthaea,** according to Varro, the name of the sibyl who offered to sell nine (or three, according to Pliny) books of prophetic oracles to Tarquin the Proud, king of Rome 534–510 BC. See also **10: sibyl.**

angels, orders of, according to Dionysius the Areopagite, there are nine orders: 1, seraphim and 2, cherubim (first circle); 3, thrones and 4, dominions (second circle); 5, virtues, 6, powers, 7, principalities, 8, archangels, and 9, angels (third circle).

archon, (in Greek antiquity) one of the nine chief judges at Athens, after the seventh century BC.

baseball team, called "nine" because of the nine defensive positions in the game.

Beatitudes, the, from Matthew 5: 3-12, namely:
1, blessed are the poor in spirit, for theirs is the Kingdom of Heaven;
2, blessed are they that mourn, for they shall be comforted;
3, blessed are the meek, for they shall inherit the earth;
4, blessed are they that hunger and thirst for righteousness, for they shall be filled;
5, blessed are the merciful, for they shall obtain mercy;
6, blessed are the pure in heart, for they shall see God;
7, blessed are the peacemakers, for they shall be called the sons of God;
8, blessed are they which are persecuted for righteousness' sake, for theirs is the Kingdom of Heaven;
9, blessed are ye, when men shall revile you and persecute you, and shall say all manner of evil against you falsely for My sake. Rejoice and be exceeding glad: for great is your reward in heaven.

bourgeois, a size of type on a nine-point body.

casting out nines, (in mathematics) a method for checking the accuracy of addition, subtraction, multiplication, or division, based on the fact that the remainder obtained when a number is divided by 9 is equal to the sum of its digits; hence, if the so-called check digit obtained for the quantities involved in a calculation is the same as the check digit for the answer derived, then the calculation was correct. However, casting out nines does not provide a check against the possibility of transposed numbers, that is, 456 instead of 465. See also **11: casting out elevens.**

cat, a cat is said to have nine lives because it displays an uncanny ability to survive all kinds of catastrophes.

cat-o'-nine-tails, a leather whip divided at its end into nine lashes, formerly used for punishment of miscreants.

Catonsville Nine, the, nine Viet Nam war protestors, joint defendants in a celebrated trial, who were charged with mutilating and destroying government property and with impeding Selective Service procedures in Catonsville, Maryland, in May, 1967, namely: 1, Daniel and 2, Philip F. Berrigan; 3, David Durst; 4, John Hogan; 5, Thomas P. Lewis; 6, Marjorie B. Melville; 7, Thomas Melville; 8, George Mische; and 9, Mary Moylan. See also **2: Berrigan Brothers.**

cubo-cubo-cube, the ninth power of a number, or the cube of the cube, as *the cubo-cubo-cube of 2 is 512* (2^9).

Curse of Scotland, the nine of diamonds, so called from its resemblance to the heraldic bearings of the Earls of Stair, one of whom was detested in Scotland as the principal author (while Master of Stair) of the massacre of Glencoe (1692).

9

Deucalion, (in Greek mythology) the son of Prometheus, and husband of Pyrrha, who, when Zeus decided to destroy all mankind by a flood, constructed a boat in which he and his wife floated for nine days before it landed on Mount Parnassus. In order to renew mankind, Zeus ordered him to throw behind them the stones from the hillside: those thrown by Deucalion became men, those by Pyrrha became women.

doctor, *Brit.* pill number 9 in the Field Medicine Chest, so called because it is prescribed so frequently.

Elaphebolion, the ninth month of the Greek calendar, from about the middle of March to the middle of April.

ennea-, a prefix in words of Greek origin, meaning 'nine.'

ennead, 1. the number nine.
2. a system of nine objects.
3. Enneads, the mystical doctrines of Plotinus (AD ?205–270), the primary and classical document of Neo-Platonism, so called because each of its six divisions contains nine books.

enneadic system, (in mathematics) a system of ten points such that on joining any one to all the rest the nine lines form an ennead.

enneadic system of numeration, a system of numbering by nines.

enneaeteric, containing or consisting of nine years, as a period.

enneagon, a plane figure having nine angles.

enneagonal number, a number of the form $\frac{1}{2}n\ (7n - 5)$, as 1, 9, 24, 46, etc.

enneahedron, *or* **enneahedria,** *pl.* **enneahedra.** a solid having nine faces.

enneander, (in botany) a plant having nine stamens.

enneapetalous, (in botany) having nine petals.

enneaphyllous, (in botany) (of a compound leaf) having nine leaflets.

enneasemic, (in ancient prosody) consisting of, having, or equal to nine morae, as an iambic or trochaic tripody.

enneasepalous, (in botany) having nine sepals.

enneaspermous, (in botany) having nine seeds.

enneastyle, (in architecture) consisting of or having nine columns.

enneasyllabic, containing or consisting of nine syllables, as a verse or word as, *interdenominationally.*

enneatical days, every ninth day of a disease.

enneatical years, every ninth year of a person's life.

Enneoctonus, a genus of shrikes, so called from the tradition that the shrike kills nine victims daily. Compare **9: nine-killer.**

Etruscan gods, the gods by which Lars Porsena swore were: **1,** Juno; **2,** Minerva; and **3,** Tinia; (the three chief); **4,** Hercules; **5,** Mars; **6,** Saturn; **7,** Summanus; **8,** Vedius; and **9,** Vulcan. [Benét, *Reader's Encyclopedia,* 2nd ed., 1965.]

flanconade, *or* **flanconnade,** (in fencing) the ninth and last thrust, usually aimed at the side.

fruits of the Holy Spirit, (in some Christian denominations) gifts that are granted by the Holy Spirit, namely: **1,** love; **2,** joy; **3,** peace; **4,** patience; **5,** graciousness; **6,** compassion; **7,** fidelity; **8,** meekness; and **9,** temperance.

gates of Hell, according to Milton, there are nine gates of Hell: three are brass, three iron, and three of adamantine rock.

glossopharyngeal nerve, either of a pair of nerves involved in the sense of taste and other senses and in the secretion of certain glands. Also called **ninth cranial nerve.**

Hydra, (in Greek mythology) a monstrous serpent or dragon of the lake of Lerna in Argolis, represented as having nine heads, each of which, being cut off, was immediately succeeded by two new ones unless the wound was cauterized. Its destruction was one of the labors of Hercules.

I, i, **1.** the ninth letter of the English alphabet.
2. the ninth item in a system where A = 1, B = 2, C = 3, etc., in which *I* is included.

innings, (in baseball) the usual number of innings, or turns at bat given each team is nine, unless the score is tied, in which case the game continues till one team has won.

iota, the ninth letter of the Greek alphabet, corresponding to the English *i.*

J, the ninth item in a system where A = 1, B = 2, C = 3, etc., in which *I* has been excluded.

L, l, the ninth consonant of the English alphabet.

look nine ways, to squint a great deal.

Louis IX, the sum of the digits of the year of his birth (1215) yield his titular number.

Lyric Poets, the nine lyric poets included in the Alexandrian Canon, namely: **1,** Alcaeus; **2,** Alcman; **3,** Anacreon; **4,** Bacchylides; **5,** Ibycus; **6,** Pindar; **7,** Sappho; **8,** Simonides; and **9,** Stesichorus.

9

mid-morn, nine o'clock in the morning.

Muses, the, nine daughters of Zeus and Mnemosyne (according to Hesiod), namely: 1, Calliope (of epic song or music); 2, Clio (of history); 3, Erato (of erotic poetry); 4, Euterpe (of lyric song); 5, Melpomene (of tragedy); 6, Polyhymnia or Polymnia (of sacred songs); 7, Terpsichore (of dance); 8, Thalia (of comedy and bucolic poetry); and 9, Urania (of astronomy). Also called **the Aganippides.** See also **3: Muses.**

nine, 1. the whole number between eight and ten.
2. one side in a game of baseball.

nine-banded armadillo, a species of armadillo having nine bands.

"nine days old," from an old nursery rime:
Pease porridge hot
Pease porridge cold,
Pease porridge in the pot
Nine days old.
Some like it hot,
Some like it cold,
Some like it in the pot
Nine days old.

Nine Days' Wonder, anything that causes a great sensation and then is quickly forgotten. [From the time it takes puppies and kittens, which are born blind, to open their eyes.] Compare **90: Ninety Days' Wonder.**

Nine Elms, an area of southwest London, since 1974 the new home of London's principal fruit and vegetable market, formerly located in Covent Garden. Battersea Power Station, a famous London landmark, is also located here. The name is thought to derive from nine huge elms which once grew on the site.

nine-eyed, having nine ('many') eyes, that is, spying, prying, etc.

ninefold, multiplied by nine; nonuple.

Nine Heavens, the, (in ancient cosmogony) the nine orbs or spheres in which the heavenly bodies were supposed to move, namely, those occupied by: 1, the moon; 2, Venus; 3, Mercury; 4, the sun; 5, Mars; 6, Jupiter; 7, Saturn; 8, the firmament (or "fixed" sphere); and 9, the crystalline. The earth was supposed to be at the center of the system.

nine-holes, (formerly) a piece of beef cut from the brisket. [So called from the holes left after the removal of the ribs.]

nine-killer, *or* **nine-murder,** a shrike, or butcher-bird, so called from its reputed killing of just nine birds a day.

nine-lived, having nine lives, as a cat.

nine men's morris, *or* **nine men's merels,** *or* **nine-penny morris,** an old game, referred to in Shakespeare's *A Midsummer Night's Dream* [II,i,98], similar to checkers, in which each side had nine pieces, or stones. Also called **fivepenny morris.**

ninepins, *or* **ninepegs,** a bowling game in which the object is to knock down as many pins as possible of the nine that are set up in a particular, usually triangular, pattern. Compare **10: tenpins.**

Nine Worthies, the, famous personages, often referred to by old writers and classed together, like the seven wonders of the world. They have been reckoned in the following manner: three Gentiles: 1, Hector; 2, Alexander; 3, Julius Caesar; three Jews: 4, Joshua; 5, David; 6, Judas Maccabaeus; and three Christians: 7, King Arthur; 8, Charlemagne; 9, Godfrey of Bouillon.

nineworthiness, a mock title formerly applied to someone as if he was one of or deserved to be ranked along with the **Nine Worthies.**

Nine Worthy Women, the, Brewer (*Reader's Handbook*) lists them as: 1, Minerva; 2, Semiramis; 3, Tomyris; 4, Jael; 5, Deborah; 6, Judith; 7, Britomart; 8, Elizabeth or Isabella of Aragon; and 9, Johanna of Naples.

ninth part of a man, a tailor: from the saying that nine tailors make a man.

nonagon, a plane figure with nine sides and nine angles.

nones, **1.** (in the ancient Roman calendar) the ninth day before the ides of the month. It was the fifth day, except in March, May, July, and October, when it was the seventh day. Compare **1: calends, 13: ides.**
2. (in the Roman Catholic and Greek churches, in religious houses, and as a devotional office in the Anglican communion) the office of the ninth hour, originally said at the ninth hour of the day (about 3 P.M.), or between noon and that hour.
3. *Obsolete.* the ninth hour after sunrise, or about three o'clock in the afternoon.

nonuple, ninefold.

noon, (according to Roman and ecclesiastical reckoning) the ninth hour of the day, namely the ninth hour from sunrise, or three o'clock in the afternoon. Compare **12: noon.**

November, the ninth month of the year in the Roman calendar, in which the New Year began on March first.

novena, (in the Roman Catholic Church) a devotion consisting of prayers said during nine consecutive days for the purpose of obtaining, through the intercession of the Virgin or of the particular saint to whom the prayers are addressed, some special blessing or mercy.

novenary, of or pertaining to the number nine.

9

novene, 1. relating to or depending on the number nine.
2. proceeding by nines.

novennial, 1. of or pertaining to a period of nine years.
2. a commemoration or celebration of an event that occurred nine years earlier; a ninth anniversary.
3. lasting for nine years; occurring once every nine years.

planet, one of the nine bodies revolving about the sun, namely, in their order from the sun: 1, Mercury; 2, Venus; 3, Earth; 4, Mars; 5, Jupiter; 6, Saturn; 7, Uranus; 8, Neptune; and 9, Pluto.

point, (in heraldry) one of the nine recognized positions on the shield which denote the locality of figures or charges. See illustration. A, dexter chief point; B, chief point; C, sinister chief point; D, honor point; E, fesse-point; F, nombril; G, base or flank point; H, dexter base point; I, sinister base point.

possession, possession is said to be nine points of the law, being equivalent to: 1, money to make good a claim; 2, patience to carry a suit through; 3, a good cause; 4, a good lawyer; 5, a good counsel; 6, good witnesses; 7, a good jury; 8, a good judge; and 9, good luck.

Prairial, the ninth month of the French revolutionary calendar, from about May 20th to June 18th.

Precious Stones, Nine, an order of knighthood, established in 1869, in Siam, for the Buddhist princes of the royal house. Also called the **Sacred Order.**

primary planet, one of the nine planets that orbit the sun, in contradistinction to the secondary planets, or satellites, which orbit the planets. See also **9: planet.**

religion and learning, house of, (in astrology) the ninth house, a cadent house. See **12: house.**

rule of nines, (in medicine) a method for estimating the percent of body surface affected by burns by allotting 9 percent to the head and to each arm, twice 9 percent to each leg, 9 percent to the back of the trunk, 9 percent to the front of the trunk, and 1 percent to the perineum.

Sabines, the nine gods of the Sabines, called the Novensiles, namely: 1, Aeneas; 2, Bacchus; 3, Esculapius; 4, Fides 'Faith'; 5, Fortuna; 6, Hercules; 7, Romulus; 8, Santa; and 9, Vesta.

September, the ninth month of the modern calendar. Compare **7: September.**

"stitch in time, a," an old proverb, "A stitch in time saves nine."

superlative, nine, "the superlative of superlatives in Eastern estimation. It is by nines that Eastern presents are given when the donor wishes to extend his bounty to the highest pitch of munificence." [Brewer, *Dictionary of Phrase & Fable,* 1894.]

Supreme Court, the highest judicial body in the United States, chiefly concerned with adjudicating matters dealing with violations of constitutional rights and as a court of last appeal in civil and criminal cases. It consists of one Chief Justice and eight Associate Justices, each of whom is appointed for life by the President with the approval of the Senate. Sometimes called **the nine old men.**

teth, the ninth letter of the Hebrew alphabet, corresponding approximately to English *t* and having a numerical value of 9.

to the nines, to perfection; fully; elaborately; generally applied to the way a person is attired, as, *She dressed to the nines.*

Valkyr, (in Scandinavian mythology) one of the nine handmaidens of Odin. Usually called *Walküre*.

Worthies of London, Nine, nine prominent men associated with London, namely: 1, Sir William Walworth, Lord Mayor, who stabbed Wat Tyler; 2, Sir Henry Pritchard, who feasted Edward III (with 5000 followers), Edward, the Black Prince, the king of Cyprus, and David, king of Scotland; 3, Sir William Sevenoke, who fought with the dauphin of France and built twenty almshouses and a free school (1418); 4, Sir Thomas White, who kept the citizens loyal to Queen Mary during the Wat Tyler rebellion (1553); 5, Sir John Bonham, commander of the army to stop Suleiman the Magnificent; 6, Christopher Croker, companion to Edward, the Black Prince, when he helped Don Pedro assume the throne of Castile; 7, Sir John Hawkwood, a knight of Edward, the Black Prince, and immortalized in Italian history as Giovanni Acuti Cavaliero; 8, Sir Hugh Caverley, who rid Poland of a monstrous bear; and 9, Sir Henry Maleverer ("Henry of Cornhill") who, as a crusader during the reign of Henry IV, became the guardian of "Jacob's well." [Benét, *Reader's Encyclopedia,* 2nd ed., 1965.]

10

animals in paradise, (in Muslim religion) ten animals are admitted into heaven, namely: 1, Kratim, the dog of the Seven Sleepers of Ephesus; 2, Balaam's ass, which scolded him; 3, the ant of Solomon, which remonstrates with the sluggard; 4, the whale of Jonah; 5, the ram of Ishmael, offered in place of sacrifice for Isaac; 6, the ass on which Jesus rode into Jerusalem; 7, the camel, miraculously produced from a rock by Saleh; 8, the cuckoo of Belkis; 9, the ox of Moses; and 10, Al Borak, the animal that bore Muhammad to heaven: it had the face of a man, the cheeks of a horse, the wings of an eagle, and spoke with a human voice.

10

Attic Orators,   the ten orators included in the Alexandrian Canon, namely: 1, Æschines; 2, Andocides; 3, Antiphon; 4, Demosthenes; 5, Dinarchus; 6, Hyperides; 7, Isaeus; 8, Isocrates; 9, Lycurgus; and 10, Lysias.

avatar, (in Hinduism) an incarnation of a god on earth as savior, especially one of the ten of Vishnu, namely: 1, Matsya (fish); 2, Kurma (tortoise); 3, Varaha (boar); 4, Narasinha (half lion, half man); 5, Vamana (dwarf); 6, Parashurama (man, as Rama with an ax); 7, Ramachandra (man, as Rama); 8, Krishna; 9, Buddha; and 10, Kalki (winged white horse).

big casino, *or* **great casino,** (in the card game of casino) the ten of diamonds.

Big Ten, the, 1. the heads of government and foreign secretaries of France, Great Britain, Italy, Japan, and the United States who met to draft the Treaty of Versailles at the Paris Peace Conference in 1919. See also **3: Big Three, the.**
2. a college athletic conference, consisting of: 1, the University of Illinois; 2, the University of Indiana; 3, the University of Iowa; 4, the University of Michigan; 5, Michigan State University, 6, the University of Minnesota; 7, Northwestern University; 8, Ohio State University; 9, Purdue University; and 10, the University of Wisconsin.

Bill of Rights,   the first ten amendments, called articles, to the Constitution of the United States which set forth fundamental legal guarantees accorded all of its citizens, namely:
I. Congress shall make no law respecting an establishment of religion, or prohibiting the free exercise thereof; or abridging the freedom of speech, or of the press; or the right of the people peaceably to assemble, and to petition the Government for a redress of grievances.
II. A well regulated Militia, being necessary to the security of a free State, the right of the people to keep and bear Arms, shall not be infringed.
III. No soldier shall, in time of peace be quartered in any house, without the consent of the Owner, nor in time of war, but in a manner to be prescribed by law.
IV. The right of the people to be secure in their persons, houses, papers, and effects, against unreasonable searches and seizures, shall not be violated, and no Warrants shall issue, but upon probable cause, supported by Oath or affirmation, and particularly describing the place to be searched, and the persons or things to be seized.
V. No person shall be held to answer for a capital, or otherwise infamous crime, unless on a presentment or indictment of a Grand Jury, except in cases arising in the land or naval forces, or in the Militia, when in actual service in time of War or public danger; nor shall any person be subject for the same offence to be twice put in jeopardy of life or limb; nor shall be compelled in any criminal case to be a witness against himself, nor be deprived of life, liberty, or property, without due process of law; nor shall private property be taken for public use, without just compensation.
VI. In all criminal prosecutions, the accused shall enjoy the right to a

speedy and public trial, by an impartial jury of the State and district wherein the crime shall have been committed, which district shall have been previously ascertained by law, and to be informed of the nature and cause of the accusation; to be confronted with the witnesses against him; to have compulsory process for obtaining witnesses in his favor, and to have the Assistance of Counsel for his defence.

VII. In Suits at common law, where the value in controversy shall exceed twenty dollars, the right of trial by jury shall be preserved, and no fact tried by a jury, shall be otherwise re-examined in any Court of the United States, than according to the rules of the common law.

VIII. Excessive bail shall not be required, nor excessive fines imposed, nor cruel and unusual punishments inflicted.

IX. The enumeration in the Constitution, of certain rights, shall not be construed to deny or disparage others retained by the people.

X. The powers not delegated to the United States by the Constitution, nor prohibited by it to the States, are reserved to the States respectively, or to the people.

Club of Ten, the, the nominal sponsor of advertisements which appeared from 1973 to 1978 in such newspapers as *The Times* [London], *The New York Times,* and the *Frankfurter Allgemeine Zeitung,* defending South Africa's policies. Readers complained that the advertisements were tasteless and misleading. Their origin was revealed in 1978, when the South African government closed the department responsible, amid investigations of financial corruption, in favor of a more orthodox public relations service.

Commandments, a woman's ten commandments are said to be the nails of her two hands with which she scratches the faces of those who offend her:

> Could I come near your beauty with my nails,
> I'd set my ten commandments in your face.
> [Shak., *2 Henry VI*, I,iii,141.]

Council of Ten, a secret tribunal instituted in the ancient republic of Venice, in 1310, and continuing to the overthrow of the republic in 1797. Composed at first of ten and later of seventeen members, it exercised unlimited power over internal and external affairs, often marked by great rigor and oppression.

count, the, (in boxing) the time it takes for the referee to count to ten (presumably, ten seconds); if a contestant who has been knocked down does not rise before the conclusion of the count, he is considered as having lost the match, and the other boxer is automatically declared the winner.

count to ten, slowly count from one to ten before reacting to a provocation, suggested as a device to those who are excitable and might do something rash and hasty if they do not take the time to think.

10

deca-, *or* **deka-,** an element in words of Greek origin, meaning 'ten.'

decacerous, having ten horns, legs, tentacles, or other processes.

decachord, a musical instrument having ten strings.

décade, (in the French revolutionary calendar) one of the three periods of ten days each into which each of the twelve months was divided, with five days left over. See also **5: Sans-culottides.**

decade, *or* **decad,** 1. the number ten.
2. *usually,* **decad.** in a Pythagorean or cabalistic sense, as an element of the universe, the tetractys or quaternary number, being $1 + 2 + 3 + 4$.
3. a set of ten objects; ten considered as a whole or unit.
3. *usually,* **decade.** a period of ten years.
4. (in music) a group of ten tones having precise acoustical relations with one another, arranged so as to explain and correct problems in harmony and modulation.
5. a division of a literary work containing ten parts or books.

decadianome, a quartic surface (dianome) having ten conical points.

decadist, one who writes a ten-part work.

decad ring, a finger ring, used like a rosary in numbering, having on its circumference ten knobs of one form for the aves, one for the pater, and sometimes a twelfth for the credo. Also called **rosary ring.**

decagon, a plane figure having ten sides and ten angles.

decagram, *or* **decagramme,** *or* **dekagram,** a weight of ten grams.

decagyn, (in botany) a plant having ten styles, or carpels.

decahedron, a solid having ten faces.

decalet, a stanza of ten lines.

decaliter, *or* **decalitre,** *or* **dekaliter,** a measure of capacity equal to ten liters.

decalitron, *pl.* **decalitra.** (in numismatics) the Syracusan name for the Attic standard didrachm (worth ten litra).

Decalogue, *or* **Ten Commandments, the** 1. the list of ten religious obligations revealed to Moses on Mt. Sinai by God (from Ex. 20:2-17 and Deut. 5:6-21). In the Jewish, Eastern Orthodox, Anglican communion, and most Protestant traditions they are: 1, I am the Lord thy God, thou shalt have no other gods before me; 2, Thou shalt not make unto thyself any graven image . . . thou shalt not bow down to them, nor serve them; 3, Thou shalt not take the name of the Lord thy God in vain; 4, Remember the Sabbath day, to keep it holy; 5, Honor thy father and thy mother; 6, Thou shalt not kill; 7, Thou shalt not commit adultery; 8, Thou shalt not steal; 9, Thou shalt not bear false witness against thy neighbor; and 10,

Thou shalt not covet thy neighbor's wife, nor thy neighbor's house. . . .
2. In the Roman Catholic and Lutheran traditions, they are listed as follows: **1**, I am the Lord thy God, Thou shalt have no other gods before me; **2**, Thou shalt not take the name of the Lord thy God in vain; **3**, Keep holy the sabbath day; **4**, Honor thy father and thy mother; **5**, Thou shalt not kill; **6**, Thou shalt not commit adultery; **7**, Thou shalt not steal; **8**, Thou shalt not bear false witness against thy neighbor; **9**, Thou shalt not covet thy neighbor's wife; and **10**, Thou shalt not covet thy neighbor's goods.

Decameron, The, (literally, 'ten days' work') a literary work by Giovanni Boccaccio (1313–75), Italian writer born in Paris. Written probably over the period 1349–51, it recounts the events of ten days of a fortnight's flight by seven women and three men from the plague in Florence in 1348. Comfortable in the countryside, each takes a turn to determine how the day is to be occupied by directing their activities, including, among other things, dances, songs, and storytelling. By the tenth day, each has told a story each day, for a total of 100 tales.

decameter, *or* **decametre,** *or* **dekameter,** a measure of length consisting of ten meters.

decanate, (in astrology) ten degrees, or a third of a zodiacal sign assigned to a planet, in which it has the least possible essential dignity.

decander, (in botany) a plant having ten stamens.

decangular, having ten angles.

decapetalous, (in botany) having ten petals.

decaphyllous, (in botany) having ten leaves.

decapod, **1.** an animal, as a crustacean, having ten legs, or, as a squid, having ten tentacles.
2. having ten arms, legs, rays, or tentacles.

Decapolis, a federation of ten Greek cities, originally in the Roman province of Syria, namely: **1**, Scythopolis; **2**, Hippos; **3**, Gadara; **4**, Abila; **5**, Dium; **6**, Pella; **7**, Gerasa; **8**, Philadelphia; **9**, Canatha, and **10**, Damascus. It was founded, probably by Pompey, about 63 BC and lasted till the 2nd century AD.

decare, a land measure equal to ten ares or one thousand square meters.

decasemic, (in ancient prosody) consisting of ten metrical units, or morae.

decasepalous, (in botany) having ten sepals.

decastere, *or* **dekastere,** a measure of volume equal to ten steres, or cubic meters.

decastich, a poem consisting of ten lines.

10

decastyle, (in architecture) **1.** having a portico with ten styles in front, or consisting of ten columns. **2.** such a portico.

decasyllabic, having ten syllables, as a verse in iambic pentameter, or a word, as *disestablishmentarianism.*

decathlon, an athletic contest for men in which competitors are awarded points for their performances in ten track and field events, namely: 1, 100-meter run; 2, 400-meter run; 3, 1500-meter run; 4, 110-meter high hurdles; 5, long jump; 6, high jump; 7, pole vault; 8, javelin throw; 9, discus throw; and 10, shot put.

December, the tenth month of the year in the Roman calendar, reckoning from March.

decemcostate, (in botany) having ten ribs or raised ridges, as certain fruits.

decemdentate, having ten teeth or points.

decemfid, divided into ten parts or lobes.

decemlocular, having ten cells, as ovaries.

decempedal, having ten feet; decapod.

decempennate, (in ornithology) having ten primaries, or flight-feathers on the pinion-bone, or manus.

decemvir, *pl.* **decemvirs, decemviri. 1.** one of ten men, the title of four differently constituted bodies in ancient Rome: **(a)** a body of magistrates elected in 451 BC for one year to prepare a system of written laws (decemviri legibus scribendis), with absolute powers of government, and succeeded by another for a second year, who ruled tyrannically under their leader, Appius Claudius, and aimed to perpetuate their own power but were overthrown in 449. The decemvirs of the first year completed ten of the celebrated twelve tables and those of the second year the remaining two, forming both a political constitution and a legal code. **(b)** a court of justice (decemviri litibus judicandis), of ancient but uncertain origin, which took cognizance of civil and, under the empire, also of capital cases. **(c)** an ecclesiastical college (decemviri sacris faciundus or decemviri sacrorum), elected for life from about 367 BC for the care and inspection of the Sibylline Books, etc.; increased to fifteen (quindecemvirs) in the first century BC. **(d)** a body of land commissioners (decemviri agris dividundis), occasionally appointed to apportion public lands among citizens. **2.** one of any official body of ten, as of the Council of Ten in Venice.

decemvirate, 1. the office or term of office of a body of decemvirs. **2.** a body of ten men in authority.

decennary[1], **1.** referring to ten, especially ten years.
2. a commemoration or celebration of an event that occurred ten years earlier; a tenth anniversary.

decennary[2], **1.** consisting of or involving ten each.
2. of or relating to a tithing.
3. (in old English law) a tithing consisting of ten freeholders and their families.

decenner, *Obsolete.* one of the freeholders of a decennary.

decennial, 1. of or pertaining to a period of ten years.
2. a commemoration or celebration of an event that occurred ten years earlier; a tenth anniversary.
3. lasting for ten years; occurring once every ten years.

decennium, a period of ten years.

decima, *pl.* **decimae.** (in music) an interval of ten diatonic degrees; an octave and a third.

decimal fraction, a fraction whose denominator is a power of 10, as $^3/_{100}$, $^7/_{1000}$, etc., usually written .03, .007, etc.

decimalism, the theory or system of a decimal notation or division, as of numbers, currency, weights, etc.

decimalization, the act of causing to conform to a decimal system, as the *decimalization* of British currency from the pound/shilling/penny system, in 1973.

decimal system, the system of writing numbers based on the powers of 10, in standard use today throughout the world. In an imperfect form, lacking the 0 (zero), it is believed to have been invented in India and is explained in the Latin geometry of Boëthius (d. *c.* AD 525). The system was disused in Europe until the invention of the zero, when it was reintroduced through the Arabs (by whom it is called the *Indian notation*), and was first explained in the work of Leonardo da Pisa, about 1200. The extension of the system to fractions was accomplished much later, by Stevinus, in 1582.

decimate, to select by lot and put to death every tenth man of (a captured army or body of prisoners or mutineers), a barbarity occasionally practised in antiquity. Compare $^1/_{100}$: **centesimation.**

decimestrial, *Rare.* consisting of or containing ten months.

decimo, (in Chile) a silver coin equivalent to ten centavos.

decimole, *or* **decuplet,** (in music) a group of ten notes which are to be played in the time of eight or four notes.

10

decuman, 1. designating the gate of a Roman military camp near which the tenth cohort of a legion was encamped.
2. large; immense, esp. of waves, because the tenth was said to be the largest.
3. (in astrology) one of the ten divisions of the ecliptic.
4. a large wave (see def. 2).

decuple, 1. tenfold; containing ten times as many.
2. a number repeated ten times.
3. to increase tenfold.

decuria, *pl.* **decuriae.** a company of ten, specifically, in Jesuit education, a minor division of a class in which chiefly memory lessons are heard.

decurion, 1. an officer in the Roman army who commanded a decury, or a body of ten soldiers.
2. any commander or overseer of ten; a tithing-man.

decury, 1. a body of ten men under a decurion.
2. the office or authority of a decurion.

decussis, *pl.* **decusses.** a large copper coin, ten times the value of an as, used in parts of Italy outside Rome during the third century BC.

dekadrachm, *or* **decadrachm,** an ancient silver coin of the value of 10 drachms, occasionally issued at Syracuse and in other parts of the Hellenic world.

dekass, a unit of mass; ten asses.

denarius, *pl.* **denarii.** the principal silver coin of the Romans under the republic and the empire, worth ten asses.

denary, *pl.* **denaries.** 1. containing ten; tenfold.
2. a division by tens; a tithing.
3. a denarius.

dicker, *Obsolete or provincial.* the number or quantity ten, especially, ten hides or skins, forming the twentieth part of a last of hides.

didecahedral, (in crystallography) (of a prism) having ten sides with pentahedral, or five-sided, bases.

dignities, house of, (in astrology) the tenth house. See **12: house.**

dime, 1. *Obsolete.* the number ten.
2. a coin of the United States, formerly of silver but more recently of silver sandwiched over baser metal alloy, worth ten cents or one tenth of a dollar.

dime a dozen, a, available cheaply; in abundant supply.

dime novel, *U. S.* **1.** *Archaic.* a story or novelette, usually lurid and sensationalistic, cheaply bound and formerly selling for a dime.
2. any cheap, sensational literary effort.

dime store, a retail shop where, formerly, many items were for sale for ten cents or less. Compare **5: five-and-ten-cent store.**

dix, a playing card with ten pips; a ten.

dizain, a poem of one or more stanzas, each of ten lines.

eagle, a former gold coin of the United States, worth ten dollars.

Fast of the Tenth Month, (in Judaism) a fast observed on the tenth day of Tebeth (the tenth month of the Jewish ecclesiastical calendar), for on that day occurred the siege of Jerusalem by Nebuchadnezzar.

hang ten, *Surfing slang.* go surfing, in facetious reference to the toes, used in gripping the board.

Hollywood Ten, the, the ten witnesses called in 1947 before the House Unamerican Activities Committee investigating Hollywood who chose to regard the Committee as unconstitutional and would not testify, namely: 1, Alvah Bessie; 2, Herbert Biberman; 3, Lester Cole; 4, Edward Dmytryk; 5, Ring Lardner, Jr.; 6, John Howard Lawson; 7, Albert Maltz; 8, Samuel Ornitz; 9, Adrian Scott; and 10, Dalton Trumbo. They were indicted and imprisoned for contempt. Also called **the Unfriendly Ten.**

Hygeia, the tenth planetoid, discovered in 1849.

instrument of ten strings, (in the Bible) a psaltery.

J, j, **1.** the tenth letter of the English alphabet.
2. the tenth item in a system where A = 1, B = 2, C = 3, etc., in which *I* is included.

K, the tenth item in a system where A = 1, B = 2, C = 3, etc., and *I* is excluded.

kappa, the tenth letter of the Greek alphabet, corresponding to the English *k.*

Knight Templar, the tenth, or highest degree of the York Rite of Masonry. See also **32: thirty-second degree Mason.**

long primer, a size of type on a ten-point body.

Lost Tribes of Israel, the ten tribes, remaining from the original twelve, under the leadership of Joshua, who led them to the Promised Land after the death of Moses, namely: 1, Asher; 2, Dan; 3, Ephraim; 4, Gad; 5, Issachar; 6, Manasseh; 7, Naphtali; 8, Reuben; 9, Simeon; and 10, Zebulon. These formed the northern Kingdom of Israel in 930 BC and were conquered by the Assyrians in 721 BC, after which they were assimi-

10

lated, or "disappeared." Legend has it that they will one day be found. See also **12: Twelve Tribes of Israel.**

M, m, the tenth consonant of the English alphabet.

Malebolge, the ten chasms into which the eighth circle of Hell is divided. See **8: Hell.**

Messidor, the tenth month of the year in the calendar of the French republic, from June 19th to July 18th.

mille, (in certain card games) a counter or chip worth ten points.

names of God, (in Christian incantatory formula) the ten divine names, namely: 1, El; 2, Elye; 3, Sabaoth; 4, Adonay; 5, Alpha; 6, Omega; 6, Messias; 8, Pastor; 9, Agnus; and 10, Fons. See also **7: names of God.**

not touch (something *or* someone) with a ten-foot pole, shun; avoid like the plague.

October, the tenth month of the year in the modern calendar, reckoning from January.

Pacific Ten, the, a college athletic conference originally composed of 1, the University of California (Berkeley); 2, the University of California at Los Angeles; 3, the University of Oregon; 4, Oregon State University; 5, the University of Southern California; 6, Stanford University; 7, the University of Washington; 8, Washington State University; and, since 1978, 9, the University of Arizona; and 10, Arizona State University. Formerly, without Arizona and Arizona State, called the **Pacific Eight.**

perfect "10," a, a thing or person that fulfills all one's ideal of rare beauty or perfection, especially a woman. [From the ultimate rating obtainable "on a scale of one to ten."]

quaternary number, the number ten, so called by the Pythagoreans because it is equal to 1 + 2 + 3 + 4.

Queen Anne's bounty, (in Roman Catholicism) the tenth part of the annual profit of every living in England, originally paid to the pope, but later transferred to the crown.

Rockefeller, John "Dime," a nickname of the multimillionaire, John D(avison) Rockefeller, 1839–1937. [From his practice of giving away dimes, especially to children, a plan conceived by his public relations counsel to improve Rockefeller's public image.]

sawbuck, *U. S. Slang.* a ten-dollar bill. [From the X shape of a sawbuck.]

Sephiroth, (in the cabala) the ten attributes or intelligences that form the Adam Kadmon ('Primordial man' = the Deity), namely: 1, Kether (the crown); 2, Hakmah (wisdom); 3, Binah (intelligence); 4, Hesed (mercy or love); 5, Pahad Geburah or Din (strength and justice); 6, Tiphereth (beauty); 7, Netsah (firmness or victory); 8, Hod (majesty or splendor); 9,

Yesod (foundation); and 10, Malkuth (kingdom). The Sephiroth divided into three categories called ʿAmudim (pillars), known together as Ets-Ḥaïm, or Tree of Life: numbers 2, 4, and 7 form the right pillar; 1, 6, 9, and 10 form the middle; and 3, 5, and 8 form the left (Pillar of Judgment). The Sephiroth are united by links forming three more triads: 1, 2, and 3 compose the first, or uppermost triad; 4, 5, and 6 form the second; and 7, 8, and 9 form the third.

sibyl, (in ancient mythology) one of ten (in common tradition) women reputed to possess special powers of prophecy or divination and intercession with the gods in behalf of those who resorted to them, namely: 1, Cimmerian; 2, Cumaean; 3, Delphian; 4, Erythrean; 5, Hellespontine or Trojan; 6, Libyan; 7, Persian or Babylonian; 8, Phrygian; 9, Samian; and 10, Tiburtine. Of these the Cumaean, named Amalthaea, was the most celebrated. According to legend, she appeared before Tarquin the Proud and offered to sell him nine books; when he refused, she burned three and offered those remaining at the same price. Again refused, she burned three more and offered the remaining three at the same price as the original nine. Tarquin bought the books, which were found to contain directions as to the worship of the gods and the policy of the Romans. These *Sibylline Books,* written in Greek hexameters, were carefully kept at Rome and consulted by oracle-keepers under direction of the senate. They were destroyed at the burning of the temple of Jupiter (83 BC). Fresh collections were made but these, too, were destroyed soon after AD 400.

stop on a dime, stop (usually a car or other vehicle) suddenly and in a very short space.

Tebeth, the tenth month of the Jewish ecclesiastical year, and the fourth month of the secular year, beginning in December (at the new moon) and ending in January.

ten, the whole number coming after nine and before eleven.

Ten, the, a group of American painters who exhibited together, formed (1898) in order to draw attention to their work, namely: 1, Frank W. Benson; 2, Joseph De Camp; 3, Thomas W. Dewing; 4, Childe Hassam; 5, Willard Leroy Metcalf; 6, Robert Reid; 7, E. E. Simmons; 8, Edmund Charles Tarbell; 9, John Henry Twachtman (replaced, on his death in 1902, by William Merritt Chase); and 10, J. Alden Weir. Also called **Ten American Painters.**

ten-dollar word, *Jocular.* a multisyllablic or unfamiliar word. Compare **5: five-dollar word.**

ten-fifteen, *U. S.* a radio code used by the police meaning 'civil disturbance.'

ten-fifty-five, *U. S.* a radio code used by the police meaning 'intoxicated driver.'

10

ten-fifty-seven, *U. S.* a radio code used by the police meaning 'hit and run accident.'

ten-fifty-two, *U. S.* a radio code used by the police meaning 'ambulance needed.'

ten-four, *U. S.* a radio code used by the police and on citizens band radio meaning 'acknowledgment; okay.'

ten-fourteen, *U. S.* a radio code used by the police meaning 'prowler report.'

ten-gallon hat, a high-crowned, broad-brimmed hat as worn by cowboys and others in the southwestern U. S., so called in facetious reference to the capacity of the crown.

ten-gauge shotgun, *or* **ten-bore shotgun,** a shotgun with a bore that accommodates a spherical bullet of a diameter such that ten weigh a pound. Compare **12: twelve-gauge shotgun.**

Ten Hours Act, an act of Parliament, in 1847, long sponsored by Lord Ashley (later, the seventh Earl of Shaftesbury), that forbade allowing children to work more than ten hours a day.

Ten Little Indians, a nursery rime. According to *The Oxford Dictionary of Nursery Rhymes,* Iona and Peter Opie, O.U.P., 1951 and 1975, in the original version, a song (1869) by Septimus Winner, the characters were ten little "Injuns"; but when the song was transferred by Frank Green to England, about 1869, reference was made to ten little niggers, negroes, darkies, or youthful Africans. There are several versions, but this is Winner's:

> Ten little Injuns standin' in a line,
> One toddled home and then there were nine;
> Nine little Injuns swingin' on a gate,
> One tumbled off and then there were eight.
> > One little, two little, three little, four little, five little Injun boys,
> > Six little, seven little, eight little, nine little, ten little Injun boys.
> Eight little Injuns gayest under heav'n,
> One went to sleep and then there were seven;
> Seven little Injuns cutting up their tricks,
> One broke his neck and then there were six.
> > One little, two little, . . . etc.
> Six little Injuns kickin' all alive,
> One kick'd the bucket and then there were five;
> Five little Injuns on a cellar door,
> One tumbled in and then there were four.
> > One little, two little, . . . etc.
> Four little Injuns up on a spree,
> One he got fuddled and then there were three;
> Three little Injuns out in a canoe,

One tumbled overboard and then there were two.
   One little, two little, . . . etc.
Two little Injuns foolin' with a gun,
One shot t'other and then there was one;
One little Injun livin' all alone,
He got married and then there were none.
   One little, two little, . . . etc.

A detective novel by Agatha Christie bore the title *Ten Little Niggers* when published in England, changed to *And Then There Were None* when it appeared in America.

"ten o'clock scholar, . . ." from an old nursery rime:
   A diller, a dollar, a ten o'clock scholar,
   What makes you come so soon?
   You used to come at ten o'clock,
   But now you come at noon.

tenpenny nail, 1. a nail that, before 1500, cost tenpence a hundred. 2. (in modern use) a nail 3 inches long.

tenpins, a bowling game played with ten pins set up in a triangular pattern at the end of an alley, the object being to knock down as many as possible with one or two rollings of a ball. Compare **9: ninepins.**

Ten-pound Act, a statute of the colony of New York (1769), replacing the Five-pound Act, that gave jurisdiction over civil cases involving sums not exceeding ten pounds to justices of the peace and to other local magistrates. Compare **5: Five-Pound Act.**

Ten Provinces, the Dominion of Canada, which is a federation of ten provinces and two territories, namely: **1,** Alberta; **2,** British Columbia; **3,** Manitoba; **4,** New Brunswick; **5,** Newfoundland; **6,** Nova Scotia; **7,** Ontario; **8,** Prince Edward Island; **9,** Quebec; **10,** Saskatchewan, and the Northwest Territories and Yukon Territory.

ten-seventy, *U. S.* a radio code used by the police meaning 'fire alarm.'

ten-seventy-eight, *U. S.* a radio code used by the police meaning 'officer needs assistance.'

tenspot, *U. S. slang.* 1. a ten-dollar bill. 2. the ten of any suit at cards.

ten-ten, *U. S.* a radio code used by the police meaning 'fight in progress.'

Tenth Avenue, a north-south street in New York City formerly associated with crime and violence, especially in the section between 40th and 55th Streets, known as "Hell's Kitchen." It is remembered in Richard Rodgers' ballet, "Slaughter on Tenth Avenue" (1936).

ten-thirty-four, *U. S.* a radio code used by the police meaning 'riot.'

10

ten-thirty-one, *U. S.* a radio code used by the police meaning 'crime in progress.'

ten-thirty-three, *U. S.* a radio code used by the police meaning 'emergency.'

ten-thirty-two, *U. S.* a radio code used by the police meaning 'man with gun.'

Ten Years' War, an abortive revolution (1868–78) of Cuban guerrillas against the corrupt Spanish colonial government in which some 200,000 died. It resulted in the abolition of slavery.

top ten, the ten most popular in a given category, as popular music recording artists, popular music records, books, best-dressed women, etc., as determined by a survey or by the expressed preferences of a selected group of professionals in the category.

tripos, (at Cambridge University) the list of successful candidates for honors in specified departments, namely, in the order of their founding: 1, mathematical; 2, classical; 3, moral sciences; 4, natural sciences; 5, theology; 6, law; 7, history; 8, Semitic languages; 9, Indian languages; and 10, medieval and modern languages.

Trojan War, the siege of Troy by the Greeks, the subject of Homer's *Iliad.* Prompted by the abduction of Helen, the wife of Menelaus, king of Sparta, by Paris, son of Priam, king of Troy, it lasted ten years, culminating in the destruction of Troy after Greek soldiers gained entrance to the city concealed in a huge wooden horse, presented as a "gift" by the Greeks.

vagus nerve, either of a pair of long nerves involved in speech, swallowing, and other functions and senses. Also called **tenth cranial nerve.**

Wicked Ten, the, in Macaulay's poem "Virginia," the epithet for the second decemvirate of ancient Rome, namely: 1, Titus Antonius Merenda; 2, Appius Claudius; 3, Marcus Cornelius Maluginensis; 4, Caeso Duillius; 5, Quintus Fabius Vibulanus; 6, Lucius Minucius; 7, Spurius Oppius Cornicen; 8, Quintus Poetelius; 9, Manius Rabuleius; and 10, Marcus Sergius. This commission of magistrates was appointed in 450 BC to hold executive power and complete the work of a similar board, which had been in office the previous year, in satisfying plebeian demands for a written code of laws. Appius Claudius was the leader of both. Legend makes the earlier group virtuous, and the second, despite the continuity of leadership and an admixture of plebeian members which the other had lacked, oppressive and arbitrary; Macaulay's poem, presented as an imaginative reconstruction of Roman balladry, incorporates this tradition.

Wilmington Ten, the, a group of nine black men and a white woman imprisoned in 1972 after racial disturbances in Wilmington, North Carolina, namely: 1, Benjamin Chavis; 2, Reginald Epps; 3, Jerry Jacobs; 4,

10

James McKoy; 5, Wayne Moor; 6, Marvin Patrick; 7, Connie Tindall; 8, Anne Shephard Turner; 9, Willi Earl Vereen; and 10, Joe Wright. They were convicted of arson, assault, and of conspiracy to commit these offenses. In 1978, after the principal witnesses for the prosecution had withdrawn their testimony, the United States government asked for the convictions to be overturned; this was done by an appeals court two years later, the prisoners having meanwhile been freed on parole.

wives, Muhammad had ten wives: 1, Ayishah, daughter of Abu Bekr, his favorite, married when she was nine years old, whom he regarded as one of the "three perfect women"; 2, Barra; 3, Hend; 4, Kadijah, his first wife and twenty-five years his senior, whose youngest child was Fatima; 5, Maimuna, who survived all of the others; 6, Rehana, who was Jewish; 7, Safiya, whose husband, Kenana, Muhammad put to death in order to marry her; 8, Souda; 9, Umm Habiba; and 10, Zainab. See also **15: concubines.**

X, the Roman numeral for 10.

yod, the tenth letter of the Hebrew alphabet, corresponding to the English *y, j,* or *i.*

11

accessory nerve, either of a pair of nerves involved in speech, swallowing, and the muscles of the head and shoulders. Also called **eleventh cranial nerve.**

casting out elevens, (in mathematics) a method for checking the accuracy of addition, subtraction, multiplication, or division, based on the fact that the difference between the sum of the digits in the odd places (counting from right to left with the units place as 1) and the sum of the digits in the even places of any number divisible by 11 is itself divisible by 11; hence, if the so-called check digit obtained for the quantities involved in a calculation is the same as the check digit for the answer derived, then the calculation was correct. See also **9: casting out nines.**

Chapter 11, a provision of the U. S. federal Bankruptcy Act for the relief of insolvent debtors and of their creditors in which, essentially, the ownership of the assets is transferred to a new entity owned by both creditors and debtor and the creditors' claims are satisfied by the future earnings of the entity rather than through liquidation of assets. Compare **7: Chapter 7, 13: Chapter 13.**

cricket team, called "eleven" because there are eleven players on each side.

Eleven, the, *or* **Hendeka,** (in ancient Athens) an administrative body of eleven persons, ten members chosen by lot and their secretary, whose duty was to superintend the prisons, receive arrested prisoners, and carry

11

out the sentences of the law. If the defendant pleaded "Not guilty," they adjudicated the case.

eleven-plus, *Brit.* an examination, formerly administered in England and Wales to 11- or 12-year-old children, to select candidates for grammar schools.

elevenses, *Brit. informal.* a light, mid-morning snack. Compare **4: fourings.**

eleventh chord, 1. a chord used in jazz, consisting of a triad (major or minor) with a superimposed seventh, ninth, and eleventh.
2. the interval between any tone and a tone of the eleventh diatonic degree above or below it; a compound fourth, or an octave and a fourth.
3. a tone distant by an eleventh from a given tone.

eleventh commandment, "Thou shalt not be found out."

eleventh hour, at the very last possible moment. Also used attributively, as *The condemned man was granted an eleventh-hour reprieve.*

football team, called "eleven" because of the eleven positions on each side.

friends and benefactors, house of, (in astrology) the eleventh house, a succedent house. See **12: house.**

hendecagon, *or* **endecagon,** a plane figure having eleven sides and eleven angles.

hendecahedron, a solid having eleven plane faces.

hendecasyllable, a metrical line having eleven syllables: *O you chorus of indolent reviewers.* [Tennyson, "Hendecasyllabics."]

jack, a playing card bearing a picture of a page and no specific numerical designation but in many games valued higher than the 10 and lower than the queen and king and (sometimes) the ace, and sometimes worth eleven. Also called **knave.**

K, k, 1. the eleventh letter of the English and of the Phoenician alphabets.
2. the eleventh item in a system where A = 1, B = 2, C = 3, etc., and *I* is included.

kaph, the eleventh letter of the Hebrew alphabet, corresponding to the Greek *kappa* and the English *k*.

L, the eleventh item in a system where A = 1, B = 2, C = 3, etc., and *I* is excluded.

lambda, the eleventh letter of the Greek alphabet, corresponding to the English *L, l*.

N, n, the eleventh consonant of the English alphabet.

11

natural, (in craps) a winning throw of dice showing a score of eleven on the first roll. See also **7: natural.**

November, the eleventh month of the modern calendar.

number eleven, *Old slang.* an umbrella.

O, the medieval Roman numeral for 11.

rule of eleven, (in bridge or whist) the principle that when the lead is of the fourth highest card of a suit, subtracting the value of that card from 11 yields the number of higher cards in the suit held by the other players.

small pica, a size of type on an eleven-point body.

soccer team, called "eleven" because there are eleven players on each side.

undecennial, 1. of or pertaining to a period of eleven years.
2. a commemoration or celebration of an event that occurred eleven years earlier; an eleventh anniversary.
3. lasting for eleven years; occurring once every eleven years.

11¼

point, one eighth of a right angle, or 11°15′, being the angle between the thirty-two adjacent points of the compass.

12

Ahijah, (in the Bible) a prophet who had a garment torn into twelve pieces, symbolizing the twelve tribes of Israel:
> And Ahijah caught the new garment that was on him, and rent it in twelve pieces: And he said to Jeroboam, Take thee ten pieces: for thus saith the Lord, the God of Israel, Behold, I will rend the kingdom out of the hand of Solomon, and will give ten tribes to thee: (But he shall have one tribe for my servant David's sake, the city which I have chosen out of all the tribes of Israel.)
> [I Kings 11:30-32.]

Arthur, battles of King, the twelve battles in which King Arthur defeated the Saxons, namely: 1, of the river Glem; 2, 3, 4, 5, (four battles) of the Duglas; 6, of Bassa; 7, of Celidon; 8, of Castle Gwenion; 9, of Caerleon (Carlisle); 10, of Trath Treroit; 11, of Agned Cathregonion (Edinburgh); and 12, of Badon Hill (Hill of Bath, now Bannerdown).

Asir, the twelve chief gods of Scandinavian mythology, namely: 1, Baldr; 2, Bragi; 3, Forseti; 4, Freya; 5, Heimdall; 6, Niord; 7, Odin; 8, Thor; 9, Tyr; 10, Ullur; 11, Vali; and 12, Vidar.

bed of Ware, a huge bed, large enough for twelve persons, on exhibit at the Victoria & Albert Museum, London.

12

boxcars, *Slang.* a pair of dice showing a score of six on each die.

daily dozen, calisthenic exercises taken informally by people at home to keep trim and fit. There may have been twelve different exercises at one time, or each exercise was to be performed twelve times; in either case, why *dozen* is unclear.

day, the number of hours during the day when daylight is available, varying according to latitude and the time of the year, but often reckoned from 8 A.M. to 8 P.M..

December the twelfth month of the modern calendar, with 31 days. Compare **10: December.**

didodecahedral, (in crystallography) (of a prism) having twelve sides with hexahedral, or six-sided, bases.

dihexagonal, twelve-sided, as a prism or pyramid or a double six-sided pyramid, or quartzoid.

dihexahedron, (in crystallography) a six-sided prism with trihedral summits.

dirhombohedron, a twelve-sided solid composed of a double hexagonal pyramid.

dodeca-, the first element in some compounds of Greek origin, meaning 'twelve.'

dodecafid, divided into twelve parts.

dodecagon, *or* **duodecagon,** a polygon with twelve sides and twelve angles.

dodecagyn, (in botany) a plant having twelve styles.

dodecahedron, *or* **duodecahedron,** a solid having twelve faces.

dodecameral, divided into twelve parts or series of twelve.

Dodecanese, a group of Greek islands in the Aegean Sea, originally consisting of twelve, namely: **1,** Astypalaia; **2,** Ikaria; **3,** Kalimnos; **4,** Karpathos; **5,** Kasos; **6,** Khalki; **7,** Leros; **8,** Lipsos; **9,** Nisiros; **10,** Patmos; **11,** Syme; and **12,** Tilos. Rhodes and Kos were added in 1912, and Kastellorizon in 1923. Also called, when under Turkish rule, **the Twelve Sporades.**

dodecapetalous, (in botany) having twelve petals.

dodecaphony, a twelve-tone system of musical composition developed and established by Arnold Schönberg (1874–1951) between 1908 and 1920.

dodecarch, one of a ruling body of twelve.

12

dodecasemic, (in prosody) consisting of twelve morae.

dodecastyle, (in architecture) **1.** (of a portico, etc.) having twelve columns in front.
2. a portico with twelve columns in front.

dodecasyllable, a word of twelve syllables, as *antidisestablishmentarianism.*

dodecatemory, a twelfth part, formerly used for a sign of the zodiac.

Dodekathlos, *or* **Twelve Labors of Hercules** (Greek **Heracles**), the twelve tasks imposed on Hercules by Eurystheus, son of Sthenelus, king of Argos, specifically (in their usual order): 1, capture of the Nemean Lion; 2, killing of the Lernean Hydra; 3, capture of the Arcadian Hind (or Stag); 4, capture of the Erymanthean Boar; 5, cleansing of the Augean Stables; 6, killing of the Stymphalian Birds; 7, capture of the Minotaur; 8, capture of the Diomedan Mares; 9, taking of the Girdle of Hippolyte, queen of the Amazons; 10, seizure of the Cattle of Geryon; 11, bringing the Golden Apples of the Hesperides; and 12, bringing up Cerberus from Hades.

douzain, a stanza of twelve lines.

douzaine, (in the Channel Islands) an administrative council of twelve elected men.

douzepere, *or* **doucepere,** one of the twelve peers (*les douze pairs*) of France.

dozen, twelve of anything.

duodecennial, **1.** of or pertaining to a period of twelve years.
2. a commemoration or celebration of an event that occurred twelve years earlier; a twelfth anniversary.
3. lasting for twelve years; occurring once every twelve years.

duodecimal, **1.** reckoning by twelves or powers of twelve; to the base 12.
2. one of a system of numerals the base of which is 12.

duodecimfid, divided into twelve parts.

duodecimo, **1.** a book size in which each leaf is one twelfth of the whole sheet of paper used in printing, usually $7\frac{2}{3}''$ × $5\frac{1}{8}''$ in the U. S. (when untrimmed), and corresponding to *crown octavo* in Britain.
2. a book with leaves of this size.
3. consisting of sheets folded into twelve leaves, measuring about $7\frac{2}{3}''$ × $5\frac{1}{8}''$. Usually written **12mo** or **12°**.

duodecim scripta, an ancient game, similar to backgammon, played with fifteen pieces on a board marked with twelve double lines.

12

duodecuple, consisting of twelves.

duodenary, 1. relating to the number twelve; twelvefold; increasing by twelves.
2. duodecimal.

duodene, (in musical theory) a group of twelve tones having precise acoustical relations with one another, arranged so as to explain and correct problems in harmony and modulation.

duodenum, (in anatomy) the first portion of the small intestine, connected directly to the stomach and extending from the pylorus to the jejunum, so called because its length is about twelve finger-breadths in man.

enemies or captivity, house of, (in astrology) the twelfth house, a cadent house. See **12: house.**

Epiphany, *or* **Twelfth Night,** January 6, the twelfth night after Christmas, commemorating the day that the infant Jesus was visited by the Magi and, later, marking the end of the Christmas season.

European Economic Community, an association of twelve western European countries formed to promote economic cooperation between its members. Founded in 1958 by 1, Belgium; 2, France; 3, Italy; 4, Luxembourg; 5, the Netherlands; and 6, West Germany; they were joined in 1973 by 7, Denmark; 8, Ireland; and 9, the United Kingdom; in 1981 by 10, Greece; and in 1986 by 11, Portugal; and 12, Spain. Also called the **Common Market, EEC, Euromarket.**

great dodecahedron, a regular solid having twelve plane faces each of which has the same boundaries as five covertical faces of an ordinary icosahedron. See also **4: Kepler-Poinsot solid.**

great stellated dodecahedron, a regular solid having twelve plane faces each of which is formed by stellating a face of the great dodecahedron. See also **4: Kepler-Poinsot solid.**

house, (in astrology) a twelfth part of the heavens, as divided by great circles drawn through the north and south points of the horizon, with six parts below the horizon and six above. Numbered round from east to south, beginning with that which lay in the east immediately below the horizon, the houses are named as follows: 1, the house of life; 2, of riches; 3, of brethren; 4, of parents; 5, of children; 6, of sickness; 7, of marriage; 8, of death; 9, of religion and learning; 10, of dignities; 11, of friends and benefactors; and 12, that of enemies and captivity. The succedent houses are the second, fifth, eighth, and eleventh. The cadent houses are the third, sixth, ninth, and twelfth.

hypoglossal nerve, either of a pair of nerves involved in swallowing and in tongue movement. Also called **twelfth cranial nerve.**

king, (in card games) a card, bearing a picture of a king and no specific numerical designation, but in many games valued higher than the queen and (sometimes) lower than the ace, sometimes given a nominal value of twelve.

Knights of the Round Table, 1. according to Dryden, the twelve knights and paladins of the Round Table were: 1, Launcelot; 2, Tristram; 3, Lamoracke; 4, Tor; 5, Galahad; 6, Gawain; 7, Gareth; 8, Palomides; 9, Kay; 10, Mark; 11, Mordred; and 12, one chosen from the following, who are shown on the frontispiece of the *History of Prince Arthur* (1470), by Sir Thomas Malory: Acolon; Ballamore; Beleobus; Belvoure; Bersunt; Bors; Ector de Maris; Ewain; Floll; Gaheris; Galohalt; Grislet; Lionell; Marhaus; Paginet; Pelleas; Percival; Sagris; Superabilis; and Turquine. 2. according to the *Mabinogion,* the twelve were: 1, Cadwr; 2, Launcelot; 3, Owain; 4, Kynon; 5, Aron; 6, Llywarch Hen; 7, Kai; 8, Trystan; 9, Gwevyl; 10, Gwalchmai; 11, Drudwas; and 12, Eliwlod.

L, l, 1. the twelfth letter of the English alphabet.
2. the twelfth item in a system where A = 1, B = 2, C = 3, etc., and *I* is included.

lamed, the twelfth letter of the Hebrew alphabet, corresponding to the Greek *lambda* and the English *L, l.* See also **30: lamed.**

M, m the twelfth item in a system where A = 1, B = 2, C = 3, etc., and *I* is excluded.

Meistersingers, *or* **Twelve Wise Masters, the,** the original corporation of the guilds of ordinary workingmen, established in the fourteenth century and continuing till the sixteenth, who cultivated music and poetry in the principal cities of Germany; it consisted of: 1, Master Altschwert; 2, Hans Blotz (barber); 3, Sebastian Brandt (jurist); 4, Hans Folz (surgeon); 5, Konrad Harder; 6, Heinrich von Mueglen; 7, Thomas Murner; 8, Master Muscablüt; 9, Master Barthel Regenbogen (blacksmith); 10, Hans Rosenblüt (armorial painter); 11, Hans Sachs (cobbler); and 12, Wilhelm Weber. Although Sachs (1494–1576) was not one of the originals, he is usually included because of his major contributions, among which were thirty-four folio volumes of manuscript containing 208 plays, 1700 comic tales, and some 450 lyric poems.

midnight, twelve o'clock at night.

mu, the twelfth letter of the Greek alphabet, corresponding to the English *m.*

Niflheim, there were twelve rivers in Niflheim, all flowing from the spring Hvergelmir. [Benét, *Reader's Encyclopedia,* 2nd ed., 1965.]

noon, midday; twelve o'clock in the daytime. Compare **9: noon.**

12

Olympians, the, the major (and original) gods and goddesses who dwelt on Mount Olympus, namely: 1, Aphrodite; 2, Apollo; 3, Ares; 4, Artemis; 5, Athena; 6, Demeter; 7, Dionysus; 8, Hephaestus; 9, Hera; 10, Hermes; 11, Poseidon; and 12, Zeus. Some sources include 13, Hades and 14, Hestia.

ordinary regular dodecahedron, a regular pentagonal dodecahedron.

ordinary year, a year of twelve months, as that of the Jewish calendar, based on a lunisolar calendar, occurring the first, second, fourth, fifth, seventh, ninth, tenth, twelfth, thirteenth, fifteenth, sixteenth, and eighteenth years of a nineteen-year cycle. Compare **13: embolismic year.**

P, p the twelfth consonant of the English alphabet.

Paladins of Charlemagne, the twelve paladins were: 1, Astolpho; 2, Ferumbras (or Fierabras); 3, Florismart; 4, Ganelon; 5, Maugris; 6, Namo or Nayme de Bavière; 7, Ogier the Dane; 8, Oliver; 9, Otuel; 10, Rinaldo; 11, Roland (or Orlando); and 12, one chosen from the following: Basin de Genevois; Geoffroy de Frises; Guerin, duc de Lorraine; Guillaume de l'Estoc; Guy de Bourgogne; Hoël, comte de Nantes; Lambert, prince de Bruxelles; Richard, duc de Normandy; Riol du Mans; Samson, duc de Bourgogne; and Thiery.

pentagonal dodecahedron, a dodecahedron whose twelve faces are each irregular pentagons. Also called **pyritohedron.**

petit jury, a body of persons charged with the function of determining the guilt or innocence of a person charged with a crime or of adjudicating the outcome of a civil case. In criminal cases, the jury is composed of twelve persons, sometimes with one or two alternates; in civil cases, there may be fewer. Compare **23: grand jury.**

pica, a size of type on a twelve-point body.

quorum of Twelve, (in the Mormon Church) the twelve apostles.

rhombic dodecahedron, a solid with twelve plane faces, each of which is a rhomb.

royal stag, a stag with twelve points on its antlers.

scalenohedron, a solid with twelve faces, each of which is a scalene triangle.

semidiurnal, accomplished in or lasting half a day (twelve hours). Compare **6: semidiurnal.**

sign, one of the twelve divisions of the zodiac, each comprising 30° of the ecliptic and marked as to position by a constellation, namely: 1, Aries (the Ram); 2, Taurus (the Bull); 3, Gemini (the Twins); 4, Cancer (the Crab); 5, Leo (the Lion); 6, Virgo (the Maid); 7, Libra (the Balance); 8, Scorpio (the Scorpion); 9, Sagittarius (the Archer); 10, Capricornus (the Goat); 11,

Aquarius (the Water-bearer); and 12, Pisces (the Fishes). Owing to the precession of the equinoxes, the signs have now moved away from the constellations from which they take their names into the next constellation to the west; (thus the sign Aries is now in the constellation Pisces, etc.).

Skirophorion, the twelfth, or last month of the ancient Attic calendar, containing 29 days and corresponding to the last part of June and first part of July.

small stellated dodecahedron, a regular solid having twelve plane faces each of which is formed by stellating each face of the ordinary dodecahedron. See also **4: Kepler-Poinsot solid.**

Titan, (in mythology) one of twelve of a race of primordial deities, children of Uranus (Heaven) and Gaea (Earth), consisting of six sons, namely: 1, Coeus; 2, Crius; 3, Hyperion; 4, Iapetus; 5, Kronos; and 6, Oceanus; and six daughters, namely: 7, Mnemosyne; 8, Phoebe; 9, Rhea; 10, Tethis; 11, Theia; and 12, Themis.

triakistetrahedron, a tetrahedron on each face of which a triangular pyramid has been erected, resulting in a solid with twelve surfaces.

Twelfth, (in Britain) the twelfth of August, the day when the grouse-shooting season opens officially.

Twelfth Avenue, the westernmost north-south street in New York City, running alongside the Hudson River from 23rd Street north. Before the construction of the West Side Highway above 72nd Street (1938), the tracks of the Hudson Division of the New York Central Railroad ran along its unprotected way, and careless or adventurous youngsters who crossed the tracks were often killed by the trains, leading it to be nicknamed "Death Avenue" by the Hearst newspapers of the day.

Twelfth-cake, a cake prepared for the festivities of Epiphany. Traditionally, a bean was introduced into its making and, when the cake was served, whoever got the piece with the bean became the bean-king and was entitled to preside over the ceremonies.

"Twelve Days of Christmas, the", a popular Christmas song in which the singer (presumably a woman) relates the gifts sent to her by her "true love," one gift the first day, two on the second (plus, again, the one from the first day) till, on the last day:

On the twelfth day of Christmas,
My true love sent to me
Twelve lords a-leaping,
Eleven ladies dancing,
Ten pipers piping,
Nine drummers drumming,
Eight maids a-milking,
Seven swans a-swimming,
Six geese a-laying,

12

> Five gold rings . . .
> Four colly birds,
> Three French hens,
> Two turtle doves,
> And a partridge in a pear tree.

Thus, counting the repetitions, the recipient receives 430 gifts in all. "Colly birds" are those that cock their heads from side to side. As this is a word generally unfamiliar today, they are usually called "calling birds."

twelve-gauge shotgun, *or* **twelve-bore shotgun,** a shotgun with a bore that accommodates a spherical bullet of a diameter such that twelve weigh a pound. Compare **10: ten-gauge shotgun.**

twelvepenny nail, 1. a nail that, before 1500, cost twelvepence a hundred.
2. (in modern use) a nail 3¼ inches long.

Twelve Tables, the tables on which were engraved and promulgated at Rome (451 and 450 BC) short statements of the rules of Roman law which were most important in daily life, first called the laws of the decemvirs. Ten were first set forth, then two were added.

Twelve Tribes of Israel, the Ten Tribes (See **10: Lost Tribes of Israel**), together with the tribes Judah and Benjamin, that constituted the original Hebrew tribes. At the time when the ten tribes established their kingdom in the north, (930 BC) Judah and Benjamin established the Kingdom of Judah in the south; they survived because they were allowed to return home after the Babylonian Exile (586 BC).

13

Archimedean solid, one of thirteen regular geometric solids, namely: 1, truncated tetrahedron (eight faces); 2, cuboctahedron (fourteen faces); 3, truncated cube (fourteen faces); 4, truncated octahedron (fourteen faces); 5, small rhombicuboctahedron (twenty-six faces); 6, great rhombicuboctahedron (thirty faces); 7, snub-cube (thirty-eight faces); 8, icosidodecahedron (thirty-two faces); 9, truncated dodecahedron (thirty-two faces); 10, truncated icosahedron (thirty-two faces); 11, small rhombicosidodecahedron (sixty-two faces); 12, great rhombicosidodecahedron (sixty-two faces); and 13, snub-dodecahedron (ninety-two faces).

baker's dozen, thirteen of anything. Originally, to avoid difficulty with the stringent laws governing short weight, bakers gave thirteen loaves, rolls, or other items for every dozen purchased.

13

bar mitzvah, 1. a Jewish boy on his thirteenth birthday, presumed to have reached the age at which he can obey the commandments and assume certain responsibilities.
2. the celebration of a Jewish boy's thirteenth birthday.

Chapter 13, a provision of the U. S. federal Bankruptcy Act for the relief of insolvent debtors and of their creditors in which, essentially, the debtor is either relieved of the obligation to pay creditors in full, or is given an extension of time in which to pay, or both. Compare **7: Chapter 7, 11: Chapter 11.**

Comic Poets, the thirteen comic poets included in the Alexandrian Canon, separated into three classes, namely: (Old Comedy): 1, Aristophanes; 2, Cratinus; 3, Epicharmus; 4, Eupolis; 5, Pherecrates; and 6, Plato; (Middle Comedy): 7, Alexis; 8, Antiphanes; (New Comedy): 9, Apollodorus; 10, Diphilus; 11, Menander; 12, Philemon; and 13, Philippides. Compare **13: Tragic Poets.**

devil's dozen, thirteen of anything, from the number of witches supposed to be present at a witches' sabbath.

embolismic year, a year of thirteen months, as that of the Jewish calendar, based on a lunisolar calendar, occurring the third, sixth, eighth, eleventh, fourteenth, seventeenth, and nineteenth years of a nineteen-year cycle. Compare **12: ordinary year.**

ides, the, the thirteenth day in the ancient Roman calendar, except in March, May, July, and October, when it was the fifteenth day. Originally, when the Roman months were strictly lunar, the ides was the day of the full moon. Compare **1: calends, 9: nones.**

koto, a Japanese musical instrument consisting of a long box over which are stretched thirteen strings of silk, each five feet in length and provided with a separate bridge. It is played by strumming, like a harp, with both hands, the tuning being effected by changing the position of the bridge.

Louis Treize, a style associated with the reign (1610–43) of Louis XIII (1601–43) in France, characterized in its early years by the influences of Italian art, especially that of Michelangelo Merisi da Caravaggio (1573–1610), a painter of realistic religious themes. Later, the paintings of Antoine (c1588–1648), Louis (c1593–1648), and Mathieu (1607–1677) Le Nain and of Peter Paul Rubens (1577–1640) were influential. The architecture of the period is represented by the Palais de Justice, at Rennes, the Palais du Luxembourg, and the church of the Sorbonne, both at Paris.

M, m, 1. the thirteenth letter of the English alphabet.
2. the thirteenth item in a system where A = 1, B = 2, C = 3, etc., and *I* is included.

13

Macheteros, los thirteen Puerto Ricans accused of robbing a Wells Fargo truck of $7.2 million, in 1985 in Hartford, Connecticut, to finance Puerto Rican independence, namely: 1, Carlos Ayes Suarez; 2, Luz María Berrios Berrios; 3, Isaac Camacho Negrón; 4, Elias Castro Ramos; 5, Luís Alfredo Colón Osorio; 6, Angel Díaz Ruíz; 7, Jorge Farinacci García; 8, Hilton Fernández Diamante; 9, Orlando González Claudio; 10, Ivón Meléndez Carrión; 11, Filiberto Ojeda Ríos; 12, Norman Ramírez Talavera; and 13, Juan Enrique Segarra III Palmer. Ann Gassin, originally charged with the others, turned government witness. Victor Gerena, an employee of Wells Fargo at the time of the robbery, is the prime suspect but was not apprehended. Also called **the Hartford Thirteen, the Hartford Fourteen, the Wells Fargo Thirteen.**

mem, the thirteenth letter of the Hebrew alphabet, corresponding to the Greek *mu* and the English letter *m*.

N, the thirteenth item in a system where A = 1, B = 2, C = 3, etc., and *I* is excluded.

nu, the thirteenth letter of the Greek alphabet, corresponding to the English *N, n*.

OPEC, an organization formed in 1960 of countries whose economies depend largely on oil exports in order to gain control over pricing and production, which had previously been determined by United States and European oil companies and by market forces, namely: 1, Algeria; 2, Ecuador; 3, Gabon; 4, Indonesia; 5, Iran; 6, Iraq; 7, Kuwait; 8, Libya; 9, Nigeria; 10, Qatar; 11, Saudi Arabia; 12, United Arab Emirates; and 13, Venezuela. Official name: **Organization of Petroleum Exporting Countries.**

original colonies, the American colonies as originally established by the British government and constituting, at the outbreak of the Revolutionary War (1775), a total of thirteen, namely: (royal colonies): 1, Georgia; 2, Massachusetts; 3, New Hampshire; 4, New Jersey; 5, New York; 6, North Carolina; 7, South Carolina; and 8, Virginia; (proprietary colonies): 9, Delaware; 10, Maryland; and 11, Pennsylvania; (corporate colonies): 12, Connecticut; and 13, Rhode Island. See also **24: Twenty-four Proprietors.**

Q, q, the thirteenth consonant of the English alphabet.

riddle of claret, thirteen bottles of claret, a magnum and twelve quarts, traditionally presented on a riddle, or sieve, with the magnum at the center surrounded by the other bottles.

stripes, there are thirteen stripes in the American flag, six white and seven red, symbolizing the thirteen original colonies, possibly adopted from the flag of the Sons of Liberty. It became official on June 14 (since celebrated as Flag day), 1777, by resolution of the Continental Congress. In 1795, when there were fifteen states, two stripes were added as well as

13

two stars; but the 13-stripe flag was returned to in 1818, by which time there were twenty states (and twenty stars).

thirteen, the whole number between twelve and fourteen, sometimes superstitiously considered unlucky because it was the number who attended the Last Supper. See also **13: triskaidekaphobia.**

Thirteen Years' War, a war between Poland and Russia, 1654–67, ending in the truce of Andruszowo.

Tragic Poets, the thirteen tragic poets included in the Alexandrian Canon, separated into two classes, namely: (first class): 1, Achaeus; 2, Æschylus; 3, Agathon; 4, Euripides; 5, Ion; and 6, Sophocles; (second class or **Tragic Pleiades**): 7, Æantides; 8, Alexander the Ætolian; 9, Homer the Younger; 10, Lycophron; 11, Philiscus of Corcyra; 12, Sosiphanes or Sosicles; and 13, Sositheus. Compare **13: Comic Poets.**

triskaidekaphobia, fear of the number 13, especially in regarding it as unlucky. See also **13: unlucky number.**

unlucky number, the superstition that it is unlucky for thirteen to sit down to dinner together at the same table, because one of them will die before a year has passed, comes from the Last Supper, attended by Christ and his twelve disciples. Christ was crucified, and Judas Iscariot hanged himself. This superstition has been extended to other applications of 13, as in hotel room numbers, numbering of floors in hotels, apartment houses, and office buildings, etc.; it also is applied, particularly, to the thirteenth day of a month that falls on a Friday, deemed as an especially unlucky time.

Veadar, the thirteenth, or intercalary month added to the Jewish year after Adar.

14

Apollo 14, a U. S. spacecraft that landed on the moon, February 5, 1971. The mission was the fourth of six manned lunar landings during the Apollo space program.

Bastille day, July 14, a national holiday in France and among French people everywhere, commemorating the fall on that date, in 1789, of the Bastille, an ancient prison in Paris that symbolized the oppression of the masses.

carbon-14, a radioactive isotope of carbon, the decay of which in deceased organisms is the basis of radiocarbon dating, a scientific means of obtaining fairly accurate estimates of the age of archaeological samples from 500 to 50,000 years old.

14

demobilization furlough, a leave of fourteen days accorded to members of the British armed services, 1918–19, on demobilization.

English, a size of type on a fourteen-point body.

Flag day, the 14th day of June, established to commemorate the adoption by Congress on that date, in 1777, of the Stars and Stripes as the official emblem of the United States.

fortnight, a period of fourteen nights; two weeks.

fourteen penn'orth of it, *Brit. Slang.* fourteen years' transportation.

fourteen-point-one continuous pocket billiards, a form of pocket billiards, played in championship competition, in which 15 numbered object balls and one white cue ball are used. Each player must identify the pocket into which he is driving each ball. When 14 object balls have been pocketed, they are again racked up. Play continues until one player has scored 150 or 125 points.

Fourteen Points, the, a pronouncement by President Woodrow Wilson, made January 8, 1918, of the war aims of the Allies in World War 1, renunciation of secret diplomacy; 2, freedom of navigation of the seas; 3, freedom of trade; 4, reduction of arms; 5, settlement accommodating the peoples of the colonies as well as the colonialist powers; 6, respect for Russia's right of self-determination; 7, restoration of Belgium; 8, German withdrawal from France and a settlement of the Alsace-Lorraine dispute; 9, readjustment of the frontiers of Italy along ethnic lines; 10, prospect of autonomy for the peoples of Austria-Hungary; 11, restoration of Romania, Serbia, and Montenegro, free access to the sea for Serbia, and international guarantees for the Balkan states; 12, prospect of autonomy for non-Turkish peoples of the Ottoman Empire and unrestricted opening of the Straits, but secure sovereignty for the Turks in their own areas; 13, independence for Poland with access to the sea; and 14, "a general association of nations" to guarantee every state's integrity. See also **4: Four Ends, 4: Four Principles, 5: Five Particulars.**

fourteenth, (in music) an interval one diatonic degree less than two octaves.

Fourteenth Amendment to the U. S. Constitution, proposed by resolution of Congress on June 13, 1866. By a concurrent resolution of Congress adopted July 21, 1868, it was declared to have been ratified by "three fourths and." Records of the Department of State show that the 14th Amendment was subsequently ratified by three more of the States. It was rejected by Kentucky and New Jersey. The Amendment forbids any state from denying or abridging the federal rights of citizens or their right to equal protection under the law. The Supreme Court applied the equal-protection principle of the Amendment in *Brown v. Board of Education of Topeka* (1954), which struck down the policy of racial segregation in schools. The text is as follows:

Section 1. All persons born or naturalized in the United States, and subject to the jurisdiction thereof, are citizens of the United States and of the State wherein they reside. No State shall make or enforce any law which shall abridge the privileges or immunities of citizens of the United States; nor shall any State deprive any person of life, liberty, or property, without due process of law; nor deny to any person within its jurisdiction the equal protection of the laws.

Section 2. Representatives shall be apportioned among the several States according to their respective numbers, counting the whole number of persons in each State, excluding Indians not taxed. But when the right to vote at any election for the choice of electors for President and Vice President of the United States, Representatives in Congress, the Executive and Judicial officers of a State, or the members of the Legislature thereof, is denied to any of the male inhabitants of such State, being twenty-one years of age, and citizens of the United States, or in any way abridged, except for participation in rebellion, or other crime, the basis of representation therein shall be reduced in the proportion which the number of such male citizens shall bear to the whole number of male citizens twenty-one years of age in such State.

Section 3. No person shall be a Senator or Representative in Congress, or elector of President and Vice President, or hold any office, civil or military, under the United States, or under any State, who, having previously taken an oath, as a member of Congress, or as an officer of the United States, or as a member of any State legislature, or as an executive or judicial officer of any State, to support the Constitution of the United States, shall have engaged in insurrection or rebellion against the same, or given aid or comfort to the enemies thereof. But Congress may by a vote of two-thirds of each House, remove such disability.

Section 4. The validity of the public debt of the United States, authorized by law, including debts incurred for payment of pensions and bounties for services in suppressing insurrection or rebellion, shall not be questioned. But neither the United States nor any State shall assume or pay any debt or obligation incurred in aid of insurrection or rebellion against the United States, or any claim for the loss or emancipation of any slave; but all such debts, obligations and claims shall be held illegal and void.

Section 5. The Congress shall have power to enforce, by appropriate legislation, the provisions of this article.

Henry IV, the number 14 is associated with him in the following ways: 1, there are 14 letters in his name (Henri de Bourbon); 2, after the extinction of the family of Navarre, he was the 14th king of France and Navarre; 3, he was born on December 14; 4, the year of his birth was 1553, whose digits total 14: 1553 was 14 centuries, 14 decades, and 14 years from the birth of Christ; 5, his first wife, Marguerite Valois, was born on May 14, also in 1553; 6, 56 years (4 × 14) after Henry II ordered the enlargement of the Rue de la Ferronnerie, on May 14, 1554, Henry IV was assassinated there; 7, he won his victory at Ivry on March 14, 1590; 8, on May 14, 1590, demonstrations were organized against him in Paris; 9, he was

14

placed under papal ban by Gregory XIV on November 14, 1592; **10,** his son, Louis XIII, was baptized on September 14, 1606; and **11,** he was assassinated on May 14, 1610.

Holy-Cross day, a festival observed by the Orthodox, Roman Catholic, and Anglican communion on September 14th, in commemoration of the exaltation of the alleged cross of Christ after its recovery from the Persians, AD 628. Also called **Holyrood day.**

Irene, the fourteenth planetoid, discovered in 1851.

line, (in typography) a line of agate type which, for various purposes, is calculated as being $1/14$ of an inch in depth, requiring 14 to make an inch. Also called **agate line.**

Louis Quatorze, a style associated with the reign (1643–1715) of Louis XIV (1638–1715) ("the Sun King") in France, characterized by formal Baroque grandeur, with gilded and veneered furniture, elaborately ornate architecture (as exemplified, especially, by the palace at Versailles, designed mainly by LeBrun), the emergence of decorative landscaping under the direction of André Le Nôtre (1613–1700), and painting chiefly under the influence of Nicolas Poussin (1594–1665), who reflected the Classicism of the Italian school, and of Giovanni Lorenzo Bernini (1598–1680). It was during Louis XIV's reign that the Gobelins factory was founded to supply lavish furnishings for the palaces and public buildings. See also next entry.

Louis XIV, king of France from 1643 until his death on September 1, 1715. The number 14 is associated with him in the following ways: **1,** he was the 14th king of France to bear his name; **2,** the first year of his reign was 1643, whose digits total 14; **3,** the year of his death was 1715, whose digits total 14; **4,** he lived for 77 years, whose digits total 14; and **5,** the year of his birth (1638) added to that of his death (1715) total 3353, whose digits total 14. See also preceding entry.

N, n, 1. the fourteenth letter of the English alphabet.
2. the fourteenth item in a system where A = 1, B = 2, C = 3, etc., and *I* is included.

Nivôse, the fourth month of the French revolutionary calendar, beginning about December 21st and ending about January 19th.

nun, the fourteenth letter of the Hebrew alphabet, corresponding to the Greek *nu* and the English *N, n*.

O, the fourteenth item in a system where A = 1, B = 2, C = 3, etc., and *I* is excluded.

Quartodeciman, a member of an early Christian community that celebrated the Paschal festival on the fourteenth day of Nissan, the same day as that on which Jews celebrate Passover, regardless of the day of the week.

quatorzain, a poem consisting of fourteen lines, or verses; a sonnet.

quatorze, (in the game of piquet) the four aces, kings, queens, jacks, or tens, so called because in the hand that holds the highest, such a group of four counts fourteen points.

R, r, the fourteenth consonant of the English alphabet.

sonnet, a poem consisting of 14 lines, usually in iambic pentameter, in one of the following patterns: a Shakespearean sonnet has a rime scheme *abab/cdcd/efef/gg*; a Petrarchan sonnet has an octave with a rime scheme *abbaabba* followed by a sestet with a rime scheme *cdecde, cdcdcd,* or of some other pattern; and a Spenserian sonnet has a rime scheme *abab/bcbc/cdcd/ee.*

stations of the cross, a series of fourteen events associated with the Passion of Christ, depicted in paintings or sculptures usually inside a church as shrines to which the devout pay their devotions: 1, condemnation by Pilate; 2, reception of the cross; 3, Christ's first fall; 4, meeting with His mother; 5, Simon of Cyrene carrying the cross; 6, Veronica wiping the face of Jesus; 7, Christ's second fall; 8, exhortation to the women of Jerusalem; 9, Christ's third fall; 10, stripping of the clothes; 11, crucifixion; 12, death; 13, descent from the cross; and 14, burial. According to the *Encyclopaedia Britannica,* 15 is sometimes added: finding of the cross by Helena.

stone, a unit of weight, equal to fourteen pounds avoirdupois, used in England in giving the weight of a person.

St. Valentine's day, February 14th, a day when friends and lovers affirm their affection for one another by greeting cards, gifts, etc.

Tantras, the Four Great, (in Tibetan Buddhism) four manuals, or sections of the Kagyur, namely: the Tantra of Activities (Kriya); the Tantra of Application (Charya); the Tantra of Perfection (Yoga); and the Tantra of Supreme Perfection (Anuttara Yoga), which are supposed to lead to realization of Supreme Truth by means of magical rites, spells, symbolisms, and meditation and through the development of intuitive wisdom and compassion.

tessaradecad, a group of fourteen items or people.

tessarescaedecahedron, a geometric solid having fourteen faces, as a cuboctahedron, truncated octahedron, or truncated cube.

tetradecapod, *or* **tetradecapodous,** having fourteen feet, as an order of crustaceans.

Trisala, 14 auspicious dreams of, (in Buddhism) Trisala was the mother of Mahavira, the latest Jaina saint of India. In Jaina tradition, like the mothers of the other twenty-three saints (Jinas), Trisala had fourteen dreams at the time of Mahavira's conception. The occurrence of all four-

14

teen dreams may portend the birth of a universal monarch (*çakravartin*), and the dreams and their interpretation are a recurrent motif in Jaina literature: representations of them are woven into rugs and hangings, illustrated in manuscripts dealing with the lives of the saints, and carved or painted on ritual objects and on the lintels of temple doors. They include: 1, a white elephant (symbolizing the universal monarch); 2, a white bull radiating light (symbolizing the spread of knowledge by the future child); 3, a leaping white lion (symbolizing the baby's conquest over his enemies and over karman, his past actions); 4, the goddess Šri, or Lakšmi, seated on a lotus anointed by a pair of elephants (symbolizing him as an anointed king); 5, a garland of fragrant flowers; 6, a white moon bearing the mark of a deer; 7, a red, radiant sun (symbolizing his dispelling of the darkness of ignorance); 8, an auspicious banner (Svetambara sect) or a pair of fishes (Digambara sect); 9, a golden water-filled ewer (symbolizing his immersion in meditation); 10, a lotus lake filled with flowers and bees; 11, a milky ocean (symbolizing the perfect knowledge he will attain); 12, a city or chariot bedecked with jewels; 13, a large, jewel-filled vase (symbolizing right knowledge, right faith, and right conduct); and 14, a fire fueled with ghee (symbolizing the illumination of the universe by his wisdom).

truncated cube, a regular solid with fourteen plane faces. Also called **tessarescaedecahedron, cuboctahedron, cubo-octahedron.** See also **13: Archimedean solid.**

truncated octahedron, a regular solid with fourteen plane faces. See also **13: Archimedean solid.**

xi, the fourteenth letter of the Greek alphabet, corresponding to the English *X, x*.

15

concubines, Muhammad had fifteen concubines; Mariyeh, the mother of his son Ibrahim, was his favorite. See also **10: wives.**

Danelagh, the counties where the Danes settled in England, namely: 1, Bedfordshire; 2, Buckinghamshire; 3, Cambridgeshire; 4, Derbyshire; 5, Essex; 6, Hampshire; 7, Hertfordshire; 8, Leicestershire; 9, Lincolnshire; 10, Middlesex; 11, Norfolk; 12, Northamptonshire; 13, Nottinghamshire; 14, Suffolk; and the territory of 15, Northumbria.

Fifteen, the, the first Jacobite rebellion (1715), when James Edward Stuart, the "Old Pretender," son of James II, failed to gain the throne of England. See also **45: Forty-five.**

fifteen decisive battles, according to Sir Edward Shepherd Creasy (*Fifteen Decisive Battles of the World,* 1851), they were: 1, defeat of 100,000 Persians by 10,000 Greeks under Miltiades at Marathon, 490 BC; 2, defeat of the Athenians at Syracuse, 413 BC; 3, defeat of Darius by Alexander at Arbela, 331 BC; 4, defeat of Hasdrubal at Metaurus, 207 BC; 5, defeat of the

Romans by the Gauls, AD 9; **6**, defeat of Attila at Chalons, AD 451; **7**, defeat of the Saracens by Charles Martel at Tours, AD 732; **8**, defeat of Harold II by William of Normandy at Hastings, 1066; **9**, victory by Joan of Arc at Orléans, 1429; **10**, defeat of the Spanish Armada, by the British, 1588; **11**, defeat of the French by Marlborough at Blenheim, 1704; **12**, defeat of Sweden by Peter the Great at Pultowa, 1709; **13**, defeat of General Burgoyne by General Gates at Saratoga, 1777; **14**, defeat of Germany by France at Valmy, 1792; and **15**, defeat of Napoleon by the Duke of Wellington at Waterloo, 1815.

"Fifteen men on a Dead Man's Chest," a song sung by the pirates in *Treasure Island* (1883), by Robert Louis Stevenson:
Fifteen men on a Dead Man's chest—
Yo-ho-ho, and a bottle of rum!
Drink and the devil had done for the rest—
Yo-ho-ho, and a bottle of rum!

fifteen-puzzle, *or* **15-puzzle,** a puzzle consisting of fifteen square numbered pieces fitted into a frame with one square left open so that the others can be slid about to form numerical patterns.

Fifteen Years' War, a series of military engagements (1591–1606) between Austria and the Ottoman Turks that blocked the Turkish expansion into Europe and helped to establish Transylvania as an eastern European power.

Louis Quinze, a style associated with the reign (1715–74) of Louis XV (1710–74) in France, characterized by the culmination of superior craftsmanship in furniture and furnishings and by elaborate Rococo decoration in which carving, inlaying of rare woods, ivory, and metal, and Chinese lacquering were brought to a fine art.

O, o, **1.** the fifteenth letter of the English alphabet.
2. the fifteenth item in a system where A = 1, B = 2, C = 3, etc., and *I* is included.

omicron, the 15th letter of the Greek alphabet, corresponding to the English *O, o.*

O's of St. Bridget, the, *or* **Fifteen O's, the** fifteen meditations on the Passion of Christ, composed by St. Bridget, each beginning with *O Jesu* or a similar invocation.

P, the fifteenth item in a system where A = 1, B = 2, C = 3, etc., and *I* is excluded.

points of a good horse, as recorded in Wynkyn de Worde (1496):
A good horse sholde have three propyrtees of a man, three of a woman, three of a foxe, three of a haare, and three of an asse. Of a *man*, bolde, prowde, and hardye. Of a *woman*, fayre-breasted, faire of heere, and easy to move. Of a *foxe*, a faire taylle, short eers,

15

with a good trotte. Of a *haare*, a grate eye, a dry head, and well rennynge. Of an *asse*, a bygge chynn, a flat legge, and a good hoof. [Brewer, *Reader's Handbook*, 1889.]

quince, a card game similar to 21, blackjack, or vingt-et-un, in which the object is to count fifteen or as near as possible to it without exceeding it.

quindecad, a set or series of fifteen.

quindecagon, a plane figure with fifteen sides and fifteen angles. Also called **pendecagon.**

quindecemvir, (in Roman antiquity) one of a body of fifteen magistrates who, at the close of the republic, had charge of the *Sibylline books*. They succeeded the decemvirs who were keepers of the books from 367 BC and who continued the functions of the duumvirs, two patricians of high rank who kept the books under the kings. See also **10: decemvir, 1, c.**

quindecennial, 1. of or pertaining to a period of fifteen years.
2. a commemoration or celebration of an event that occurred fifteen years earlier; a fifteenth anniversary.
3. lasting for fifteen years; occurring once every fifteen years.

quinzain, *or* **quinzaine,** the fifteenth day after a feast-day, including the feast-day.

rugby, the number of players on each side in the sport of rugby football is fifteen.

S, s, the fifteenth consonant of the English alphabet.

samekh, the fifteenth letter of the Hebrew alphabet, corresponding to the Greek *sigma* and the English *S, s.*

song of degrees, a title given to 15 psalms, cxx to cxxxiv.

St. Swithin's day, July 15th, a festival in honor of St. Swithin, bishop of Winchester, 852–862. See also **40: St. Swithin's day.**

XV, the Roman numeral for the number fifteen.

16

arhat, (in Buddhism) one of the sixteen apostles whose duty it is to spread the teachings of Buddha. Compare **18: lohan.**

ayin, the sixteenth letter of the Hebrew alphabet.

dioctahedral, (in crystallography) having the form of an octahedral prism with tetrahedral summits.

hexadecimal, 1. pertaining to sixteen.
2. a quantity represented in the hexadecimal system, in which 16 is the base, or radix. Figures used are 0 through 15, or, more commonly, 0

through 9 and A through F. The quantity represented by 140 in the decimal system (base 10) would be shown as 8C in the hexadecimal system: (8 × 16^1) + (12 × 16^0). Compare **2: binary number; 8: octal number.**

in sixteens,   (in typography and bookbinding) describing a form of imposed type or plates containing sixteen pages, or a book having sixteen pages in each signature.

Louis Seize,   a style associated with the reign (1774–93) of Louis XVI (1754–93) in France, characterized by a return to Classicism (in Neoclassicism) in the rejection of the Rococo excesses of the preceding period. The relatively brief era was marked by the emergence of the Neoclassicist painter Jacques-Louis David (1748–1825) (in preference to the Rococo styles of Jean-Honoré Fragonard (1732–1806)) and of the sculptor Jean-Antoine Houdon (1741–1828); it was a time when the porcelain factories of Sèvres and Meissen became fully developed and the French and German cabinetmakers excelled in the design and manufacture of simple, though elegant furniture.

P, p,   1. the sixteenth letter of the English alphabet.
2. the sixteenth item in a system where A = 1, B = 2, C = 3, etc., and *I* is included.

pi,   the sixteenth letter of the Greek alphabet, corresponding to the English, *P, p.*

Psyche,   the sixteenth planetoid, discovered in 1852.

Q,   the sixteenth item in a system where A = 1, B = 2, C = 3, and *I* is excluded.

sexadecimal,   sixteenth; relating to sixteen.

sexto-decimo,   a book in which the signatures are formed by folding each sheet into sixteen leaves. Also **16mo, 16°, sixteenmo.**

sixteen,   the whole number coming after fifteen and before seventeen.

Sixteen,   the central committee of the Catholic Holy League, in France, which set up a Committee of Public Safety in Paris (1588) and established a reign of terror that lasted for several years.

Sixteen Kingdoms, the,   the many short-lived non-Chinese dynasties that ruled various parts (or all) of northern China from about 316 till about 589 AD and were ultimately Sinicized. They included members of nomadic tribes like the Mongolians, the Huns, and the Tibetans.

sixteenpenny nail,   1. a nail that, before 1500, cost sixteenpence a hundred.
2. (in modern use) a nail 3½ inches long.

sweet sixteen,   when a girl reaches the age of sixteen, it is a tradition to have a party of her friends for her, referred to as a "sweet sixteen party."

16

T, t, the sixteenth consonant of the English alphabet.

two-line brevier, a size of type on a sixteen-point body.

17

haiku, a popular Japanese verse form, also used in English, that consists of seventeen syllables and is characterized by pithy allusions and similes.

pe, the seventeenth letter of the Hebrew alphabet, corresponding to Greek *pi* and the English *P, p*.

Q, q, 1. the seventeenth letter of the English alphabet.
2. the seventeenth item in a system where A = 1, B = 2, C = 3, etc., and *I* is included.

R, r, the seventeenth item in a system where A = 1, B = 2, C = 3, etc., and *I* is excluded.

rho, the seventeenth letter of the Greek alphabet, corresponding to the English, *R, r*.

salute, *U. S.* a 17-gun salute is fired on the arrival and departure of any general or admiral and of the Assistant Secretaries of Defense, Army, Navy, or Air Force. It is fired only on the arrival of the Chairman of a Congressional Committee. Compare **19: salute, 21: salute.**

seventeen, the whole number between sixteen and eighteen; a prime number.

Seventeenth Amendment to the Constitution of the United States, an amendment, ratified on May 31, 1913, that provided for the number of senators to be elected from each State (two), the length of their term (six years), and the procedures to be followed in the event of vacancies.

Seventeenth Parallel, a provisional line of demarcation, which became the political boundary between North and South Vietnam, established (1954) in Vietnam by the Geneva Accords. It only approximated the line of latitude.

seventeen-year locust, a locust, or cicada, the nymphs of which remain in the earth for seventeen years (in northern climates) and for thirteen years (in warmer climates) before emerging, often in huge numbers. It is native to the northeastern United States. Also called **periodical cicada.**

St. Patrick's day, March 17th, the day observed by the Irish in honor of St. Patrick, bishop and patron saint of Ireland, who is supposed to have died in 460.

V, v, the seventeenth consonant of the English alphabet.

18

eighteenmo, 1. a size of book of which each signature is made up of eighteen folded leaves, making thirty-six pages, usually written **18mo** or **18°**. In the United States, the usual untrimmed size is 4" × 6⅓". Also **18°, 18mo, octodecimo.**

"eighteenth of April, . . . " Henry Wadsworth Longfellow's (1807-82) poem, "Paul Revere's Ride" (1863), begins:
Listen my children, and you shall hear
Of the midnight ride of Paul Revere,
On the eighteenth of April, in Seventy-five;
Hardly a man is now alive
Who remembers that famous day and year.

eighteen-wheeler, a large trailer truck and its prime mover, having eighteen wheels in all, the prime mover with two at the front and eight at the rear, the trailer with eight at the rear.

great primer, a size of type on an eighteen-point body.

lohan, (in Chinese Buddhism) one of the eighteen supernatural guardians of the faith. Compare **16: arhat.**

Louis XVIII, the sum of the digits of the year of his birth (1755) yields his titular number.

octodecimal, (of a number system) based on the number eighteen. Compare **2: binary number, 10: decimal system.**

Prohibition, the law forbidding the manufacture, export, or import of alcoholic beverages in, from, or to the United States as set forth in the XVIIIth amendment to the Constitution, ratified by Congress on January 16, 1919, effective one year later, namely:
Section 1. After one year from the ratification of this article the manufacture, sale, or transportation of intoxicating liquors within, the importation thereof into, or the exportation thereof from the United States and all territory subject to the jurisdiction thereof for beverage purposes is hereby prohibited.
Section 2. The Congress and the several States shall have concurrent power to enforce this article by appropriate legislation.
Section 3. This article shall be inoperative unless it shall have been ratified as an amendment to the Constitution by the legislatures of the several States, as provided in the Constitution, within seven years from the date of the submission hereof to the States by the Congress. See also **21: Repeal.**

R, r, 1. the eighteenth letter of the English alphabet.
2. the eighteenth item in a system where A = 1, B = 2, C = 3, etc., and *I* is included.

18

S, s, the eighteenth item in a system where A = 1, B = 2, C = 3, etc. and *I* is excluded.

sadi, the eighteenth letter of the Hebrew alphabet.

sigma, the eighteenth letter of the Greek alphabet, corresponding to the English *S, s.*

Treaty of the Eighteen Articles, a treaty (June 26, 1831) regulating the relationship between Belgium and the Netherlands. See also **24: Treaty of the Twenty-four Articles.**

W, w, the eighteenth consonant of the English alphabet, sometimes called a semi-vowel.

19

decennoval, relating to or involving the number nineteen.

Decennovium, a nineteen-mile-long section of the Appian Way, an ancient road between Rome and southern Italy; the section, constructed about 250 BC, went through the Pontine marshes, which were partly drained to accommodate it.

koph, the nineteenth letter of the Hebrew alphabet. See also **20: koppa.**

nineteenth hole, the bar at a golf club, where the players often retire after having played a full round (of 18 holes).

nineteen to the dozen, superabundantly; redundantly; rapidly; excessively: *The children keep on talking nineteen to the dozen, so you can't get a word in edgewise.*

S, s, 1. the nineteenth letter of the English alphabet.
2. the nineteenth item in a system where A = 1, B = 2, C = 3, etc., and *I* is included.

salute, *U. S.* a 19-gun salute is fired on the arrival of various officials, namely: 1, the Vice President of the United States; 2, the Speaker of the House; 3, any Cabinet Member; 4, the Chairman of the Joint Chiefs of Staff; 5, the President of the Senate pro tempore; 6, any Governor of a State; 7, the Chief Justice of the United States; 8, a General of the Army; 9, the Chief of Naval Operations; 10, the Air Force Chief of Staff; 11, the Marine Commandant; 12, a General of the Air Force; 13, a Fleet Admiral; 14, an American Ambassador; 15, a Foreign Ambassador; or 16, the Premier or Prime Minister of a Foreign Country. It is also fired on the departure of the Secretary of Defense, Army, Navy, or Air Force; the Chairman of the Joint Chiefs of Staff; the Army Chief of Staff; the Chief of Naval Operations; the Air Force Chief of Staff; the Marine Commandant; a General of the Army; a General of the Air Force; or a Fleet Admiral. Compare **17: salute, 21: salute.**

19

suffrage, "The right of citizens of the United States to vote shall not be denied or abridged by the United States or by any State on account of sex": the XIXth amendment to the Constitution.

T, t, the nineteenth item in a system where A = 1, B = 2, C = 3, etc., and *I* is excluded.

tau, the nineteenth letter of the Greek alphabet, corresponding to the English *T, t*.

X, x, the nineteenth consonant of the English alphabet.

20

decimal vigesimal system, a number system based on both the number 10 and the number 20, used in many parts of the world, especially formerly; traces of it can be found in French *quatre-vingt* 'four twenties' (80), *quatre-vingt-dix* 'four twenties and ten' (90), *soixante-dix* 'sixty and ten' (70), etc., which are said to have originated among the Basque. Compare **20: vigesimal system.**

Doric column, the usual number of channels in a classical Doric column is twenty.

great icosahedron, a regular solid having twenty plane faces. See also **4: Kepler-Poinsot solid.**

icosahedron, *or* **icosihedron,** a solid having twenty plane triangular faces.

icosian, of or pertaining to twenty.

Inauguration day, January 20th, since 1933 the day on which the President elect of the United States takes the oath of office. See also **4: Inauguration day.**

koppa, the twentieth letter of the original Greek alphabet, k, analogous in form and corresponding in position and use to the Phoenician and Hebrew *koph* and the Latin *g*. The *kappa* was substituted for it in the words in which it had been used, but the symbol was retained as a numeral with its ancient value of 90.

Odysseus, (in Greek mythology) the hero who wandered for twenty years, trying to return home to Ithaca after the Trojan War.

paragon, a size of type on a twenty-point body.

resh, the twentieth letter of the Hebrew alphabet, corresponding to the Greek *rho* and the English *R, r*.

Rip Van Winkle, a character in a short story, "Rip Van Winkle," that appeared in a collection, *The Sketch Book of Geoffrey Crayon, Gent.* (1819–20), by Washington Irving. Rip, who lived in the Catskill Moun-

20

tains just before the Revolutionary War, falls asleep after drinking a magical potion; upon awakening, he discovers that he has slept for twenty years, and much of the rest of the tale concerns his subsequent adventures and his astonishment at the many changes that have taken place during his absence.

Roaring Twenties, in the 1920s, a nickname applied in America to the decade of flappers, bootlegging, and Murder Incorporated.

score, twenty.

T, t, 1. the twentieth letter of the English alphabet.
2. the twentieth item in a system where A = 1, B = 2, C = 3, etc., and *I* is included.

teeth, the number of deciduous teeth in the normal human mouth is 20. Compare **32: teeth.**

Twentieth Century Limited, a luxurious Pullman train on the New York Central Railroad between New York City and Chicago. Service began in 1905; it was discontinued in the 1970s.

twenty-mule team, a team of twenty mules, formerly used for drawing wagons of borax in Death Valley, where it was mined. It was widely known through a 1930s radio show, "Death Valley Days," which was sponsored by a cleaning preparation called *20-Mule Team Borax.*

twenty questions, an oral game in which one player thinks of a specific object, person, title of a book, film, piece of music, song, or the like and the others try to guess what it is from Yes or No replies given to a maximum of twenty questions.

twenty-twenty hindsight, *Facetious.* the ability to determine the nature of an event after it has occurred. Also called **Monday-morning quarterbacking.**

twenty-twenty vision, (in ophthalmology) the condition of having normal visual acuity, needing no corrective lenses.

U, u, the twentieth item in a system where A = 1, B = 2, C = 3, and *I* is excluded.

upsilon, the twentieth letter of the Greek alphabet, corresponding to the English *Y, y* or *U, u.*

vicennial, 1. of or pertaining to a period of twenty years.
2. a commemoration or celebration of an event that took place twenty years earlier; a twentieth anniversary.
3. lasting for twenty years; occurring once every twenty years.

vigesimal, twentieth.

20

vigesimal system, a number system based on the number 20, used in many parts of the world, especially formerly; it originated with the digits on both hands and feet, and traces of its use can be found in Mexico and Central America. Compare **20: decimal vigesimal system.**

vintiner, the commander of an old division of English infantry, numbering twenty men and called a *twenty*.

Y, y, the twentieth consonant of the English alphabet, sometimes called a semivowel.

zwanziger, an Austrian silver coin of the nineteenth century, equivalent to twenty kreutzers.

21

free, white, and 21, an old-fashioned catch-phrase, rarely heard because of its racial overtones, used to describe a person, especially a young woman, from the standpoint of independence, youth, and opportunity.

midsummer, the time of the summer solstice, the 21st of June. [From the custom, in Great Britain, of considering summer as beginning on May 1st.]

midwinter, the middle or depth of winter, the time of the winter solstice, the 21st of December. [From the custom, in Great Britain, of considering winter as beginning on November 1st.]

phi, the twenty-first letter of the Greek alphabet, corresponding to the English *F, f,* and *ph*.

Repeal, the repeal, effective December 5, 1933, of the XVIIIth amendment to the Constitution of the United States as set forth in the XXIst amendment. See also **18: Prohibition.**

salute, a twenty-one-gun salute is fired on the arrival and departure of the President, an ex-President, and a President-elect of the United States and of the sovereign or Chief of State of a foreign country, or a member of a reigning royal family. It is also accorded to the national flag of the United States. Compare **17: salute, 19: salute.**

shin, the twenty-first letter of the Hebrew alphabet, corresponding to the English *sh*.

twenty-first rule, a rule adopted by the U. S. House of Representatives in 1840 providing that no petitions for the abolition of slavery should be received by that body. It was dropped in 1844.

21

Twenty-one Years' War, a war, 1701–21, between Russia, under Peter I ("the Great"), and Sweden, under Charles XII, concluded at the peace of Nystad, August 30, 1721, in which Russia gained access to the open-water ports of the Baltic Sea.

U, u, 1. the twenty-first letter of the English alphabet.
2. the twenty-first item in a system where A = 1, B = 2, C = 3, etc., and *I* is included.

V, v, the twenty-first item in a system where A = 1, B = 2, C = 3, etc., and *I* is excluded.

Z, z, the twenty-first consonant of the English alphabet.

22

catch-22, a dilemma in which the choices offered have equally disastrous results. [From the title of Joseph Heller's novel, *Catch-22,* 1961, in which the protagonist is continually faced with such situations.]

chi, the twenty-second letter of the Greek alphabet, corresponding to English *ch.*

major arcana, picture cards of which the original tarot deck consisted; they have representations on them for use in divination that correspond to the twenty-two letters of the Hebrew alphabet, namely: 1, (aleph) Magician or Juggler; 2, (beth) High Priestess; 3, (gimel) Empress; 4, (daleth) Emperor; 5, (heh) Pope or Hierophant; 6, (vav) Lovers; 7, (zayin) Chariot; 8, (cheth) Justice; 9, (teth) Hermit; 10, (yod) Wheel; 11, (kaph) Strength; 12, (lamed); Hanged Man; 13, (mem) Death; 14, (nun) Temperance or Alchemist; 15, (samekh) Devil; 16, (ayin) Tower; 17, (pe) Star; 18, (tsade) Moon; 19, (koph) Sun; 20, (resh) Judgment; 21, (shin) World; and 22, (tav) Fool. Fool sometimes appears as aleph and sometimes bears no number or letter; other letters and numbers are occasionally changed, as well. See also **78: tarot.**

śin, the twenty-second letter of the Hebrew alphabet.

two-line small pica, a size of type on a twenty-two-point body.

V, v, 1. the twenty-second letter in the English alphabet.
2. the twenty-second item in a system where A = 1, B = 2, C = 3, etc., and *I* is included.

W, w, the twenty-second item in a system where A = 1, B = 2, C = 3, etc., and *I* is excluded.

23

grand jury, a body of twenty-three persons charged with the function of deciding whether "probable cause" exists to believe that a person has committed a crime and, if so, to return an indictment for same. Compare **12: petit jury.**

lesser sanhedrin *or* **sanhedrim,** *or* **provincial sanhedrin** *or* **sanhedrim,** a local authority, consisting of twenty-three members having jurisdiction over minor civil and criminal cases. Appointed by the great sanhedrin, it included a physician, a scribe, and a schoolmaster. See also **70: great sanhedrin.**

psi, the twenty-third letter of the Greek alphabet, corresponding to the English *ps.*

tav, the twenty-third letter of the Hebrew alphabet, corresponding to the Greek *tau* and the English *T.*

Thalia, the twenty-third planetoid, discovered by Hind in 1852.

twenty-three skidoo, *or* **twenty-three skiddoo,** an almost meaningless expression popular in the 1920s and 1930s.

W, w, 1. the twenty-third letter of the English alphabet.
2. the twenty-third item in a system where A = 1, B = 2, C = 3, etc., and *I* is included.

X, x, the twenty-third item in a system where A = 1, B = 2, C = 3, etc., and *I* is excluded.

Bartholomew day, the 24th day of August, on which is held a festival in honor of St. Bartholomew, one of the twelve apostles, and which is noted in history as: 1. the day in 1572 on which the great massacre of French Protestants (the St. Bartholomew massacre) was begun in Paris by order of the king, which order was executed in other towns on its receipt, the last being in Bordeaux on October 3rd.
2. the day in 1662 on which the penalties of the English Act of Uniformity came into force.
3. the day on which a great fair (Bartholomew fair) was held at Smithfield, in London, from 1133 to 1855.

day, the number of hours in a day, twenty-four reckoned from midnight to midnight.

diploid, (in crystallography) a solid with 24 trapezoidal planes. Also called **didodecahedron, dyakis-dodecahedron.**

discalenohedron, *pl.* **discalenohedra, discalenohedrons.** a double twelve-sided (dihexagonal) pyramid, whose faces belong to two complementary scalenohedra.

dyakis-dodecahedron, 1. See **2: diploid.**
2. a solid having twenty-four faces, resembling the deltoidal icositetrahedron.

24

"four-and-twenty blackbirds," a nursery rime that goes:
Sing a song of sixpence,
A pocketful of rye,
Four and twenty blackbirds
Baked in a pie.
When the pie was opened,
The birds began to sing.
Wasn't that a dainty dish
To set before a king?

"four-and-twenty sailors . . . ," from an old nursery rime:
I saw a ship a-sailing,
A-sailing on the sea,
And, oh, but it was laden
With pretty things for thee.
There were comfits in the cabin
And kisses in the hold
The sails were made of silk,
And the masts were all of gold.
The four-and-twenty sailors,
That stood upon the decks
Were four-and-twenty white mice
With chains about their necks.
The captain was a duck
With a packet on his back;
And when the ship began to move,
The captain said "Quack! Quack!"

hexatetrahedron, a solid bounded by twenty-four scalene triangles; the inclined hemihedral form of the hexoctahedron, sometimes found in diamonds.

hour, the number of hours in a day is twenty-four.

House of Keys, the lower branch of the legislature (Court of Tynwald) of the Isle of Man, constituted of a body of twenty-four representatives.

icositetrahedron, a solid having twenty-four similar quadrilateral faces; a tetragonal trisoctahedron; a trapezohedron.

Manhattan, Peter Minuit, the first director of the New Netherlands (1624–33), is said to have purchased the island of Manhattan (New York City) from its Indian inhabitants for $24 worth of trinkets and trading goods.

nycthemeron, the whole natural day, consisting of twenty-four hours.

omega, the 24th, and last letter of the Greek alphabet, corresponding to the English long *o*; used metaphorically to mean the end or last of something. Compare **1: alpha.**

poll tax, "The right of citizens of the United States to vote in any primary or other election for President or Vice President, for electors for President or Vice President, or for Senator or Representative in Congress, shall not be denied or abridged by the United States or any State by reason of failure to pay any poll tax or other tax": the XXIVth amendment to the Constitution. (It should be noted that although the individual States may have complementary laws, the Constitution is silent in the matter of statewide, county, and municipal elections.)

quire, twenty-four sheets of paper; the twentieth part of a ream.

tetrahexahedron, a geometric solid having twenty-four equal triangular faces, four corresponding to each face of the cube. Also called **tetrakishexahedron.**

Treaty of the Twenty-four Articles, a treaty (October 15, 1831) regulating the relationship between Belgium and the Netherlands, replacing the Treaty of the Eighteen Articles and finally accepted by William II of the Netherlands on March 14, 1838.

triakisoctahedron, an octahedron on each face of which a triangular pyramid has been erected, resulting in a solid with twenty-four surfaces.

Twenty-four Proprietors, the, twenty-four Quakers who bought East Jersey in 1682. In 1674, a group of Quakers headed by Edward Byllynge had bought what is now the state of New Jersey from Lord John Berkeley and Sir George Carteret who had been given it by James, Duke of York. In 1676, the land was divided into two sections—West and East Jersey. West Jersey, under Byllynge, became the first Quaker colony in the United States. In 1702, the owners having given up their claim to the land, England united the two colonies into a single royal colony.

two-line pica, a size of type on a twenty-four-point body.

X, x, 1. the twenty-fourth letter of the English alphabet.
2. the twenty-fourth item in a system where A = 1, B = 2, C = 3, etc., and *I* is included.

Y, y, the twenty-fourth item in a system where A = 1, B = 2, C = 3, etc., and *I* is excluded.

pony, (in British sporting slang) the amount of twenty-five pounds, usually the amount of a wager.

Twenty-five Articles, the, the thirty-nine articles of the Church of England, with the omission of the third, eighth, thirteenth, fifteenth, seventeenth, eighteenth, twentieth, twenty-first, twenty-third, twenty-sixth, thirty-third, thirty-fourth, and thirty-seventh, that were adopted in 1784 as the doctrinal base of the Methodist Episcopal Church.

25

wedding, the silver wedding anniversary is celebrated at the conclusion of the twenty-fifth year of marriage.

Y, y, 1. the twenty-fifth letter of the English alphabet; a semivowel.
2. the twenty-fifth item in a system where A = 1, B = 2, C = 3, etc., and *I* is included.

Z, z, the twenty-fifth item in a series where A = 1, B = 2, C = 3, etc., and *I* is excluded.

26

rhombicuboctahedron, a regular solid having twenty-six plane faces.

small rhombicuboctahedron, a regular solid with twenty-six plane faces. See also **13: Archimedean solid.**

Z, z, 1. the twenty-sixth letter of the English alphabet.
2. the twenty-sixth item in a system where A = 1, B = 2, C = 3, etc., and *I* is included.

27

"twenty-seven different wigs," from the old nursery rime:
Gregory Griggs, Gregory Griggs,
Had twenty-seven different wigs.
He wore them up, he wore them down,
To please the people of the town;
He wore them east, he wore them west,
But he never could tell which he loved the best.

28

dominoes, a game played with 28 tiles.

twenty-eight, the Australian yellow-collared parakeet, so named from its call, which is sometimes repeated several times.

two-line English, a size of type on a twenty-eight-point body.

30

Battle of the Thirty, a battle, on March 27, 1351, in the continuing conflict over the succession to the duchy of Brittany between Charles of Blois, supported by John II, king of France, and John of Montfort, supported by Edward III, king of England. Jean de Beaumanoir, marshal of Brittany, sent a challenge to John Bramborough, captain of the English forces, to send thirty selected champions, knights and squires, to engage the same number from his own forces in a pitched battle near Ploërmel. The battle, fought with lances, swords, daggers, and maces, was won by Beaumanoir; one of his champions, Guillaume de Montauban, overcame seven of the English champions singlehandedly. Bramborough was among those slain.

great rhombicuboctahedron, a regular solid with thirty plane faces. See also **13: Archimedean solid.**

lamed, the twelfth letter of the Hebrew alphabet, having a numerical value of 30.

Memorial day, the 30th of May, formerly called Decoration day, originally established to honor the U. S. soldiers killed 1861–65; now celebrated to honor the dead of all U. S. wars.

pieces of silver,
Then one of the twelve, called Judas Iscariot, went unto the chief priests, and said unto them, What will ye give me, and I will deliver him unto you? And they covenanted with him for thirty pieces of silver. [Matt. 26.14-15.] Later, Judas tried to repent by returning the pieces of silver to the chief priests and elders, but they would not have them, so he cast them down in the temple and departed and hanged himself. As the priests could not return the blood money to the treasury, they used it to buy "the potter's field [Aceldama], to bury strangers in." [Matt. 27.7.]

rouge-et-noir, a gambling card game in which the object is to achieve a total count of values, from adding the pips on the cards, of not less than thirty, and not more than forty. Also called **trente-et-quarante.**

thirty, the numeral "30," a holdover from telegraphese, typed at the end of a story by a journalist; rarely used today.

"Thirty days hath September . . ." an old counting rime to tell the number of days in a month:
Thirty days hath September,
April, June, and November.
All the rest have thirty-one
Excepting February alone,
And that has twenty-eight days clear
And twenty-nine in each leap year.

thirtypenny nail, **1.** a nail that, before 1500, cost thirty pence a hundred.
2. (in modern use) a nail $4\frac{1}{2}$ inches long.

Thirty Tyrants, the, the governors appointed by Lysander of Sparta to administer Athens (404 BC) They were in power for only eight months, when Thrasybulos deposed them and restored the republic. According to Xenophon, "The Thirty" put more people to death in eight months of peace than the enemy had done in a war of thirty years. [Brewer, *Reader's Handbook,* 1889.]

30

Thirty Years' Truce, a treaty, concluded in 445 BC, between Athens on one side and Sparta and her allies on the other. Followed by about fifteen years of peace and Athenian prosperity under Pericles, it was terminated by the outbreak of the Peloponnesian War (431–404).

Thirty Years' War, the, 1. the Peloponnesian war between Athens and Sparta (404–431 BC).
2. the wars between the Protestants and Catholics of Germany (1618–48).

trental, a collection of thirty of anything, especially (in Roman Catholicism) a service of thirty Masses for a deceased person. Also, **trigintal.**

triacontahedral, a plane geometric figure having thirty sides.

triaconter, (in Greek antiquity) a vessel with thirty oars.

tricennial, 1. of or pertaining to a period of thirty years.
2. a commemoration or celebration of an event that occurred thirty years earlier; a thirtieth anniversary.
3. lasting for thirty years; occurring once every thirty years.

32

double bourdon, the lowest stop in an organ, of 32-feet pitch.

dyocaetriacontahedron, a solid having thirty-two faces. Also **dyokai-triakontahedron.**

écarté, a card game for two players in which thirty-two cards are used, the cards from two to six being excluded.

four-line brevier, a size of type on a thirty-two-point body.

icosidodecahedron, a solid having thirty-two plane faces, twenty of which are triangular and twelve pentagonal. See also **13: Archimedean solid.**

piquet, a card game played with thirty-two cards, the twos through the sixes having been removed.

point of the compass, one of the thirty-two points on a compass, being, in clockwise order from north (with degrees in parentheses following), in a list called *boxing the compass:* north by east (11¼°), north northeast (22½°), northeast by north (33¾°), northeast (45°), northeast by east (56¼°), east northeast (67½°), east by north (78¾°), east (90°); east by south (101¼°), east southeast (112½°), southeast by east (123¾°), southeast (135°), southeast by south (146¼°), south southeast (157½°), south by east (168¾°), south (180°); south by west (191¼°), south southwest (202½°), southwest by south (213¾°), southwest (225°), southwest by west (236¼°), west southwest (247½°), west by south (258¾°), west (270°); west by north (281¼°), west northwest (292½°), northwest by

32

west (303¾°), northwest (315°), northwest by north (326¼°), north northwest (337½°), north by west (348¾°), north (360°).

teeth, the number of permanent teeth in the normal human mouth is 32. Compare **20: teeth.**

thirty-second degree Mason, the highest degree a man can advance to in the Scottish Rite of Masonry. See also **33: thirty-third degree Mason.**

truncated dodecahedron, a regular solid with thirty-two plane faces. See also **13: Archimedean solid.**

truncated icosahedron, a regular solid with thirty-two plane faces. See also **13: Archimedean solid.**

33

Bear Party, *or* **Bears, the,** a group of thirty-three Americans who seized what is now Sonoma County, California, from the Mexicans in 1846, so-called because they depicted a grizzly bear on the flag they designed for the territory they named the California Republic.

thirty-third degree Mason, an honorary degree awarded by the Scottish Rite of Masonry for outstanding service to Masonry, the community, or the nation. See also **32: thirty-second degree Mason.**

36

face, (in astrology) one of the thirty-six parts of the zodiac, formed by dividing each sign into three equal parts.

thirty-six righteous men, (in Jewish legend) the minimum number of virtuous individuals in each generation to whom the world owes its continued existence; their identities are known only to God.

three-line pica, a size of type on a thirty-six-point body.

38

snub-cube, *or* **octocaetriacontahedron,** a regular solid with thirty-eight plane faces. See also **13: Archimedean solid.**

39

Moses's law, a term used by pirates for the laying on of thirty-nine lashes on the naked back, given as "Lay on forty, less one." See also **39: forty stripes.**

Thirty-nine Articles, the, the thirty-nine points of doctrine maintained by the Church of England, first framed in forty-two articles (1552), then revised and set forth in thirty-nine (1562–63).

40

"Ali Baba and the Forty Thieves," one of the tales from *The Thousand and One Nights.*

Borden, Lizzie, (1860–1927) a figure in a celebrated American murder case, she was found not guilty (1893) of murdering her mother and father with an ax the year before. Public opinion as to her innocence was divided, as demonstrated by this anonymous jingle, which became popular at the time:

Lizzie Borden took an ax
And gave her mother forty whacks.
When she saw what she had done,
She gave her father forty-one.

double paragon, a size of type on a forty-point body.

F, a medieval Roman numeral with a value of forty.

fair, fat, and forty, an old-fashioned catch phrase, presumably describing the person who, less than twenty years before, was "free, white, and 21." See also **40: fat, fair, and forty, 21: free, white, and 21.**

Fat, fair and forty, from *The Irish Mimic; or Blunders at Brighton* (1795), by John O'Keefe (1747–1833): "Fat, fair and forty were all the coasts of the young men." [*Bartlett's Familiar Quotations,* 15th ed., 1980.]

Field of the Forty Footsteps, a former field, at the rear of the British Museum, where, according to tradition, during the Monmouth rebellion two brothers who took different sides fought to the death. Forty impressions of their feet were traceable for years afterwards. [Brewer, *Reader's Handbook.*]

fortescue, a fish found in Australian waters, the name being an adaptation of the form, *forty-skewer,* in allusion to its many sharp spines.

forty, (in the Bible) a number frequently used, for example: the number of years the Israelites wandered in the wilderness [Ex. 16.35]; the number of days Christ spent on earth after the Resurrection [Acts 1.3]; or the number of days required for embalming [Gen. 1.3].

Forty, the, 1. (in ancient Attica) a body of magistrates who adjudicated the trial of small causes in rural demes.
2. (in the Venetian republic) the name (with qualifying terms) of two appellate tribunals and a criminal court.
3. Also called **the Forty Immortals.** the forty members of the French Academy.

forty and eight, a jocular name given by soldiers in the American Expeditionary Forces (1917–1918) to French railway freight cars, which were actually marked for "40 men or 8 horses."

forty days and forty nights, (in the Bible) a period of time commonly referred to, for example: the time that Noah spent fasting on Mt. Sinai [Ex. 25.18]; that Elijah spent on Mt. Horeb without food [1 Kings 19.8]; or that Jesus spent fasting in the wilderness [Matt. 4.2]. See also **40: St. Swithin's Day.**

forty hours, (in the Roman Catholic Church) a continuous exposition of the Eucharist for forty hours.

forty-legs, an Australian myriapod.

fortypenny nail, 1. a nail that, before 1500, cost forty pence a hundred. 2. (in modern use) a nail 5 inches long.

forty-rod lightning, *Obsolete U. S. slang.* whisky so strong that it could kill at a distance of forty rods.

forty-spot, an Australian shrike.

forty stripes, 1. according to Mosaic law, Jews were forbidden to inflict more than forty stripes on an offender, hence used a scourge with three lashes thirteen times, that is, "forty save one." 2. *Facetious.* **Forty stripes save one,** the thirty-nine articles of the Anglican communion. See also **39: Moses's law.**

forty winks, *Informal.* a brief nap; a catnap.

Lent, *or* **Quadragesima,** (in the Christian Church) an annual fast of forty days, beginning with Ash Wednesday and ending on Easter, in commemoration of Christ's forty-day fast (Matt. 4.2). Although the days between Ash Wednesday and Easter number forty-six, the intervening Sundays (called *Sundays in Lent*) are not counted, because Sunday is always a feast-day.

Life begins at 40, a catch-phrase offering solace to those, forty and older, who should be optimistic because their best years are still ahead.

mem, the Hebrew alphabetic character with a numerical value of 40.

ombre, *or* **omber,** a card game for from two to five persons playing with a deck of forty cards, the eights, nines, and tens having been removed.

quadragenarian, a person between the ages of forty and forty-nine, inclusive.

quadragenarious, consisting of forty; being forty years old.

quadragene, a papal indulgence for forty days; a remission of the temporal punishment due to sin, corresponding to the forty days of the ancient canonical penance.

Quadragesima Sunday, the first Sunday in Lent. Compare **50: quinquagesima, 60: Sexagesima Sunday; 70: Septuagesima Sunday.**

40

quarantine, 1. a period of forty days, specifically, Lent.
2. a term, originally of forty days but now of varying length determined by the individual circumstances, during which a ship arriving in port and known or suspected to be infected with a malignant contagious disease is obliged to remain isolated from the inhabitants at the place of arrival.
3. the enforced isolation of individuals suffering from or recuperating from acute contagious disease or of a dwelling, town, or area in which such a disease exists.
4. to put under quarantine.

quorum, in the British House of Commons forty persons are required for a quorum.

Roaring Forties, the, the region between 40° and 50° south or north latitude in the north Atlantic or, especially, in the south Pacific ocean that is known for its rough seas and strong winds.

St. Swithin's day, When St. Swithin, bishop of Winchester, 852–62, was canonized, within the next century, the monks desired to transfer his remains from the churchyard at Winchester, where he had at his own request been buried, to the cathedral, and selected July 15th as the date. Heavy rains lasting for forty days delayed the transfer: hence the popular saying that if it rains on St. Swithin's day, it is sure to rain continuously for forty days. See also **15: St. Swithin's day.**

41

Forty-one, the, (formerly) the forty-one Venetian councillors who elected the Doge.

42

Forty-second Street, a street in New York City associated with honky-tonk, "adult" movie theaters, amusement arcades, and the kinds of people who patronize and operate such businesses, chiefly in the section between 6th (the Avenue of the Americas) and 9th Avenues.

Forty-Second Street thief, a pickpocket who works only in one locale.

44

'coffee,' (in short-order restaurants, soda-fountains, etc., on the west coast of the U. S. in the 1940s) "forty-four" was used to designate 'a cup of coffee.'

four-line small pica, a size of type on a forty-four-point body.

45

Forty-five, the, the second Jacobite rebellion (1745), when "Bonnie Prince Charlie," also known as the "Young Pretender," grandson of James II, failed to gain the throne of England and fled to France. See also **15: Fifteen.**

midangle, an angle of 45°, half of a right angle.

octant, the eighth part of a circle, or 45°.

47

Pythagorean proposition, the forty-seventh proposition of the first book of Euclid's *Elements,* that the square of the hypotenuse of a right triangle is equal to the sum of the squares of the other two sides, said to have been discovered by Pythagoras.

48

Forty-Eight, the, the forty-eight preludes and fugues included in *The Well-Tempered Clavier,* by Johann Sebastian Bach.

forty-eighter, a participant in the German revolution of 1848.

forty-eightmo, a book signature containing forty-eight leaves, or ninety-six pages, approximately 2½″ × 4″. Also, **48mo, 48°.**

four-line pica, a size of type on a forty-eight-point body.

hexoctahedron, *or* **hexakisoctahedron,** a solid bounded by forty-eight equilateral triangles. Also called **adamantoid,** because it is a common form of the diamond.

pinochle deck, a pack of playing cards used in pinochle, consisting of two each of the 9, 10, jack, queen, king, and ace in each suit, or 48 cards in all. Compare **52: playing cards.**

49

Forty-Niners, the, those who went to California, during the Gold Rush, to seek their fortunes by mining gold after its discovery at Sutter's Mill, in 1849.

50

Aegyptus, the twin brother of Danaüs. His fifty sons married Danaüs's fifty daughters. See also **50: Danaides.**

Danaides, the fifty daughters of Danaüs, king of Argos, who married the fifty sons of his twin brother Ægyptus, king of Arabia and Egypt. At Danaüs' instigation, all but one, Hypermnestra (whose husband was

50

Lynceus), slew their husbands on their wedding night, for which they were condemned in Hades to pour water everlastingly into sieves.

Dirty Half-Hundred, the, "the Queen's Own," formerly, the 50th Foot, so called from men wiping their faces with their black cuffs. [Brewer, *Dictionary of Phrase and Fable,* 1894.]

Fifty Decisions, the final disposition by Justinian of fifty questions concerning which the authorities on Roman law were not agreed. They were made AD 529–30 and were embodied in the revised Code of Justinian.

fifty-fifty, equally shared between two people, parties, interests, etc.

fiftypenny nail, 1. a nail that, before 1500, cost fiftypence a hundred. 2. (in modern use) a nail 5½ inches long.

go fifty-fifty, *or* **50/50,** divide equally with someone else. Also, **split** *or* **divide fifty-fifty** *or* **50/50.**

L, the Roman numeral for the number fifty.

Nereid, one of fifty or, according to other sources, one hundred beautiful nymphs of the sea, daughters of Nereus and Doris.

nun, the fourteenth letter of the Hebrew alphabet, having a numerical value of 50.

penteconter, an ancient Greek trading ship that was propelled by fifty oars.

Pentecost, 1. (in the Old Testament) a Jewish harvest festival observed on the fiftieth day after Passover (the 14th of Nissan). It is called the feast of weeks in the Old Testament (Deut. 16.10ff), Hebrew Shabuoth. While primarily concerned with the celebration of the completion of the harvest, in the minds of later Jews it seems also to have been associated with the giving of the Ten Commandments by God to Moses on Mt. Sinai on the fiftieth day after the Exodus from Egypt.
2. the feast of Whitsunday, a festival of the Christian Church, observed annually in remembrance of the descent of the Holy Ghost upon the apostles during the feast of the Pentecost. As the Jewish Pentecost is associated with the celebration of the first fruits of the earth, so the Christian Pentecost is associated with the celebration of the first fruits of the Spirit. Formerly, *Pentecost* referred both to Whitsunday and to the whole fifty-day period ending with Whitsunday.

pentecoster, (in ancient Greece) a commander of fifty men.

pentecostys, (in ancient Greece) a company of fifty men.

quinquagenarian, a person between the ages of fifty and fifty-nine, inclusive.

50

quinquagesima, a period of fifty days, used especially in Quinquagesima Sunday, the fiftieth day before Easter and the last Sunday before Lent. Also called **Shrove Sunday.** Compare **40: Quadragesima Sunday, 60: Sexagesima Sunday, 70: Septuagesima Sunday.**

quinquagesimal, pertaining to the number fifty or to the fiftieth.

semicentennial, 1. of or pertaining to a period of fifty years.
2. a commemoration or celebration of an event that occurred fifty years earlier; a fiftieth anniversary.
3. lasting for fifty years; occurring once every fifty years.

United States of America, a nation composed originally of thirteen, now of fifty states, all of which are on the continent of North America except for Hawaii, listed in the order in which they gained statehood, namely: 1, Delaware (7 December 1787); 2, Pennsylvania (12 December 1787); 3, New Jersey (18 December 1787); 4, Georgia (2 January 1788); 5, Connecticut (9 January 1788); 6, Massachusetts (6 February 1788); 7, Maryland (28 April 1788); 8, South Carolina (23 May 1788); 9, New Hampshire (21 June 1788); 10, Virginia (26 June 1788); 11, New York (26 July 1788); 12, North Carolina (21 November 1789); 13, Rhode Island (29 May 1790); 14, Vermont (4 March 1791); 15, Kentucky (1 June 1792); 16, Tennessee (1 June 1796); 17, Ohio (1 March 1803); 18, Louisiana (30 April 1812); 19, Indiana (11 December 1816); 20, Mississippi (10 December 1817); 21, Illinois (3 December 1818); 22, Alabama (14 December 1819); 23, Maine (15 March 1820); 24, Missouri (10 August 1821); 25, Arkansas (15 June 1836); 26, Michigan (26 January 1837); 27, Florida (3 March 1845); 28, Texas (29 December 1845); 29, Iowa (28 December 1846); 30, Wisconsin (29 May 1848); 31, California (9 September 1850); 32, Minnesota (11 May 1858); 33, Oregon (14 February 1859); 34, Kansas (29 January 1861); 35, West Virginia (20 June 1863); 36, Nevada (31 October 1864); 37, Nebraska (1 March 1867); 38, Colorado (1 August 1876); 39, North Dakota (2 November 1889); 40, South Dakota (2 November 1889); 41, Montana (8 November 1889); 42, Washington (11 November 1889); 43, Idaho (3 July 1890); 44, Wyoming (10 July 1890); 45, Utah (4 January 1896); 46, Oklahoma (16 November 1907); 47, New Mexico (6 January 1912); 48, Arizona (14 February 1912); 49, Alaska (3 January 1959); 50, Hawaii (21 August 1959). See also **13 original colonies.**

wedding, the golden wedding anniversary is celebrated at the conclusion of the fiftieth year of marriage.

51

fifty-first state, the, (in the U. S.) an epithet of Puerto Rico (since Hawaii gained statehood, in 1959), a commonwealth whose citizens have U. S. citizenship.

52

playing cards, the number of cards in an ordinary deck is fifty-two, consisting of four suits with thirteen cards each, numbered from ace to ten with three face-cards, the jack (or knave), the queen, and the king. In some games the ace counts as one or eleven, in others it ranks above the king, and in others the choice is determined by the player or the circumstances of play. The deck may contain fewer for certain games, as a pinochle deck which contains forty-eight, made up by removing all cards from the 2 through the 8 and doubling the number remaining.

week, the number of weeks in a year, approximately fifty-two.

54

"Fifty-four Forty or Fight," the rallying cry of U. S. expansionist supporters of the presidential campaign of James K. Polk (1844), who wished to exclude Great Britain from the Oregon Territory, the northern boundary of which (54°40′N latitude) was contiguous with Alaska, then in Russian possession.

57

Heinz, the fifty-seven varieties of canned foods for many years offered by the Heinz Food Company.

60

Babylonian system, the number 60 was the base of the Babylonian number system, remnants of which are preserved in the division of the angular degree into sixty minutes.

like sixty, *U. S. informal.* very fast; furiously fast.

sexagenarian, a person between the ages of sixty and sixty-nine, inclusive.

sexagenary, *or* **sexagenal,** 1. referring to the number sixty, especially sixty years.
2. a commemoration or celebration of an event that occurred sixty years earlier; a sixtieth anniversary.

Sexagesima Sunday, the second Sunday before Lent. [It is not clear why the fifty-sixth day before Easter should be called by this name.] Compare **40:Quadragesima Sunday, 50:quinquagesima, 70:Septuagesima Sunday.**

sextile, (in astrology) noting the aspect or position of two planets that are sixty degrees, or two signs, distant from one another, considered a favorable aspect.

60

sixtypenny nail, 1. a nail that, before 1500, cost sixty pence a hundred. 2. (in modern use) a nail 6 inches long.

triakisicosahedron, an icosahedron on each face of which a triangular pyramid has been erected, resulting in a solid with sixty surfaces.

61

lurch, (in cribbage) the position of a player when his opponent has won every point (61 holes) before he himself has made 30 holes.

62

great rhombicosidodecahedron, a regular solid with sixty-two plane faces. See also **13: Archimedean solid.**

rhombicosidodecahedron, a regular solid having sixty-two plane faces.

small rhombicosidodecahedron, a regular solid with sixty-two plane faces. See also **13: Archimedean solid.**

64

sixty-four-dollar question, *or* **$64 question,** the key, most difficult, or crucial question. Also, **sixty-four-thousand-dollar question** *or* **$64,000 question.** [From **The $64 Question,** a U. S. "double-or-nothing" radio quiz show of the 1940s; later, a television quiz show on which the top prize was $64,000.]

sixty-fourmo, a book size, approximately $2\frac{1}{4}'' \times 3\frac{1}{4}''$, of signatures formed by folding a sheet into sixty-four leaves. Also, **64mo, 64°.**

66

Route 66, a highway, the main road between Chicago and Los Angeles before the construction of the interstate highway system, made famous in a song, "(Get Your Kicks On) Route 66," by Bob Troup (1946.)

69

soixante-neuf, *or* **sixty-nine** a sexual activity in which the partners perform simultaneous cunnilingus or fellatio, or cunnilingus and fellatio on one another.

70

great sanhedrin, *or* **great sanhedrim,** the supreme council and highest ecclesiastical and judicial tribunal of the ancient Jewish nation, consisting of seventy chief priests, elders, and scribes, said to have had its origin in the appointment by Moses of seventy elders to assist him as mag-

70

istrates and judges (Num. 11.16). The name comes from Greek, which suggests that the council originated during the Macedonian supremacy in Palestine. The name fell into disuse under Gamaliel IV (AD 270–300), and the institution became extinct in 425, on the death of its last president, Gamaliel VI. Also called **the Seventy.** See also **23: lesser sanhedrin.**

S, the medieval Roman numeral 70. See also **7: S.**

septuagenarian, a person between the ages of seventy and seventy-nine, inclusive.

septuagenary, 1. referring to seventy, especially seventy years.
2. a commemoration or celebration of an event that occurred seventy years earlier; a seventieth anniversary.

septuagesima, a period of seventy days.

Septuagesima Sunday, the third Sunday before Lent; Lost Sunday; Alleluia Sunday. [It is not clear why the sixty-third day before Easter should be called by this name.] Compare **40: Quadragesima Sunday, 50: quinquagesima, 60: Sexagesima Sunday.**

Septuagint, a Greek translation of the Hebrew Scriptures made by the Seventy, that is, seventy (or more) persons, completed by the second century BC.

Seventy, the, 1. See **70: great sanhedrin.**
2. the body of disciples appointed by Christ to preach the Gospel and heal the sick. [Luke 10.]
3. the body of scholars who, according to tradition, were the translators who produced the Septuagint. See **70: Septuagint.**
4. See **70: seventy disciples.**

seventy disciples, the, (in the Mormon Church) a body of men who rank in the hierarchy after the Twelve Apostles. It is their duty "to travel into all the world and preach the Gospel and administer its ordinances." Also called **the Seventy.**

three score and ten, seventy, the normal lifespan of man, according to the Bible:
"The days of our years are threescore years and ten; and if by reason of strength they be fourscore years, yet is their strength labour and sorrow; for it is soon cut off, and we fly away." [Ps. 90.10.]

74

seventy-four, a warship rated as carrying 74 guns.

75

French "75," a drink made with lemon juice, sugar, gin, and champagne, iced and served in a tall glass with a slice of lemon or orange and a cherry.

75

wedding, the diamond wedding anniversary is celebrated at the conclusion of the seventy-fifth year of marriage.

77

Charter 77, a document signed by more than 200 Czechoslovakian citizens and published in Prague in January 1977. It complained that articles in two international agreements on human rights, to which Czechoslovakia had recently acceded, were not being observed by that country's government (notably in such areas as freedom of expression, association, and religious confession); the government responded with a crackdown on signatories of Charter 77, some prominent intellectuals among them receiving prison terms.

78

rule of 78, a method of calculating the payment of interest on a one-year loan by which $12/78$ of the total interest amount is paid (along with $1/12$ of the principal) in the first month, $11/78$ is paid in the second month, and so forth, till the entire loan is retired. [From the sum of the numerators, 12 + 11 + 10 . . . + 1.]

tarot, *or* **tarok,** a deck of seventy-eight cards, developed in the fourteenth century from the original picture cards, or major arcana, by adding numbered cards; used for divination and, later, for games, they are divided into four suits (clubs, diamonds, hearts, and spades), each of which has ten cards (numbered from ace to ten) called the minor arcana, and four face cards (king, queen, knight, and knave). See **22: major arcana.**

80

Eighty Years' War, a war between Spain and the Netherlands extending, with intermissions, from about 1568 to 1648, when Spain recognized Dutch independence.

fourscore, eighty; four twenties.

octogenarian, 1. a person between eighty and eighty-nine years of age, inclusive.
2. of or pertaining to the number 80.

pe, the seventeenth letter of the Hebrew alphabet, having a numerical value of 80.

R, the medieval Roman numeral for 80.

81

'water,' (in short-order restaurants, soda-fountains, etc., in the U. S. in the 1940s) "eighty-one" used to designate 'a glass of water.' Still used—especially as 81 'one glass of water,' 82 'two glasses of water,' and so on through 85. See **86: 'all gone.'**

86

'all gone,' 'supply exhausted,' (in short-order restaurants, soda-fountains, etc., in the U. S.) "eighty-six" to designate 'There isn't/aren't any more,' usually heard from the chef in response to an order from a waiter.

87

four score and seven, the eighty-seven years between 1775 and 1863, cited in President Abraham Lincoln's address of that year to commemorate those who fell at the Battle of Gettysburg:
"Fourscore and seven years ago our fathers
brought forth on this continent a new nation
conceived in liberty and dedicated to the
proposition that all men are created equal. . . ."

88

Devil's Own, the, a name jocosely given to the 88th regiment of foot in the British army on account of its bravery in the Peninsular War (1808–14).

eighty-eight, *or* **88,** a piano. [From the eighty-eight keys of a standard pianoforte.]

Stuarts, the "fatal number" of the Stuarts was said to be eighty-eight: James III was killed in 1488; Mary Stuart was beheaded 1588; James II of England was dethroned 1688; and Charles Edward died 1788.

90

enneacontahedron, a polyhedron having ninety faces.

Gay Nineties, the decade of the 1890s, a term not in use till long after 1900.

N, the medieval symbol for 90.

Ninety Days' Wonder, *U. S. Army slang.* (in World War II) a second lieutenant who received his commission after only ninety days' training, a common occurrence in 1943–45. [A facetious reference to **9: Nine Days' Wonder.**]

90

nonagenarian, 1. a person between ninety and ninety-nine years of age, inclusive.
2. of or pertaining to the number 90.

quartile, (in astrology) noting the aspect or position of two planets that are distant from each other the fourth part of the zodiac, or 90°; considered an unfavorable aspect.

strontium-90, a radioactive isotope of strontium, found in the atmosphere and in many foods as a result of the fallout following the explosion of an atomic bomb.

92

snub-dodecahedron, a regular solid with ninety-two plane faces. See also **13: Archimedean solid.**

95

Theses, 95, a collection of ninety-five remonstrations against the abuse of indulgence by the Roman Catholic church, posted on the door of the Castle church at Wittenberg by Martin Luther on October 31, 1517; a copy was sent, with a strongly worded letter of protest, to the archbishop.

97

'to go,' (in short-order restaurants, soda-fountains, etc., esp. on the west coast of the U. S., chiefly in the 1940s) "ninety-seven" appended to a food order 'to take out,' as 'Two rare burgers, two Cokes, 97.'

100

C, the Roman numeral for one hundred. [Taken to stand for Latin *centum* 'hundred.']

cantred, a territorial division containing a hundred townships.

cent, the hundredth part of a dollar, rupee, or florin.

cental, 1. pertaining to or consisting of a hundred; proceeding or reckoned by the hundred.
2. a weight of one hundred pounds avoirdupois, formerly used in Great Britain.

centavo, a hundredth part of a dollar or peso, as in Manila, Chile, and some other Spanish-speaking countries.

centenarian, noting or pertaining to a person who has lived for one hundred years.

100

centenarius, the president of the court of one hundred in the Salic and other Teutonic legal systems.

centenary, 1. referring to a hundred, especially a hundred years.
2. a commemoration or celebration of an event that occurred a hundred years earlier; a hundredth anniversary.
3. lasting for a hundred years; occurring once every hundred years.
4. a centenarian.

centennial, 1. of or pertaining to a period of a hundred years.
2. a commemoration or celebration of an event that occurred a hundred years earlier; a hundredth anniversary.
3. lasting for a hundred years; occurring once every hundred years.

centennium, *pl.* **centennia.** one hundred years; a century.

centigrade, graduated into one hundred divisions or parts, as the centigrade, or Celsius, temperature system, in which water freezes at 0° and boils at 100°.

centipitous, having a hundred heads.

centoculated, having a hundred eyes.

centuple, multiply by a hundred or multiplied by a hundred.

centuplicate, centuple; a hundredfold.

centuria, (in ancient Rome) a measure of land, said to have been originally a hundred times the quantity Romulus distributed to each citizen.

centurial, 1. of or pertaining to a century or centuries; existing for a century or for centuries.
2. completing a century.

centurion, (in ancient Rome) a military officer in command of a company of infantry, or century.

century, 1. a hundred, especially a period of a hundred years.
2. (in ancient Rome) **(a)** a division of the people, originally and later approximately a hundred in number. **(b)** a subdivision of a legion, nominally of a hundred men. **(c)** an allotment of land varying in size according to region and period; such an allotment given to soldiers in a conquered country.

century plant, a species of agave, *Agave americana*, that blooms after ten to fifteen years. [So called because the flowering interval, which is long, was once thought to be a hundred years.] Also called **agave, aloe.**

centussis, (in ancient Rome) a unit of weight consisting of one hundred asses.

100

Chiltern Hundreds, a hilly district of Buckinghamshire, England, formerly owned by the crown. It has a nominal office attached to it, of which the holder is called the Steward of the Chiltern Hundreds. Because a member of the House of Commons who has not been disqualified cannot resign his seat directly, any member who wishes to resign may accept stewardship of the Chiltern Hundreds, which being under the crown, vacates the seat. The stewardship is under the control of the Chancellor of the Exchequer, and the recipient usually resigns immediately after his appointment.

C-note, a hundred-dollar bill.

dollar, the monetary standard of the United States, containing 100 cents.

drachma, a coin of modern Greece, equivalent to 100 lepta.

hecatomb, (in classical antiquity) a sacrifice of a hundred oxen or other beasts of one kind.

hecatompedon, a building 100 feet long or wide, especially the cella of the temple of Athena, the Parthenon, at Athens.

Hecatoncheires, the, according to Vergil's *Aeneid* (25 BC?), three giants who had fifty heads and 100 arms and hands, namely: Cottus; Briareus; and Gyges (or Gyes), who conquered the Titans. According to Homer, Briareus was so called by the gods; men called him Aegeon.

hecatonstylon, a building with a hundred columns.

hecatontarchy, government by a hundred.

hectare, a land measure equal to 100 ares, 10,000 square meters, or 2.471 acres.

hectogram, *or* **hectogramme,** a weight of 100 grams.

hectoliter, *or* **hectolitre,** a unit of capacity of 100 liters.

hectometer, *or* **hectometre,** a unit of length of 100 meters.

hectostere, a unit of volume of 100 steres, or 100 cubic meters.

honors, hundred, (in bridge) a declarer's hand with any four trumps from among the following: ten, jack, queen, king, ace. See also **150: honors.**

hundred, 1. (in early Teutonic history) a territorial or administrative district, especially, in southern and central England, a division or subdivision of a county (corresponding to a *wapentake* in northern England). [Of uncertain origin, possibly derived from bodies each consisting of a hundred warriors, from a hundred hides of land, from groups of a hundred families, etc.]
2. a similar division introduced into the colonies of Pennsylvania, Virginia, Maryland, and Delaware.

100

hundred-court, (in England) a court having jurisdiction in a hundred. See also **100: hundred, 1.**

Hundred Days, the, 1. the period from approximately March 13th, 1815, when Napoleon, after his escape from exile at Elba, was joined by Ney with his army, to June 22nd, the date of his abdication after Waterloo. No calculation yields an interval of exactly 100 days, and the period generally accepted is taken as from March 20th, when he entered Paris, to June 22nd, when he abdicated.
2. the period from March 9 to June 16, 1933, when Congress, called into special session by President Franklin Delano Roosevelt, passed most of the reform and recovery laws that he presented to them to end the Depression.

hundred-legs, a centipede.

hundredweight, 1. (in England) an avoirdupois denomination of weight equal to 112 pounds.
2. (in the U. S., Canada) an avoirdupois denomination of weight equal to 100 pounds.
3. a centiweight; abbr.: *cwt.*

Hundred Years' War, an intermittent struggle between England and France that lasted from 1337 to 1457, originating from the huge territories held in southern France by the Plantagenet kings, through Eleanor of Aquitaine (who in 1152 had married Henry Plantagenet, who became Henry II of England). It was ended in 1457 by a truce between the Lancastrian party and Louis XI.

I. Q., a person's intelligence quotient, representing one hundred times the mental age divided by the chronological age, or 100 for the average person.

Old Hundred, *or properly* **Old Hundredth,** a celebrated tune set in England about the middle of the 16th century to Kethe's version of the 100th Psalm. The earliest extant copy is in the Genevan psalter of 1554, where it is set to Beza's version of the 134th Psalm. There is evidence that it is older, and was originally a popular tune set to words of a light, gay character.

one hundred percent, complete; perfect.

per cent, *or* **percent,** in or by the hundred. Formerly, **per centum.**

quintal, a weight of one hundred pounds, varying locally by country.

101

Dirty Shirts, the, the Royal Bengal Fusileers, formerly, the 101st Foot, which fought at Delhi (1857) in shirtsleeves. [Brewer, *Dictionary of Phrase and Fable,* 1894.]

105

centumvir, *pl.* **centumvirs, centumviri.** (in ancient Rome) one of a body of 105 (called in round numbers 100) judges, three from each of thirty-five tribes, appointed to decide common causes among the people.

120

dyakis-hexacontahedron, a solid having 120 faces, reciprocal to the great rhombicosidodecahedron.

great hundred, *or* **long hundred,** six score, or 120, formerly a legal counting measure in England for balks, deals, eggs, spars, stones, etc., but not for men, money, pins, etc.

120 in the water bag, *Australian slang.* an extremely hot day.

135

trioctile, (in astrology) an aspect of two planets which, when viewed from earth, are three octants, or 135°, distant from each other.

144

gross, (in commerce) twelve dozen of a product, usually considered as a wholesale quantity.

post, a pile of 144 sheets of handmade paper fresh from the mold, arranged alternately with pieces of felt ready to be placed in a screw-press. Also called **felt-post.** Compare **144: white post.**

white post, a pile of 144 sheets of handmade paper after the interleaved felts have been removed. Compare **144: post.**

150

honors, hundred and fifty, (in bridge) a declarer's hand with all the following trumps: ten, jack, queen, king, ace; or, in no-trump, all four aces. See also **100: honors.**

Psalms, a book of the Old Testament, following Job and preceding Proverbs, that contains 150 psalms and hymns; the Book of Psalms.

psalter, (in the Roman Catholic Church) **1.** a series of 150 devout utterances in honor of certain mysteries, as the sufferings of Christ. **2.** a large chaplet or rosary with 150 beads, one for each of the Psalms.

Round Table, a huge round table, made at Carduel by Merlin for Uther Pendragon, King Arthur's father, who gave it to King Leodegraunce of Cameliard. It was given to King Arthur upon his marriage to Guinevere (the daughter of Leodegraunce). The table could accommodate 150 knights; three seats of it were reserved: two were for honor; one, the

150

Siege Perilous, was fatal to all who sat in it but the knight destined to find the Holy Grail, Sir Galahad.

sesquicentennial, 1. of or pertaining to a period of one and a half centuries, or one hundred fifty years.
2. a commemoration or celebration of an event that occurred one hundred-fifty years earlier; a one-hundred-fiftieth anniversary.
3. lasting for one hundred fifty years; occurring once every one hundred fifty years.

Y, the medieval Roman numeral for 150.

160

quarter-section, (in the U. S. and Canada) a square tract of land equal to one quarter of a square mile in area, or 160 acres; one fourth of a section.

T, the symbol for the medieval Roman numeral 160.

180

one-eighty, do a *Slang.* to reverse direction; go the opposite way (180° from the original direction); reverse oneself; change one's mind. Also, **do a 180° turn.**

200

bicentennial, 1. of or pertaining to a period of two hundred years.
2. a commemoration or celebration of an event that occurred two hundred years earlier; a two-hundredth anniversary.
3. lasting for two hundred years; occurring once every two hundred years.

Dutch 200, (in bowling slang) a score of 200 points, achieved by bowling alternate strikes and spares.

H, the medieval symbol for 200.

238

uranium-238, a radioactive isotope of uranium, used in making an atomic bomb.

250

E, a medieval Roman numeral for the number 250.

K, the medieval symbol for the number 250.

300

bees,   three hundred golden bees were found in the tomb of Childeric, when it was opened in 1653. Hence, the bee was adopted as the symbol of the French empire.

tercentenary, *or* **tricentenary,**   **1.** referring to three hundred, especially three hundred years.
2. a commemoration or celebration of an event that occurred three hundred years earlier; a three-hundredth anniversary.

tercentennial,   **1.** of or pertaining to a period of three hundred years.
2. a commemoration or celebration of an event that occurred three hundred years earlier; a three-hundredth anniversary.
3. lasting for three hundred years; occurring once every three hundred years.

350

tricentquinquagenary,   a commemoration or celebration of an event that occurred three hundred and fifty years earlier; a three-hundred-fiftieth anniversary. [Coined by Dr. Frank R. Abate in connection with the 350th anniversary of the founding of Old Saybrook, Connecticut, 1985.]

365

abraxas,   a word which, in numerology, represents the number 365, being the sum of: a = 1; b = 2; r = 100; a = 1; x = 60; a = 1; and s = 200. The significance of 365 lies in the number of days in a year, the number of eons ruling the year, the order of spirits, and the number of heavens. [Jobes, *Dictionary of Mythology, Folklore and Symbols.*]

common year,   a year of 365 days. Compare **366: leap year.**

vague year,   a year of the Egyptian calendar, so called because it "wandered" throughout the seasons in the course of a period of 1507 years.

365¼

Julian year,   a period of 365¼ days.

366

leap year,   a year of 366 days, that is, one containing an intercalary day added every fourth year (except those divisible by 400) to make the calendar year conform more closely to the solar year. Compare **365: common year.**

400

Baltadjii, the 400 halberdiers who attended the royal prince and princess of Constantinople. Their colonel is called the "Kizlar-agasi." The name means 'hatchet-bearers.'

Four Hundred, the, 1. an assembly, in 411 BC, of four hundred Athenians empowered to rule the city with absolute authority. After three months they were deposed and the democracy was restored.
2. *U. S.* the highest society in New York City, as characterized by Ward McAllister in 1892. [From the number of guests invited by Mrs. William Astor to an annual ball, as her ballroom could accommodate no more.]

G, the medieval Roman numeral for 400.

P, a medieval symbol for the number 400.

quadricentennial, 1. of or pertaining to a period of four hundred years.
2. a commemoration or celebration of an event that occurred four hundred years earlier; a four-hundredth anniversary.
3. lasting for four hundred years; occurring once every four hundred years. See also **400: quater-centenary.**

quadrigenarious, consisting of four hundred.

quater-centenary, 1. referring to four hundred, especially four hundred years.
2. a commemoration or celebration of an event that occurred four hundred years earlier; a four-hundredth anniversary. See also **400: quadricentennial**

451

Fahrenheit 451°, the temperature at which books are said to burn.

480

mill-ream, a package of handmade paper containing 480 sheets of which the two outer quires (48 sheets) are imperfect. A ream of 480 sheets of perfect paper is called a *ream of insides.*

490

seventy times seven, (in the Bible) an infinite number, an allusion to the number of times one must be willing to forgive others. [Matt. 18.22.]

500

agathoergi, (in ancient times) the five hundred knights who attended kings of Sparta as bodyguards in time of war.

500

Council of Five Hundred, an assembly of 500 members forming the second branch of the Legislative Body (the first being the Council of Ancients) in France during the government of the Directory, 1795–99.

D, the Roman numeral for the number 500.

five hundred rummy, a kind of rummy, usually for two players, in which the winner is the first to score 500 points.

Fortune 500, an annual listing, by *Fortune* magazine, of the five hundred companies in the United States with the highest gross income.

Indy 500, an annual 500-mile automobile race held on Memorial Day at the Speedway in Indianapolis, Indiana.

Q, the medieval symbol for the number 500.

quincentenary, 1. referring to five hundred, especially five hundred years.
2. a commemoration or celebration of an event that occurred five hundred years earlier; a five-hundredth anniversary.

ream by convention, five hundred sheets of paper.

Standard & Poor's 500 Index, a statistical report, compiled at five-minute intervals, on the behavior of the shares of 500 companies (400 industrial firms, 40 utilities, 40 financial institutions, and 20 transportation firms) on the New York Stock Exchange. The Index is calculated relative to average prices in 1941–43.

string of cash, a sum of 500 or 1,000 cash, the only coin formerly in use by the Chinese (called *tsien* by them), strung together by means of a hole at the center.

600

Light Brigade, on October 25, 1854, 600 men of the Light Brigade, under the command of the Earl of Cardigan, charged a Russian force of 5000 cavalry and six battalions of infantry, at Balaklava, on the Black Sea, during the Crimean War. Fewer than 200 survived. They are immortalized in "The Charge of the Light Brigade," composed by Alfred, Lord Tennyson, after he read a newspaper account of the event:
>   "Forward, the Light Brigade!"
> Was there a man dismayed?
> Not though the soldier knew
>   Someone had blundered.
> Theirs not to make reply,
> Theirs not to reason why,
> Theirs but to do and die.
> Into the valley of death
>   Rode the six hundred. [*excerpt*]

600

sexcentenary, 1. referring to six hundred, especially six hundred years. 2. a commemoration or celebration of an event that occurred six hundred years earlier; a six-hundredth anniversary.

606

salvarsan, dioxydiamido-arsenobenzol, a remedy for syphilis and other spirochetal diseases. [After the 606th compound tested by Paul Ehrlich (1854–1915), German bacteriologist, in his search for a specific remedy for syphilis.] See also **914: neosalvarsan.**

700

septicentennial, 1. of or pertaining to a period of seven hundred years. 2. a commemoration or celebration of an event that occurred seven hundred years earlier; a seven- hundredth anniversary. 3. lasting for seven hundred years; occurring once every seven hundred years.

707

year of confusion, the year of 445 days, before the introduction of the Julian calendar, the 707th year of the Roman era ending with 47 BC.

800

octocentennial, 1. of or pertaining to a period of eight hundred years. 2. a commemoration or celebration of an event that occurred eight hundred years earlier; an eight-hundredth anniversary. Also called **octingentenary, octocentenary.** 3. lasting for eight hundred years; occurring once every eight hundred years.

Valhalla, (in Norse mythology) the number of warriors in Valhalla is 800.

914

neosalvarsan, an improved version of salvarsan, developed by Dr. Paul Ehrlich and named for the 914th formula tested. See also **606: salvarsan.**

1000

chiliad, one thousand, especially the period of a thousand years.

chiliagon, a plane figure with a thousand angles and a thousand sides.

chiliarch, (in ancient Greece) a military commander or chief in charge of a thousand men.

chiliarchy, a body of a thousand men.

1000

chiliasm, the doctrine, suggested by the twentieth chapter of Revelations, of a visible and corporeal government of Christ and the saints on Earth in the last days, continuing for a thousand years. Compare **1000: millenarian, 2.**

chiliomb, a sacrifice of a thousand oxen. Compare **100: hecatomb.**

grand, a thousand dollars.

Guineas, 1000, a classic one-mile stakes horserace for fillies, established in 1814 and run at Newmarket, in Cambridgeshire, England. Compare **2000: Guineas.**

K, a symbol for 1000, taken from the Greek *kilo* 'one thousand.'

kilo-, a prefix taken from Greek, where it meant 'one thousand,' via French, where it was adopted for use in describing metric units of measurement, as *kilogramme, kilomètre,* etc.

M, the Roman numeral for 1000.

mahayugas, the one thousand great ages in Hindu mythology, each 4,320,000 years in length and equal to one *Day of Brahma.*

millenarian, 1. relating to a thousand, especially to an expected period of righteousness on earth to last a thousand years.
2. one who believes in the millennium, especially one who believes that Christ will visibly reign on earth with his saints for a thousand years or for an indefinite period before the end of the world; a chiliast.

millenary, 1. referring to a thousand, especially a thousand years.
2. a commemoration or celebration of an event that occurred a thousand years earlier; a thousandth anniversary.

millenary petition, a petition presented by about a thousand Puritan ministers to James I on his progress to London in April, 1603, asking for certain changes in ceremony, etc.

milleniad, *or* **milliad,** a period of one thousand years.

millennium, 1. a period of a thousand years.
2. (in Christian eschatology) a period of a thousand years during which the kingdom of Christ will be established on earth, based on a literal interpretation of the phrase "a thousand years" in Rev. 20.1-5. Compare **360,000: millennium.**

milli-, a word element, taken from Latin, meaning 'one thousand,' as in *millipede,* or 'one thousandth,' as in *millisecond.*

milliary, of or pertaining to the ancient Roman mile of a thousand paces, or five thousand Roman feet.

1000

millier, (in the metric system) a weight equal to one thousand kilograms, or 2204.6 pounds avoirdupois, being the weight of one cubic meter of water at 4°C.

milreis, a Portuguese or Brazilian unit of money equal to 1,000 reis.

permillage, the ratio of a certain part to the whole when the latter is taken at one thousand; the rate per thousand.

thousand, ten hundreds in quantity, 1000.

thousand days, the number of days served in the presidency of the United States by John F. Kennedy, between January 4, 1961, and November 22, 1963, when he was assassinated.

thousand-legged table, an old-fashioned folding or extension table that appears to have a large number of supports.

thousand-legger, an insect with a large number of pairs of legs, seemingly as many as a thousand.

1001

Arabian Nights, or Thousand and One Nights, a collection of Arabian short stories related by Scheherezade, daughter of the vizier, who has arranged her marriage to king Shahryar, who, to revenge himself on all women because of an unfaithful wife, takes a new wife every day and then kills her. Scheherezade saves herself by telling the king a different story every night but withholds the ending till the following night, if she survives. The king, eager to learn the conclusion of each tale, spares her life from day to day and finally abandons his murderous plan. It is the source of the stories about "Sindbad the Sailor," "Aladdin and the Wonderful Lamp," "Ali Baba and the Forty Thieves," and others.

1200

due-cento, the thirteenth century, in Italian art characterized by a blending of romanesque and gothic influences and by the proto-renaissance of Frederick II (1215–50), who showed unusual appreciation of classic art. The chief master of the period is Niccola Pisano, the sculptor and architect (*c.* 1207–78).

1300

trecento, the thirteen hundreds (fourteenth century), especially as applied to Italian art and literature and their distinguishing styles and characteristics.

1400

"Fourteen Hundred," a warning called out in the London Stock Exchange, in use in the mid seventeenth century, that a stranger had entered the premises.

quattrocentist, an Italian of the fifteenth century, especially an artist of the style called quattrocento.

quattrocento, 1. the fifteenth century considered as an epoch of art or literature, especially in Italy.
2. of the style of the fifteenth century.

1500

Cinquecentesti, the Italian worthies who flourished in the 1500s, especially the poets Ariosto, Tasso, and Rucellai; the painters Raphael, Titian, and Michelangelo; as well as Machiavelli, Luigi Alamanni, Bernardo Baldi, and others. Compare **1600: Seicentisti.**

cinque-centist, a writer or an artist of the sixteenth century or one who imitates sixteenth-century style.

cinque-cento, 1. the sixteenth century in Italy, especially the fine arts of that period.
2. designed or executed in the sixteenth century, referring especially to the decorative art and architecture of that period; renaissance.

1600

secentismo, *or* **seicentismo,** the literary style of Italian writers of the seventeenth century, regarded as a period of decadence in Italian literature. See also **1600: Secentisti.**

Seicentisti, the Italians who flourished in the 1600s and were regarded as contributing to *Seicentista*, or degraded art, characterized by bombastic writing, Rococo-style art, etc. Its chief representatives were the poet Giambattista Marini, the painter Michelangelo Merisi da Caravaggio, the sculptor Giovanni Lorenzo Bernini, and the architect Francesco Borromini. Compare **1500: Cinquecentesti.**

1961

upside-down year, the January 1961 issue of *Mad Magazine* celebrated "the last upside-down year."

2000

Guineas, 2000, a classic one-mile stakes horserace for three-year-olds, established in 1809 and run at Newmarket, in Cambridgeshire, England. Compare **1000: Guineas.**

2000

Z, the medieval Roman numeral for 2000.

5000

D̄, the Roman numeral for the number 5000.

6000

dicast, *or* **dikast,** (in ancient Athens) one of 6000 citizens chosen by lot annually to sit as judges.

7000

S̄, the medieval Roman numeral 7000.

10,000

myriad, the number ten thousand.

myriarch, a commander of ten thousand men.

Ten Thousand, a Greek force of ten thousand men who were in the service of the Persian prince Cyrus and who, after his death, returned to the Greek city of Trapezus, on the Black Sea via Kurdistan and Armenia, in 400 BC. Their adventures are recounted in Xenophon's *Anabasis.*

Ten Thousand Immortals, the main part of the Achaemenian army (640–330 BC), so called because their number was restored to that total at once if any loss was sustained.

20,000

Twenty Thousand Leagues Under the Sea, (*Vingt mille lieues sous les mers*) a novel (1869), by Jules Verne, about the adventures of Captain Nemo, a mysterious figure who travels the world in the *Nautilus,* a submarine he designed.

40,000

F̄, a medieval Roman numeral with a value of 40,000.

50,000

L̄, the Roman numeral for 50,000.

80,000

R̄, the medieval Roman numeral for 80,000.

87,000

Colosseum, the amphitheater in ancient Rome could accommodate an estimated 87,000 spectators.

90,000

$\overline{N}$, the medieval symbol for 90,000.

160,000

$\overline{T}$, the symbol for the medieval Roman numeral 160,000.

200,000

$\overline{H}$, the medieval symbol for 200,000.

250,000

$\overline{E}$, a medieval Roman numeral for the number 250,000.

$\overline{K}$, the medieval symbol for 250,000.

360,000

millennium, a period of 360,000 days during which the kingdom of Christ will be established on earth, based on Rev. 20.1-5, and on the principle that the Jewish year contained 360 days. Compare **1000: millennium.**

400,000

$\overline{G}$, the medieval Roman numeral for 400,000.

$\overline{P}$, a medieval numeral with a value of 400,000.

1,000,000

"Bet-a-million" Gates, (properly, "Bet-you-a-million" Gates) John Warne Gates (1855–1911), a speculator on the stock exchange, promoter, and manufacturer whose fame spread from his successes as a salesman of barbed wire in Texas. After an unsuccessful clash with J. P. Morgan, he retired to Texas where he helped develop Spindletop and became the founder of the Texas Company.

feel like a million dollars, to feel marvelously healthy and euphoric; to be in the best of spirits. Also, **feel like a million.**

$\overline{M}$, the Roman numeral for 1,000,000.

mega-, *or* **meg-,** a prefix meaning 'one million.'

1,000,000

million, the number one thousand thousand, or ten hundred thousand. See also **10⁹: milliard.**

millionaire, a person worth a million dollars, pounds, francs, etc.

3,000,000

three-million bill, a bill passed in the U. S. Congress in 1847 appropriating three million dollars for the purchase of land from Mexico. Introduced in the House of Representatives with the Wilmot proviso as a rider, it was passed by the Senate after rejection of the rider, which prohibited slavery in any property purchased by the U. S. from Mexico.

50,000,000

"Fifty million Frenchmen can't be wrong," a saying popular with American soldiers during World War I. [*Bartlett's Familiar Quotations,* 15th ed., 1980.]

10^9

billion, (in the American system) ten to the ninth power, or a thousand followed by 6 zeros. See also **10^{12}: billion.**

milliard, (formerly, in the British system) ten to the ninth power; ten followed by 9 zeros. See also **1,000,000: million.**

10^{12}

billion, (formerly, in the British system) ten to the twelfth power, or a million to the second power; ten followed by 11 zeros. See also **10^9: billion.**

trillion, (in the American system) ten to the twelfth power, or a thousand followed by nine (3 × 3) zeros. See also **10^{18}: trillion.**

10^{15}

quadrillion, (in the American system) ten to the fifteenth power, or a thousand followed by twelve (4 × 3) zeros. See also **10^{24}: quadrillion.**

10^{18}

quintillion, (in the American system) ten to the eighteenth power, or a thousand followed by fifteen (5 × 3) zeros. See also **10^{30}: quintillion.**

trillion, (in the British system) ten to the eighteenth power, or a million to the third power; ten followed by 17 zeros. See also **10^{12}: trillion.**

10^{21}

sextillion, (in the American system) ten to the twenty-first power, or a thousand followed by eighteen (6 × 3) zeros. See also 10^{36}: **sextillion.**

10^{24}

quadrillion, (in the British system) ten to the twenty-fourth power, or a million to the fourth power; ten followed by 23 zeros. See also 10^{15}: **quadrillion.**

septillion, (in the American system) ten to the twenty-fourth power, or a thousand followed by twenty-one (7 × 3) zeros. See also 10^{42}: **septillion.**

10^{27}

octillion, (in the American system) ten to the twenty-seventh power, or a thousand followed by twenty-four (8 × 3) zeros. See also 10^{48}: **octillion.**

10^{30}

nonillion, (in the American system) ten to the thirtieth power, or a thousand followed by twenty-seven (9 × 3) zeros. See also 10^{54}: **nonillion.**

quintillion, (in the British system) ten to the thirtieth power, or a million to the fifth power; ten followed by 29 zeros. See also 10^{18}: **quintillion.**

10^{33}

decillion, (in the American system) ten to the thirty-third power, or a thousand followed by thirty (10 × 3) zeros. See also 10^{60}: **decillion.**

10^{36}

sextillion, (in the British system) ten to the thirty-sixth power, of a million to the sixth power; ten followed by 35 zeros. See also 10^{21}: **sextillion.**

undecillion, (in the American system) ten to the thirty-sixth power, or a thousand followed by thirty-three (11 × 3) zeros. See also 10^{66}: **undecillion.**

10^{39}

duodecillion, (in the American system) ten to the thirty-ninth power, or a thousand followed by thirty-six (12 × 3) zeros. See also 10^{72}: **duodecillion.**

10^{42}

septillion, (in the British system) ten to the forty-second power, or a million to the seventh power; ten followed by 41 zeros. See also 10^{24}: **septillion.**

tredecillion, (in the American system) ten to the forty-second power, or a thousand followed by thirty-nine (13 × 3) zeros. See also 10^{78}: **tredecillion.**

10^{45}

quattuordecillion, (in the American system) ten to the forty-fifth power, or a thousand followed by forty-two (14 × 3) zeros. See also 10^{84}: **quattuordecillion.**

10^{48}

octillion, (in the British system) ten to the forty-eighth power, or a million to the eighth power; ten followed by 47 zeros. See also 10^{27}: **octillion.**

quindecillion, (in the American system) ten to the forty-eighth power, or a thousand followed by forty-five (15 × 3) zeros. See also 10^{90}: **quindecillion.**

10^{51}

sexdecillion, (in the American system) ten to the fifty-first power, or a thousand followed by forty-eight (16 × 3) zeros. See also 10^{96}: **sexdecillion.**

10^{54}

nonillion, (in the British system) ten to the fifty-fourth power, or a million to the ninth power; ten followed by 53 zeros. See also 10^{30}: **nonillion.**

septendecillion, (in the American system) ten to the fifty-fourth power, or a thousand followed by fifty-one (17 × 3) zeros. See also 10^{102}: **septendecillion.**

10^{57}

octodecillion, (in the American sytstem) ten to the fifty-seventh power, or a thousand followed by fifty-four (18 × 3) zeros. See also 10^{108}: **octodecillion.**

10^{60}

decillion, (in the British system) ten to the sixtieth power, or a million to the tenth power; ten followed by 59 zeros. See also 10^{33}: **decillion.**

novemdecillion, (in the American system) ten to the sixtieth power, or a thousand followed by fifty-seven (19 × 3) zeros. See also 10^{114}: **novemdecillion.**

10^{63}

vigintillion, (in the American system) ten to the sixty-third power, or a thousand followed by sixty (20 × 3) zeros. See also 10^{120}: **vigintillion.**

10^{66}

undecillion, (in the British system) ten to the sixty-sixth power, or a million to the eleventh power; ten followed by 65 zeros. See also 10^{36}: **undecillion.**

10^{72}

duodecillion, (in the British system) ten to the seventy-second power, or a million to the twelfth power; ten followed by 71 zeros. See also 10^{39}: **duodecillion.**

10^{78}

tredecillion, (in the British system) ten to the seventy-eighth power, or a million to the thirteenth power; ten followed by 77 zeros. See also 10^{42}: **tredecillion.**

10^{84}

quattuordecillion, (in the British system) ten to the eighty-fourth power, or a million to the fourteenth power; ten followed by 83 zeros. See also 10^{45}: **quattuordecillion.**

10^{90}

quindecillion, (in the British system) ten to the ninetieth power, or a million to the fifteenth power; ten followed by 89 zeros. See also 10^{48}: **quindecillion.**

10^{96}

sexdecillion, (in the British system) ten to the ninety-sixth power, or a million to the sixteenth power; ten followed by 95 zeros. See also 10^{51}: **sexdecillion.**

10^{102}

septendecillion, (in the British system) ten to the one hundred and second power, or a million to the seventeenth power; ten followed by 101 zeros. See also 10^{54}: **septendecillion.**

10^{108}

octodecillion, (in the British system) ten to the one hundred and eighth power, or a million to the eighteenth power; ten followed by 107 zeros. See also 10^{57}: **octodecillion.**

10^{114}

novemdecillion, (in the British system) ten to the one hundred and fourteenth power, or a million to the nineteenth power; ten followed by 113 zeros. See also 10^{60}: **novemdecillion.**

10^{120}

vigintillion, (in the British system) ten to the one hundred and twentieth power, or a million to the twentieth power; ten followed by 119 zeros. See also 10^{63}: **vigintillion.**

10^{303}

centillion, (in the American system) ten to the three hundred and third power, or a thousand followed by three hundred (100 × 3) zeros. See also 10^{600}: **centillion.**

10^{600}

centillion, (in the British system) ten to the six hundredth power, or a million to the hundredth power; ten followed by 599 zeros. See also 10^{303}: **centillion.**

10^{1000}

googol, a large number, one followed by a hundred zeros.

$10^{10^{10}}$

googolplex, a very large number, one followed by a googol of zeros.

∞

infinity, a concept of limitlessness, once thought to obtain in space but now confined to mathematics as a quantity used to express the limit of an increasing sequence or series that has no bound.

MISCELLANEOUS

deficient number, a number whose aliquot parts add up to less than itself: $1 \times 2 \times 4 = 8$; $1 + 2 + 4 = 7$.

manque, (in roulette) one of the numbers from 1 to 18. Compare **Misc.: passe.**

masculine number, an odd number. See also **2: feminine number.**

masculine sign, the first, third, fifth, etc., signs of the zodiac. See also **2: feminine sign.**

passe, (in roulette) one of the numbers from 19 to 36. Compare **Misc.: manque.**

APPENDIX

Latin

Cardinal	Ordinal
1. unus, una, unum	primus
2. duo, duae, duo	secundus
3. tres, tria	tertius
4. quattuor	quartus
5. quinque	quintus
6. sex	sextus
7. septem	septimus
8. octo	octavus
9. novem	nonus
10. decem	decimus
11. undecim	undecimus
12. duodecim	duodecimus
13. tredecim	tertius decimus
14. quattuordecim	quartus decimus
15. quindecim	quintus decimus
16. sedecim	sextus decimus
17. septemdecim	septimus decimus
18. duodeviginti	duodevicesimus
19. undeviginti	undevicesimus *or* nonus decimus
20. viginti	vicesimus *or* vigesimus
21. unus et viginti *or* viginti unus	unus et vicesimus *or* vicesimus primus
30. triginta	trigesimus
40. quadraginta	quadragesimus
50. quinquaginta	quinquagesimus
60. sexaginta	sexagesimus
70. septuaginta	septuagesimus
80. octoginta	octogesimus
90. nonaginta	nonagesimus
100. centum	centesimus
1,000. mille	millesimus
1,000,000. only expressed through multiplication.	

French

	Cardinal	Ordinal
1.	un, une	premier, première
2.	deux	second or deuxième
3.	trois	troisième
4.	quatre	quatrième
5.	cinq	cinquième
6.	six	sixième
7.	sept	septième
8.	huit	huitième
9.	neuf	neuvième
10.	dix	dixième
11.	onze	onzième
12.	douze	douzième
13.	treize	treizième
14.	quatorze	quatorzième
15.	quinze	quinzième
16.	seize	seizième
17.	dix-sept	dix-septième
18.	dix-huit	dix-huitième
19.	dix-neuf	dix-neuvième
20.	vingt	vingtième
21.	vingt et un	vingt-et-unième
22.	vingt-deux	vingt-deuxième
30.	trente	trentième
40.	quarante	quarantième
50.	cinquante	cinquantième
60.	soixante	soixantième
70.	soixante-dix	soixante-dixième
80.	quatre-vingts	quatre-vingtième
90.	quatre-vingt-dix	quatre-vingt-dixième
100.	cent	centième
1,000.	mille	millième
1,000,000.	million	millionième

Italian

Cardinal	Ordinal
1. uno	primo
2. due	secondo
3. tre	terzo
4. quattro	quarto
5. cinque	quinto
6. sei	sexto
7. sette	settimo
8. otto	ottavo
9. nove	nono
10. dieci	decimo *or* decima
11. undici	decimoprimo *or* undicesimo
12. dodici	dodicesimo
13. tredici	tredicesimo
14. quattordici	quattordicesimo *or*decimoquarto
15. quindici	decimoquinto *or* quindicesimo
16. sedici	sedicesimo
17. diciassette *or* diciasette	decimosettimo *or*diciassettisimo
18. diciotto	diciottesimo
19. diciannove *or* dicianove	diciannovesimo *or*decimonono
20. venti	ventesimo
21. ventuno	ventunesimo
30. trenta	trentesimo
40. quaranta	quarantesimo
50. cinquanta	cinquantesimo
60. sessanta	sessantesimo
70. settanta	settantesimo
80. ottanta	ottantesimo
90. novanta	novantesimo
100. cento	centesimo
1,000. mille	millesimo
1,000,000. milione	milionesimo

Spanish

	Cardinal	Ordinal
1.	un, uno, una	primero, primera
2.	dos	segundo, segunda
3.	tres	tercero, tercer
4.	cuatro	cuarto
5.	cinco	quinto
6.	seis	sexto
7.	siete	séptimo
8.	ocho	octavo
9.	nueve	noveno *or* nono
10.	diez	décimo
11.	once	undécimo
12.	doce	duodécimo
13.	trece	décimotercio
14.	catorce	catorceno
15.	quince	décimoquinto
16.	diez y seis	décimosexto
17.	diez y siete	décimo séptimo
18.	diez y ocho	décimooctavo
19.	diez y nueve	décimonono
20.	veinte	vigésimo
21.	veintiuno	vigésimo primero
30.	treinta	trigésimo
40.	cuarenta	cuadragésimo
50.	cincuenta	quincuagésimo
60.	sesenta	sexagésimo
70.	setenta	septuagésimo
80.	ochenta	octogésimo
90.	noventa	nonagésimo
100.	cien, ciento	centésimo
1,000.	mil	milésimo
1,000,000.	millón	millonésimo

German

	Cardinal	Ordinal
1.	eins	erste
2.	zwei	zweite
3.	drei	dritte
4.	vier	vierte
5.	fùnf	fùnfte
6.	sechs	sechste
7.	sieben	siebente
8.	acht	achte
9.	neun	neunte
10.	zehn	zehnte
11.	elf	elfte
12.	zwòlf	zwòlfte
13.	dreizehn	dreizehnte
14.	vierzehn	vierzehnte
15.	fùnfzehn	fùnfzehnte
16.	sechzehn	sechzehnte
17.	siebzehn	siebzehnte
18.	achtzehn	achtzehnte
19.	neunzehn	neunzehnte
20.	zwanzig	zwanzigste
21.	einundzwansig	einundzwansigste
30.	dreissig	dreissigste
40.	vierzig	vierzigste
50.	fùnfzig	fùnfzigste
60.	sechzig	sechzigste
70.	siebzig	siebzigste
80.	achtzig	achtzigste
90.	neunzig	neunzigste
100.	hundert	hundertste
1,000.	tausend	tausendste
1,000,000.	eine million	millionste

INDEX

INDEX

A

A, a: 1
Aaron   7: Seventh Heaven
Ab   5: Fast of the Fifth Month
abdominal   3: trisplanchnic
abducens nerve:6
Abednego   3: Shadrach
Abhidharma   3: Tripitaka
Abila   10: Decapolis
abolition of slavery   21: twenty-first
   rule
Abraham   3: patriarch; 6: apostle
abraxas:365
absolute zero:−273
Absolution   3: trisacramentarian
abstract body   5: Buddha, 1
Abu Bekr   10: wives
Abu Dhabi   7: United Arab Emirates
'abundant'   10: dime a dozen
Abyla   2: Pillars of Hercules
accessory nerve:11
Aceldama   30: pieces of silver
Achaean   3: Ionian
Achaeus   13: Tragic Poets
Acheron   5: rivers of Hell; 8: Hell
Acolon   12: Knights of the Round
   Table, 1
acoustic nerve   8: vestibulocochlear
   nerve
action, right   8: Noble Eightfold
   Path
Activities, Tantra of   14: Tantras
Adam   6: apostle; 7: Seventh
   Heaven, "Sindbad the Sailor"
adamantoid   48: hexoctahedron
Adam Kadmon   10: Sephiroth
Adar   13: Veadar
addition   4: four species, 1
à deux:2

Adilard   7: Pleiad, 2
Aditi   3: aum
administration   1: M. O. 1
admiral   17: salute
admonish sinners   7: spiritual works
   of mercy
Adonai   4: Tetragrammaton;
   7: names of God
Adonay   10: names of God
Adoptive Emperors:5
adoration of the Magi   7: Joys of
   Mary
Adrastus   7: Seven Against Thebes
Advent   7: church year
Aeacus   3: Judges of Hades
Aeantides   13: Tragic Poets
Aegean   7: Seven Seas
Aegean Sea   12: Dodecanese
Aegeon   100: Hecatoncheires
Aegyptus:50
Aegyptus   50: Danaides
Aelius Lampridius   6: Augustae
   Historiae Scriptores
Aelius Spartianus   6: Augustae
   Historiae Scriptores
Aello   2: Harpies; 3: Harpies
Aeneas   9: Sabines
Aeneid   100: Hecatoncheires
Aeolian   3: Ionian
Aeschines   5: Philosophers; 10: Attic
   Orators
Aeschylus   2: deuteragonist;
   13: Tragic Poets
Africans, judge of   3: Judges of
   Hades
Aganippides   9: Muses
agate:5½
agate line   14: line
agathoergi:500
Agathon   13: Tragic Poets
agave   100: century plant

237

'aggressive' 2:two-fisted
Aglaia 3:Graces
Agned Cathregonion 12:Arthur
Agni:3; :8
Agnus 10:names of God
'agreed' 10:ten four
agretae:9
Ahijah:12
Ahmed, Eqbal 6:Harrisburg Six
ai:2
air 4:elemental quality
Air Force 5:five-star general;
17:salute; 19:salute
Air Force Chief of Staff 19:salute
airy trigon ⅓:trigon
Ajman 7:United Arab Emirates
Akasagarbha 8:Dhyanibodhisattva
Aksobhya 5:Dhyanibuddha,
Herukabuddha
Alabama 50:United States of
America
"Aladdin and the Wonderful Lamp"
1001:*Arabian Nights*
Alamanni,
Luigi 1500:Cinquecentesti
Alaska 50:United States of
America; 54:"Fifty-four Forty"
alauda:5
Alberta 10:Ten Provinces
"Albinus" 7:Pleiad, 2
Al Borak 10:animals in paradise
Alcaeus 4:Lesbian Poets; 9:Lyric
Poets
Alchemist 22:major arcana
alchemy:7
Alcman 9:Lyric Poets
Alcuin 7:Pleiad, 2
Alcyone 7:Pleiades
Aldebaran 7:Hyades, 1
Alecto 3:Furies
aleph:1
aleph 22:major arcana
Alexander 9:Nine Worthies;
15:fifteen decisive battles
Alexander I 3:Three Emperors
Alexander II 3:League of the Three
Emperors
Alexander the Aetolian 13:Tragic
Poets
Alexandria 7:Seven Wonders of the
Middle Ages

Alexandria, Pharos at 7:Wonders
of the Ancient World
Alexandria, Pleiad of 7:Pleiad, 1
Alexandrian Canon 3:Iambic Poets;
4:Elegiac Poets; 5:Epic Poets,
Philosophers; 9:Lyric Poets;
10:Attic Orators; 13:Comic Poets,
Tragic Poets
Alexis 13:Comic Poets
Al Fujairah 7:United Arab Emirates
Algeria 13:OPEC
Alhambra 2:Hermanas
"Ali Baba and the Forty
Thieves":40
"Ali Baba and the Forty Thieves"
1001:*Arabian Nights*
alif:1
Alkoremmi:5
Alleluia Sunday 70:Septuagesima
Sunday
All Fools' day:1
'all gone':86
Alliance of the Three Kings:3
Alliklik 4:Serrano
'all right' 10:ten four
aloe 100:century plant
alpha:1
Alpha 10:names of God
Alsace-Lorraine
dispute 14:Fourteen Points
alternatives 2:between two fires,
horns of the dilemma
Altschwert, Master
12:Meistersingers
Amadis de Jamyn 7:Pleiad, 3
Amalthea:9
Amalthea 10:sibyl
ambiguity 2:dilogy
ambivalent:2
'ambulance needed' 10:ten-fifty-
two
America firster:1
American Ambassador 19:salute
American Painters, Ten 10:Ten
Amerus 3:Magi
Amherst College 3:Little Three
Amitabha 5:Dhyanibuddha,
Herukabuddha
Amitayus 5:Dhyanibuddha
Amoghasiddhi 5:Dhyanibuddha,
Herukabuddha

Amphiaraus   7: Seven Against
   Thebes
amphigonic   2: duoparental
amphora   2: diota
ᶜAmudim   10: Sephiroth
Anabasis   10,000: Ten Thousand
Anacreon   9: Lyric Poets
anadiplosis   2: epanastrophe
ancestor   2: diphyletic
Ancients, Council of   500: Council
   of Five Hundred
Anderson Co, Arthur   8: Big Eight, 2
Andocides   10: Attic Orators
Andrews, Laverne   3: Andrews
   Sisters
Andrews, Maxine   3: Andrews
   Sisters
Andrews, Patty   3: Andrews Sisters
Andrews Sisters: 3
Andruszowo, truce of   13: Thirteen
   Years' War
*And Then There Were
   None*   10: Ten Little Indians
Angel of Death   7: Seventh Heaven
Angel of Tears   7: Seventh Heaven
angels   9: angels
angels, circles of   9: angels, orders
   of
angels, orders of: 9
anger   7: sins; 8: Hell
Angilbert   7: Pleiad, 2
Anhwei, Four Masters of   4: Four
   Masters of Anhwei
Animalia: 3
Animalia   3: Primalia, Vegetabilia
animal kingdom   3: kingdom
animals in paradise: 10
Annunciation   7: Joys of Mary
anointed king   14: Trisala
ant   10: animals in paradise
antenna   2: dicerous, dipole
Antenorides   6: Troy
Anthesterion: 8
Antimachus   5: Epic Poets
Antiochus IV   7: menorah
Antiochus Epiphanes   8: Hanukkah
Antiphanes   13: Comic Poets
Antiphon   10: Attic Orators
Antoine de Baïf   7: Pleiad, 3
Antonius Pius   5: Adoptive
   Emperors

Antrim   6: Six Counties
Anuttara Yoga   14: Tantras
Aoide   3: Muses
Apachnas   6: Hyksos
apartment   2: duplex; 3: triplex
Apellius   3: Magi
Aphrodite   3: Golden Apples;
   12: Olympians
Apocalypse, seven lamps
   of   7: seven-branched candlestick
Apocrypha: 3
Apollo   4: Pythian games;
   12: Olympians
Apollo 14: 14
Apollodorus   13: Comic Poets
Apollonios Rhodios   7: Pleiad, 1
Apophis   6: Hyksos
Apostemus   4: Fast of the Fourth
   Month
apostle: 6
Apostles   3: Christian Creeds;
   12: quorum of Twelve
apostolic fathers: 6
Apostolic Hours of Prayer: 3
Appian Way   19: Decennovium
Appius Claudius   10: decemvir,
   Wicked Ten
Application, Tantra of   14: Tantras
Appolonia   5: Pentapolis
April: 4
Aquarius   ⅓: trigon; 12: sign
Aquila   6: Hexapla
***Arabian Nights*: 1001**
Arabian Nights' Entertainments
   7: "Sindbad the Sailor"
Arabian Peninsula   7: United Arab
   Emirates
Arabian Sea   7: Seven Seas
Arab-Israeli War   6: Six-Day War
Aralius   6: Hyksos
Aramis   3: *Three Musketeers*
Aratos   7: Pleiad, 1
Arbela   15: fifteen decisive battles
Arcadian Hind   12: Dodekathlos
Arcadian Stag   12: Dodekathlos
archangels   7: Menorah; 9: angels
Archer   12: sign
arches, temporal   2: diapsid
Archilochus   3: Iambic Poets
Archimedean solid: 13
archon: 9

Arctic Ocean   7:Seven Seas
are $^1/_{1000}$:milliare; $^1/_{100}$:centiare;
   $^1/_{10}$:deciare; 10:decare; 100:hectare
areostyle:4
Ares   12:Olympians
Arges   3:Cyclopes
Argolis   9:Hydra
Argos   7:Homer, Seven Against
   Thebes; 50:Danaides
arhat:16
Ariadne   7:seven youths
Aries $^1/_3$:trigon; 12:sign
Arion   4:Lesbian Poets
Ariosto   1500:Cinquecentesti
Aristophanes   13:Comic Poets
Aristotle   3:tritagonist; 4:elemental
   quality, secondary;
   5:Philosophers, quintessence
arithmetic   4:quadrivium; 7:seven
   arts
Arizona   50:United States of
   America
Arizona, University of   10:Pacific
   Ten
Arizona State University   10:Pacific
   Ten
Arkansas   50:United States of
   America
arm   3:tribrachial; 5:pentact;
   10:decapod
armadillo   6:poyou
Armagh   6:Six Counties
Arminians   5:Five Articles and the
   Five Points
Armour   5:Big Five Packers
arms, reduction of   14:Fourteen
   Points
Armstrong, Olive   8:Grand Jury
   Eight
Army Chief of Staff   19:salute
Arnold, Henry H.   5:five-star
   general
Aron   12:Knights of the Round
   Table, 2
arrow   4:quarrel
arrows, set of   3:pair
Arsinë   5:Pentapolis
Artemis   3:Hecate; 12:Olympians
Artemis, temple of   7:Wonders of
   the Ancient World
Arthur, battles of King:12

Arthur, King   9:Nine Worthies;
   150:Round Table
as $^1/_6$:sextans; $^1/_4$:quadrans;
   2:dupondius; $2^1/_2$:sestertius;
   3:triens, 1; 5:dekass, quinarius;
   10:decussis, dekass, denarius;
   100:centussis
Asani   8:Agni
Ascension   7:church year, Sorrows
   of Mary
Ascension day   3:Rogation days
ascetic body   5:Buddha, 1
Asgard   3:Norn, Yggdrasil
Ashcan School   8:Eight
Ashdod   5:Philistines
Asher   10:Lost Tribes of Israel
Ashley, Lord   10:Ten Hours Act
Ash Wednesday:1
Ash Wednesday   3:tripos; 40:Lent
Asia   4:perfect women
Asians, judge of   3:Judges of Hades
Asir:12
Askalon   5:Philistines
ass   6:Lucifera; 10:animals in
   paradise
Asseth   6:Hyksos
Assistant Secretary of Air Force,
   Army, Defense, Navy 17:salute
Associated Counties:7
Assumption   7:Joys of Mary
Assyria   6:Hyksos
Asterope   7:Pleiades
Astolpho   12:Paladins of
   Charlemagne
astrology   3:Hermes Trismegistus
astronomy   4:quadrivium; 7:seven
   arts; 9:Muses
Astypalaia   12:Dodecanese
Atalanta   3:Golden Apples of the
   Hesperides
at first blush:1
at first hand:1
Athanasian   3:Christian Creeds
Athena   12:Olympians
Athena, priestesses of   9:agretae
Athens   4:tetralogy; 7:Homer,
   Seven Cities; 30:Thirty Tyrants,
   Thirty Years' Truce, Thirty Years'
   War; 100:hecatompedon;
   400:Four Hundred
Athos   3:*Three Musketeers*

Atlantic Ocean   7:Seven Seas
Atlas   7:Pleiades
'at loose ends'   6:at sixes and
  sevens
'at odds'   6:at sixes and sevens
'at once'   2:two shakes of a lamb's
  tail
Atonement   5:Fundamentalism
atonement, substitutionary
  7:Fundamentalism
Ator   3:Magi
Atropos   3:Fates
at second hand:2
at sixes and sevens:6
Attic Orators:10
Attila   15:fifteen decisive battles
Augean Stables   12:Dodekathlos
Augsburg, Diet of   4:Tetrapolitan
Augsburg
  Confession   4:Tetrapolitan
August:8
Augustae Historiae Scriptores:6
"Augustine"   7:Pleiad, 2
Augustinian   4:four orders
Augustus   2:dupondius
aum:3
Auric, Georges   6:Six
Austerlitz   2:Napoleon; 3:War of
  the Third Coalition
Austerlitz, Battle of   3:Three
  Emperors
Austin Friars   4:four orders
Australian Air Force   6:"I've got
  sixpence, . . ."
Austria   1:War of the First
  Coalition; 2:War of the Second
  Coalition; 3:League of the Three
  Emperors, Trias; 7:Seven Weeks'
  War, Seven Years' War; 15:Fifteen
  Years' War
Austria-Hungary   2:Dual Alliance, 2;
  3:Triple Alliance, 3
Austria-Hungary, autonomy of
  14:Fourteen Points
Austro-Hungarian
  monarchy   2:Dual Monarchy
autocracy   1:monocracy
Avalokitesvara   5:Manusibuddha;
  8:Dhyanibodhisattva
Avarice   6:Lucifera
avarice   7:sins

avaricious   8:Hell
avatar:10
ave   10:decad ring
Avenging Angel   7:Seventh Heaven
Aventine   7:Rome
'awkward'   2:(have) two left feet
axis   2:dyad, 4; 3:triaxial
Ayes Suárez, Carlos   13:Macheteros
ayin:16
ayin   22:major arcana
Ayishah   10:wives
Azrael   7:Seventh Heaven

B

B:7
B, b:1; :2
Babartsky, Al   7:seven blocks of
  granite
Baby Bear   3:"Goldilocks and the
  Three Bears"
Babylon   7:Homer, Seven Cities
Babylon, hanging gardens of
  7:Wonders of the Ancient World
Babylonian Exile   12:Twelve Tribes
  of Israel
Babylonian Sibyl   10:sibyl
Babylonian system:60
Bacchus   4:tetralogy, 1; 7:Hyades,
  2; 9:Sabines
Bacchylides   9:Lyric Poets
Bach, Johann Sebastian   48:Forty-
  Eight
backstall   3:ugly man
backwater   1:one-horse town
Badon Hill   12:Arthur
Baïf, Antoine de   7:Pleiad, 3
Baker, Bernard L.   7:Watergate
  Seven
baker's dozen:13
Balaam's ass   10:animals
Balakirev, Mily A.   5:St. Petersburg
  Five
Balaklava   600:Light Brigade
Balance   12:sign
Baldi,
  Bernardo   1500:Cinquecentesti
Baldr   12:Asir
Balkan states, guarantees for
  14:Fourteen Points

be on all fours with:4
Berenice   5:Pentapolis
Berkeley, Lord John   24:Twenty-
    four Proprietors
Bernard, Jacqueline   8:Grand Jury
    Eight
Bernini, Giovanni
    Lorenzo   14:Louis Quatorze;
    1600:Seicentisti
Berrigan, Daniel   2:Berrigan
    Brothers; 9:Catonsville Nine
Berrigan, Philip F.   2:Berrigan
    Brothers; 6:Harrisburg Six;
    9:Catonsville Nine
Berrigan Brothers:2
Berrios, Luz María Berrios
    13:Macheteros
Bersunt   12:Knights of the Round
    Table, 1
Besser, Joe   3:Three Stooges
Bessie, Alvah   10:Hollywood Ten
best-ball foursome:4
beta:2
"Bet-a-million" Gates:1,000,000
beth:2
beth   22:major arcana
betrayal   7:Sorrows of Mary
better half:½
between two fires:2
Beza   100:Old Hundred
Bezer   6:Cities of Refuge
Bhaisajyaguru   5:Buddha, 2
Bhava   8:Agni
bi-:2
biarchy:2
Bias   7:Seven Sages of Ancient
    Greece, Seven Wise Men
biathlon:2
Biberman, Herbert   10:Hollywood
    Ten
Bible   3:Apocrypha; 7:sacred books
Bible, preservation of
    5:Fundamentalism;
    7:Fundamentalism
bibles   7:sacred books
bicameral:2
bicentennial:200
bicorn:2
bicycle race, six-day   6:six-day
    bicycle race
bicycle ride   3:triathlon

biennial:2
bifarious:2
bifid:2
bifocal:2
bifrons:2
bifurcate:2
bifurcated   2:bifurcate
bifurcous   2:bifurcate
biga:2
bigamy:2
Big Bad Wolf   3:"Three Little Pigs"
big casino:10
Big Dipper   7:muni
Big Eight:8
Big Five:5
Big Five Packers:5
Big Four:4
Big Ten:10
Big Three:3
bilateral:2
bilingual:2
biliteral:2
billion:10^9; :10^{12}
Bill of Rights:10
Bill of Rights   7:Seventh
    Amendment
bimanous:2
bimensal:2
bimestrial:2
bimetallism:2
Binah   10:Sephiroth
binary:2
binary digit   2:bit
binary name:2
binary number:2
binary star:2
binary system:2
binaural:2
binocular:2
binoculate   2:binocular
binomial:2
binominal   2:binomial
biped:2
bipennis:2
bipod:2
birdie:−1
bird in the hand is worth two in
    the bush, a:1
birefringent:2
bireme:2
birth, previous   5:Dhyanibuddha

birth cry   1:primal scream
birthwort   4:diatessaron
bisexual:2
bisexual   2:digenous
Bishamonten   7:Shichi Fukujin
bishops, case of the seven   7:seven
   bishops
bisulcate:2
bit:0; :2
bit   1:word; 4:nibble; 8:byte
bivalve:2
Black Friars   4:four orders
blackjack   2:double down;
   15:quince
'blackness'   3:Harpies
Black Prince   3:Prince of Wales's
   feathers
Black Sea   7:Seven Seas; 600:Light
   Brigade
blasphemers   8:Hell
Blenheim   15:fifteen decisive
   battles
Blitzen   8:Santa's reindeer
Blois, Charles of   30:Battle of the
   Thirty
Bloods, Five   5:Five Bloods
bloom   3:Graces
Blotz, Hans   12:Meistersingers
blue   3:primary color; 7:perfect
   color
Bluebeard   7:Fatima
boar   10:avatar
bob-tail flush:4
Boccaccio   3:Triumvirate, 3;
   10:Decameron
bodies, legislative   2:bicameral
body   2:disomatous; 3:trichotomy
body, king of the   5:Five Great
   Kings
body: abstract, meditative, mortal,
   ascetic, heavenly   5:Buddha, 1
body and matter   3:aum
body parts   4:four-letter word
Boëdromion:3
Boëthius   10:decimal system
bogey:1
bolt   4:quarrel
Bonds, Howard   8:New York Eight
Bonham, Sir John   9:Worthies of
   London

"Bonnie Prince Charlie"   45:forty-
   five
Book of Changes   6:hexagram, 5
Book of Psalms   150:Psalms
Books of Discipline:2
bootlegging   20:Roaring Twenties
Borden, Lizzie:40
'boreal'   7:septentrional
Borodin, Aleksandr P.   5:St.
   Petersburg Five
Borromini, Francesco
   1600:Seicentisti
Bors   12:Knights of the Round
   Table, 1
Boston Port Bill   5:Intolerable Acts
Boston Tea Party   5:Intolerable
   Acts
Bourbons   3:tricolor
bourgeois:9
Bow bell   5:"five farthings"
'bowel movement'   2:number two
boxcars:12
boxing the compass   32:point of the
   compass
brace:2
bract   3:tribracteate
Braddock, General   2:unlucky
   number
Bradley, Omar N.   5:five-star
   general
'braggart'   4:fourflusher
Bragi   12:Asir
Brahma   3:Trimurti
Brahma, Day of   1000:mahayugas
Brakefield, William   5:Fort Dix Five
Bramborough, John   30:Battle of
   the Thirty
branch   5:pentact
Brandt,
   Sebastian   12:Meistersingers
breadth   3:tridimensional;
   4:dimension
Brecht, Bertolt   3:*Threepenny
   Opera*
brethren   12:house
brethren, house of:3
brevier:8
Brgya-byin   5:Five Great Kings
Briareus   100:Hecatoncheires
brilliance   3:Graces
brilliant:3½

brilliant light   **5:Dhyanibuddha**
British Columbia   **10:Ten Provinces**
British Museum   **40:Field of the Forty Footsteps**
British Petroleum   **7:Seven Sisters, 2**
Britomart   **9:Nine Worthy Women**
Brittany   **30:Battle of the Thirty**
Brontë:3
Brontes   **3:Cyclopes**
brood   **2:digoneutic; 3:trigoneutic**
brother/brother relationship   **5:five social relations**
Brown, Cecil   **8:Murrow's Boys**
Brown University   **8:Ivy League**
Brown v. Board of Education of Topeka   **14:Fourteenth Amendment**
Brumaire:2
Bryn Mawr   **7:Seven Sisters, 1**
Buckinghamshire   **15:Danelagh; 100:Chiltern Hundreds**
bucolic poetry   **9:Muses**
Buddha:5
Buddha   **3:aum; 10:avatar; 16:arhat**
Buddha and dharma, union of   **3:aum**
Buddhas, celestial   **5:Buddha, 2**
Buddhas, creative   **5:Manusibuddha**
Buddhas, earthly   **5:Buddha, 1**
Buddhas, meditative   **5:Dhyanibuddha**
Buddhism   **3:Threefold Refuge; 4:Noble Truths; 14:Trisala**
Bulgaria   **3:League of the Three Emperors**
Bull   **7:Hyades, 1; 12:sign**
'bungling, bungled'   **½:half-assed**
Burdett, Winston M.   **8:Murrow's Boys**
Burgoyne, General   **15:fifteen decisive battles**
burial   **14:stations of the cross**
burn: first-degree   **1:first-degree burn**
burn one:1
burn: second-degree   **2:second-degree burn**
burn: third-degree   **3:third-degree burn**

bury the dead   **7:corporal works of mercy**
business double   **2:double, 8b**
Byllynge, Edward   **24:Twenty-four Proprietors**
Byng, Admiral   **2:unlucky number**
Byron   **3:terza rima**
byte:8
byte   **4:nibble**

C

C:100
C   **6:M. I. 6**
C, c:2; :3
Cabinet Member   **19:salute**
cadent   **12:house**
Cadwr   **12:Knights of the Round Table, 2**
Caeline   **7:Rome**
Caerleon   **12:Arthur**
Caesar   **1:First Triumvirate**
Caeso Duillius   **10:Wicked Ten**
çakravartin   **14:Trisala**
calends:1
calf   **4:Evangelists**
Calhoun, John C.   **3:Great Triumvirate**
California   **50:United States of America**
California at Los Angeles, University of **10:Pacific Ten**
California (Berkeley), University of **10:Pacific Ten**
California Republic   **33:Bear Party**
calisthenics   **12:daily dozen**
Callimachos   **7:Pleiad, 1**
Callimachus   **4:Elegiac Poets**
Callinus   **4:Elegiac Poets**
Calliope   **9:Muses**
Calpe   **2:Pillars of Hercules**
Calvinists   **5:Five Articles and the Five Points**
calyx   **2:dichlamydeous**
Camacho Negrón, Isaac **13:Macheteros**
Cambridge   **7:Associated Counties**
Cambridgeshire   **15:Danelagh; 1000:Guineas; 2000:Guineas**

Cambridge University   3: tripos, 2;
    10: tripos
camel   6: Lucifera; 10: animals in
    paradise
Camoëns   6: sestina
Canada   7: Group of Seven, 1
Canada, Dominion of   10: Ten
    Provinces
Canadian Group of
    Painters   7: Group of Seven, 2
Canatha   10: Decapolis
Cancer   ⅓: trigon; 12: sign
candelabrum: 7
candlestick   2: dicerion; 3: tricerion;
    7: menorah
candlestick of the tabernacle
    7: seven-branched candlestick
canine   1: unicuspid
Canisius College   3: Little Three
cannon   4: four-pounder
canon, second   2: deuterocanonical
    books
canonical hour   1: prime
Canonical Hours: 7
canter   5: five-gaited
cantred: 100
Capaneus   7: Seven Against Thebes
Capitoline   7: Rome
cap quarto   ¼: quarto
Capricornus   ⅓: trigon; 12: sign
captivity   12: house
Caravaggio, Michelangelo Merisi da
    13: Louis Treize; 1600: Seicentisti
carbon-14: 14
Cardigan, Earl of   600: Light
    Brigade
cardinal point: 4
carlino, half   ½: demicarlino
Carlisle   12: Arthur
Carmelites   4: four orders
Carmichael, Franklin   7: Group of
    Seven, 2
Carmichael, Mary   4: Four Marys
carnal and sinful love   8: Hell
carpel   2: digyn; 10: decagyn
carré: 4
Carter, Ruth   8: New York Eight
Carteret, Sir George   24: Twenty-
    four Proprietors
case of the seven bishops   7: seven
    bishops

Cassiel   7: menorah
caste, in India   2: Duija
casting out elevens: 11
casting out nines: 9
Castor   2: Gemini, Twins
Castro Ramos, Elias   13: Macheteros
cat: 9
catacombs   7: Seven Wonders of the
    Middle Ages
catamaran: 2
catch-22: 22
cater: 4
Catholic Holy League   16: Sixteen
Catlow, Thomas   5: Fort Dix Five
catnap   40: forty winks
cat-o'-nine-tails: 9
Catonsville Nine: 9
Cattle of Geryon   12: Dodekathlos
cause   9: possession
Cause of Pain, Noble Truth of the
    4: Noble Truths
causeuse: 2
Cavaliero, Giovanni Acuti
    9: Worthies of London
Caverley, Sir Hugh   9: Worthies of
    London
cavity   2: dicoelous
Cayuga   5: Five Nations
Celeno   3: Harpies; 7: Pleiades
celibacy   6: Six Articles
Celidon   12: Arthur
Celion, Mount   7: Seven Sleepers
cell   2: dispermatous;
    10: decemlocular
Celsius   100: centigrade
censorship   1: M. O. 1
cent: 100
cent   ¹⁄₁₀₀: penny; 100: dollar
cental: 100
centavo: 100
centavo   10: decimo
centenarian: 100
centenarius: 100
centenary: 100
centennial: 100
centennium: 100
centesimal: ¹⁄₁₀₀
centesimation: ¹⁄₁₀₀
centiare: ¹⁄₁₀₀
centigrade: 100
centigram: ¹⁄₁₀₀

centigramme $^1/_{100}$: centigram
centiliter: $^1/_{100}$
centilitre $^1/_{100}$: centiliter
centillion: 10^{303}; :10^{600}
centimeter: $^1/_{100}$
centimetre $^1/_{100}$: centimeter
centipede   100: hundred-legs
centipitous: 100
centoculated: 100
Central Pacific Railroad   4: Big Four
centrosome   2: diplosome
centumvir: 105
centuple: 100
centuplicate: 100
centuria: 100
centurial: 100
centurion: 100
century: 100
century plant: 100
centussis: 100
Cephalonia   7: Ionian Islands
Cerberus: 3
Cerberus   8: Hell; 12: Dodekathlos
Ceres   3: Hecate
Cervantes   6: sestina
Cessation of Pain, Noble Truth of the
   4: Noble Truths
Ceto   3: Graeae
Chairman of a Congressional
   Committee 17: salute
Chairman of the Joint Chiefs of Staff
   19: salute
Chalons   15: fifteen decisive battles
Champions of Christendom: 7
change-ringing   5: cinques,
   grandsire
Chang Hsu   8: Eight Immortals of
   the Wine Cup
Chang Ling   5: Five Pecks of Rice
Chang Lu   5: Five Pecks of Rice
Chang Shuming   6: Six Idlers of the
   Bamboo Streams
channel   2: diglyph; 3: triglyph
chanterelle   5: quint
Chanukah   8: Hanukkah
chaplet   150: psalter, 2
Chapter 7: 7
Chapter 11: 11
Chapter 13: 13
"Charge of the Light Brigade"
   600: Light Brigade

chariot   2: biga; 4: quadriga;
   14: Trisala
Chariot   22: major arcana
charioteer   4: faction
Charites   3: Graces
charity   7: corporal works, spiritual
   works, virtues
Charlemagne: 7
Charlemagne   3: pawnbroker;
   7: Pleiad, 2;   9: Nine Worthies
Charlemagne, Paladins of
   12: Paladins of Charlemagne
Charlemagne, Pleiad of   7: Pleiad, 2
Charles II   2: unlucky number
Charles XII   21: Twenty-one Years'
   War
Charles Edward   2: unlucky
   number; 88: Stuarts
Charles of Blois   30: Battle of the
   Thirty
Charles's Wain   7: septentrional,
   stars
Charlotte Three: 3
Charnouch   7: Seven Sleepers
Charter Oath: 5
Charter 77: 77
Charya   14: Tantras
Chase, William Merritt   10: Ten
Ch'a Shih-piao   4: Four Masters of
   Anhwei
chastity   6: Six Articles
Chaucer   3: Triumvirate, 2; 7: rime-
   royal
Chavis, Benjamin   10: Wilmington
   Ten
'cheap'   6: sixpenny; 10: dime a
   dozen
'cheat'   4: fourflusher
checkering file: 2
Cheekye-Dunsmuir   5: Vancouver
   Five
Ch'en   6: Six Dynasties
Chernogorsk   7: Siberian Seven
Cherokee   5: Five Civilized Tribes
cherubim   9: angels
cheth: 8
cheth   22: major arcana
chi: 22
Chiang Kai-shek   5: Five-Power
   Constitution
Chicago   66: Route 66

Chicago Eight   7:Chicago Seven
Chicago Seven:7
Chickasaw   5:Five Civilized Tribes
Chico   3:Marx Brothers
Chief Justice of the United States
   19:salute
Chief of Naval
   Operations   19:salute
chief point   9:point
child, first   1:primipara, unipara
child, fourth   4:quadripara
child, second   2:secundigravida,
   secundipara
Childeric   300:bees
child labor law   10:Ten Hours Act
children   12:house
children, house of:5
chiliad:1000
chiliagon:1000
chiliarch:1000
chiliarchy:1000
chiliasm:1000
chiliomb:1000
Chilo   7:Seven Sages of Ancient
   Greece
Chilon   7:Seven Wise Men
Chiltern Hundreds:100
Chimurenga, Coltrain   8:New York
   Eight
China   5:Big Five, 1, Five-Power
   Constitution, Five-Year Plan;
   7:Seven Wonders of the Middle
   Ages; 16:Sixteen Kingdoms
Chinese fire-drill   3:three-ring
   circus
Chinese immigrants   6:Six
   Companies
Ch'ing dynasty   4:Four Masters of
   Anhwei
Chios   7:Homer
Chio Sui   8:Eight Immortals of the
   Wine Cup
Chmykhalova, Maria   7:Siberian
   Seven
Chmykhalova, Timofey   7:Siberian
   Seven
Choctaw   5:Five Civilized Tribes
Chomo Ling   4:Four Lings
chorus, entrance of   2:epiparados
Christ, nature of   1:Monophysitism;
   2:Diphysitism, dualism

Christ, Second Coming
   of   2:Second Coming; 5:Fifth
   Monarchy men; 7:Fundamentalism
Christ, wills of   1:Monothelitism;
   2:Dyothelitism
Christian Creeds:3
Christian name   1:first name
Christians   7:Hell
Christie, Agatha   10:"Ten Little
   Indians"
Christmas   7:church year
Christmas, Twelve Days of
   12:Twelve Days of Christmas
Christmas season,
   end   12:Epiphany
Christ on the cross:7
Christ on the cross   3:three hours
Christ's fast   40:Lent
Christ's first fall   14:stations of the
   cross
Christ's human genealogy
   4:Evangelists
Christ's royalty   4:Evangelists
Christ's sacrifice or sacerdotal office
   4:Evangelists
Christ's second fall   14:stations of
   the cross
Christ's third fall   14:stations of the
   cross
Christ to work miracles
   5:Fundamentalism
chronic degree:$^1/_{360}$
chronological age   100:I. Q.
Chu Chin   8:Eight Immortals of the
   Wine Cup
Churchill, Winston   3:Big Three, 2
Church of England   39:Thirty-nine
   Articles
church year:7
Cibola, Seven Golden Cities of:7
cicada   17:seventeen-year locust
Cicero   4:Eloquence
"cigar, good five-cent"   5:"What
   this country needs . . ."
Cimmerian Sibyl   10:sibyl
cinquain:5
cinque:5
Cinquecentesti:1500
cinque-centist:1500
cinque-cento:1500
cinquefoil:5

248

cinque-pace:5
Cinque Ports:5
cinques:5
cinque-spotted:5
cinquième:⅕
circular number:5; :6
Cities of Refuge:6
city 14:Trisala
'civil disturbance' 10:ten-fifteen
Civil War, England 7:Associated
 Counties
Civil War, U. S. 7:Seven Days'
 Battle
'clairvoyance' 2:second sight
clam 2:bivalve
Clarisses 2:Second Order of St.
 Francis
Clark, Barney 7:Jarvik-7
classical 10:tripos
Classicism 14:Louis Quatorze;
 16:Louis Seize
classic orders:8
Clay, Dorie 8:Grand Jury Eight
Clay, Henry 3:Great Triumvirate
Clay, Omowale 8:New York Eight
Clemenceau, Georges 3:Big Three,
 1
Clement of Rome 6:apostolic
 fathers
Cleobulus 7:Seven Sages of Ancient
 Greece
Cleostratus of
 Tenedos 8:octaëteris
clergy 4:Fourth Estate
"climacteric years" 7:lease
Clio 9:Muses
Cliobulus 7:Seven Wise Men
clothe the naked 7:corporal works
 of mercy
Clotho 3:Fates
clover, five-leafed 5:cinquefoil
Club of Ten:10
clubs 78:tarot
clubs, ace of 3:matador
clubs, four of 4:Devil's four-poster
clubs, two of 2:manille
Cluchette, John W. 3:Soledad
 Brothers
C-note:100
coach and four:4
coal 6:Six Dynasties

Coalition, War of the First 1:War
 of the First Coalition
Coalition, War of the
 Second 2:War of the Second
 Coalition
Coalition, War of the Third 3:War
 of the Third Coalition
cocoon 2:dupion, 1
Cocteau, Jean 6:Six
Cocytus 5:rivers of Hell; 8:Hell
Coercive Acts 5:Intolerable Acts
Coeus 12:Titan
'coffee':44
'coffee' 1:draw one
cohort, tenth 10:decuman, 1
cold 4:elemental quality,
 secondary
Cole, Lester 10:Hollywood Ten
Cole, Old King 3:"fiddlers three"
Coleman, Glenn 8:Eight
Collet, Henri 6:Six
Collingwood, Charles 8:Murrow's
 Boys
colon 2:dicolon
Colón Osorio, Luís Alfredo
 13:Macheteros
colonies, corporate 13:original
 colonies
colonies, proprietary 13:original
 colonies
colonies, royal 13:original
 colonies
colonies, settlement
 of 14:Fourteen Points
colony, royal 24:Twenty-four
 Proprietors
Colophon 7:Homer
Colorado 50:United States of
 America
Colorado, University of 8:Big
 Eight, 1
color blindness 2:deuteranopia,
 dichromat, dichromism
Colosseum:87,000
Colosseum 7:Seven Wonders of the
 Middle Ages
Colossus of Rhodes 7:Wonders of
 the Ancient World
Columbia University 8:Ivy League
Columbus, Christopher:3

column   2:dipteral; 4:tetrastyle;
    6:hexaplar, 1; 8:octastyle;
    9:enneastyle; 10:decastyle;
    12:dodecastyle; 20:Doric column;
    100:hecatonstylon
comedy   9:Muses
Comet   8:Santa's reindeer
comfort the sorrowful   7:spiritual
    works of mercy
Comic Poets:13
commandment   7:commit the
    seventh
commandment, eleventh
    11:eleventh commandment
Commandments:10
Commandments,
    Ten   10:Decalogue
Commire   7:Pleiad, 4
committee of one:1
Committee of Public Safety
    16:Sixteen
commit the seventh:7
Common Market   12:European
    Economic Community
common people   4:Fourth Estate
Commons   3:Parliament
common wit   5:five wits
common year:365
communion   6:Six Articles
compass, boxing the   32:point of
    the compass
compassion   9:fruits of the Holy
    Spirit
compass points   32:point of the
    compass
complementary color:2
'complete'   100:one hundred
    percent
compline   7:Canonical Hours
Composite   8:classic orders
Conant, Colonel   1:"One if by land,
    and two if by sea, . . . "
concentration, right   8:Noble
    Eightfold Path
concubines:15
condemnation by Pilate   14:stations
    of the cross
condyle   2:dicondylian
Confederate Army   7:Seven Days'
    Battles

confession   3:third degree, 2; 6:Six
    Articles
Confirmation   7:Holy Sacrament
'confused'   6:at sixes and sevens
confusion, year of   707:year of
    confusion
Connaught   5:Five Bloods
Connecticut   13:original colonies;
    50:United States of America
conquest   14:Trisala
'consider carefully'   2:think twice
consolidated threes:3
consonant   2:double consonant;
    3:triconsonantal; 4:quadriliteral
Constance   4:Tetrapolitan
Constantine   7:Seven Sleepers
Constantinople   4:faction; 7:Seven
    Cities, Seven Wonders of the
    Middle Ages; 400:Baltadjii
Constitution   2:double jeopardy;
    10:Bill of Rights; 18:Prohibition;
    19:suffrage; 21:Repeal; 24:poll tax
contraplex   2:duplex telegraphy
Coopers & Lybrand   8:Big Eight, 2
copper   7:alchemy
Corbet, Sir John   5:Darnel's Case
Corfu   7:Ionian Islands
Corinthian   8:classic orders
Cornell University   8:Ivy League
corolla   2:dichlamydeous
Coronado, Francisco Vásquez de
    7:Cibola, Seven Golden Cities of
corporal works of mercy:7
corporate colonies   13:original
    colonies
Cos   9:agretae
Cottus   100:Hecatoncheires
Council of Five Hundred:500
Council of Ten:10
Council of Ten   10:decemvir
counsel   7:seven gifts of the Holy
    Ghost; 9:possession
counsel the doubtful   7:spiritual
    works of mercy
count:10
counterintelligence   1:M. O. 1; 5:M.
    I. 5
count to ten:10
couple:2
Court of Tynwald   24:House of
    Keys

Covent Garden   9:Nine Elms
covetousness   7:sins
cow-shark   7:sevengills
Crab   12:sign
cranial   3:trisplanchnic
cranial nerve, eighth
   8:vestibulocochlear nerve
cranial nerve,
   eleventh   11:accessory nerve
cranial nerve, fifth   5:trigeminal
   nerve
cranial nerve, first   1:olfactory
   nerve
cranial nerve, fourth   4:trochlear
   nerve
cranial nerve, ninth
   9:glossopharyngeal nerve
cranial nerve, second   2:optic nerve
cranial nerve, seventh   7:facial
   nerve
cranial nerve, sixth   6:abducens
   nerve
cranial nerve, tenth   10:vagus nerve
cranial nerve, third   3:oculomotor
   nerve
cranial nerve, twelfth
   12:hypoglossal nerve
Crassus   1:First Triumvirate
Cratinus   13:Comic Poets
Creasy, Sir Edward Shepherd
   15:fifteen decisive battles
Creation:6
credo   10:decad ring
Creed   3:Christian Creeds
Creek   5:Five Civilized Tribes
crescent   4:quadricrescentic
Cretaceous   3:Triassic
Crete   7:seven youths
cretic   2:diagyíos, paeon diagyíos
cricket team:11
Crimean War   600:Light Brigade
'crime in progress'   10:ten-thirty-
   one
Crius   12:Titan
Crocker, Charles   4:Big Four
Croker, Christopher   9:Worthies of
   London
Cromwell, Oliver   8:"Tell me this
   riddle while I count eight"
cronebane   ½:halfpenny

cross   2:double fitché; 3:treble
   fitché; 4:erminé, tetragammadion
Cross, Exaltation of   4:Ember Days
cross-country skiing   2:biathlon
cross-file   2:double-half-round file
crossroads   4, 2:quadrivium
Crowley, Jim   4:Four Horsemen
crown   10:Sephiroth
crown octavo   12:duodecimo
Crucifixion   7:Sorrows of Mary;
   14:stations of the cross
crystalline   9:Nine Heavens
Cuba   10:Ten Years' War
cube   2:ditesseral; 5:Platonic solid
cube, duplication of   2:Delian
   problem
cubic   2:quantic
cubic equation:3
cuboctahedron   13:Archimedean
   solid; 14:truncated cube
cubo-cubo-cube:9
cubo-octahedron   14:truncated
   cube
cuckoo   10:animals in paradise
Cudahy   5:Big Five Packers
Cui, César A.   5:St. Petersburg Five
Cumaean Sibyl   10:sibyl
cup, double   2:depas
   amphikypellon
Cupid   8:Santa's reindeer
Curse of Scotland:9
cusp   1:unicuspid; 3:tritubercular;
   4:quadricuspidate
Cybele   3:Hecate
Cyclopes:3
Cyclops   7:"Sindbad the Sailor"
cyme   2:dichasium
Cyrenaica, Pentapolis of
   5:Pentapolis
Cyrene   5:Pentapolis
Cyrus   10,000:Ten Thousand
Cythera   7:Ionian Islands
Czechoslovakia, human rights in
   77:Charter 77

D

<u>D</u>:2; :3; :500
D̄:5000
D, d:3; :4

Dacia   6:Six Islands
dactyl:3
Daedalus   7:seven youths
Daikoku   7:Shichi Fukujin
daily double:2
daily dozen:12
daleth:4
daleth   22:major arcana
"Damaetas"   7:Pleiad, 2
Damascus   3:Magi; 10:Decapolis
Dan   10:Lost Tribes of Israel
Dana, Richard Henry   2:*Two Years Before the Mast*
Danaides:50
Danaüs   50:Danaides
dance   9:Muses
Dance of the Seven Veils:7
Dancer   8:Santa's reindeer
Danelagh:15
Daniel, Arnaut   6:sestina
Dante   3:terza rima, Triumvirate, 3; 6:sestina; 8:Hell
Dardan   6:Troy
Darius   15:fifteen decisive battles
darkness, dispelling of the   14:Trisala
Darnel's Case:5
D'Artagnan   3:*Three Musketeers*
Dartmouth College   8:Ivy League
Dasher   8:Santa's reindeer
"David"   7:Pleiad, 2
David   9:Nine Worthies
David, Jacques-Louis   16:Louis Seize
David, king of Scotland   9:Worthies of London
David, Star of   6:hexagram, 4
Davies, Arthur B.   8:Eight
Davis, Rennie   7:Chicago Seven
'dawn'   1:first light
day:7; :12; :24
day   24:nycthemeron
'daybreak'   1:first light
day of rest:7
dead, bury the   7:corporal works of mercy
deadly sins   7:sin
dearborn:4
death   4:Four Horsemen of the Apocalypse, last things; 12:house; 14:stations of the cross

Death   22:major arcana
death, house of:8
'death'   8:theta
"Death Avenue"   12:Twelfth Avenue
Death Valley   20:twenty-mule team
death-watch beetle:7
Debermouch   7:Seven Sleepers
Deborah   9:Nine Worthy Women
DeBussy, Claude   6:Six
deca-:10
decacerous:10
decachord:10
decad   10:decade
decade:10
décade:10
decadianome:10
decadist:10
decadrachm   10:dekadrachm
decad ring:10
decagon:10
decagram:10
decagramme   10:decagram
decagyn:10
decahedron:10
decalet:10
decaliter:10
decalitre   10:decaliter
decalitron:10
Decalogue:10
Decameron:10
decameter:10
decametre   10:decameter
De Camp, Joseph   10:Ten
decanate:10
decander:10
decangular:10
decapetalous:10
decaphyllous:10
decapod:10
Decapolis:10
decare:10; :1000
decasemic:10
decasepalous:10
decastere:10
decastich:10
decastyle:10
decasyllabic:10
decathlon:10
'deceitful'   2:Janus-faced, two-faced

December:10; :12
decemcostate:10
decemdentate:10
decemfid:10
decemlocular:10
decempedal:10
decempennate:10
decemvir:10
decemvir   2:duumviri;
  15:quindecemvir
decemvirate:10
decemvirate   10:Wicked Ten
decemviri agris dividundis
  10:decemvir
decemviri legibus scribendis
  10:decemvir
decemviri litibus judicandis
  10:decemvir
decemviri sacris faciundus
  10:decemvir
decemviri sacrorum   10:decemvir
decemvirs, laws of the   12:Twelve
  Tables
decennary[1]:10
decennary[2]:10
decenner:10
decennial:10
decennium:10
decennoval:19
Decennovium:19
deci-:$\frac{1}{10}$
deciare:$\frac{1}{10}$
decigram:$\frac{1}{10}$
decigramme $\frac{1}{10}$:decigram
decil:$\frac{1}{10}$
deciliter:$\frac{1}{10}$
decilitre $\frac{1}{10}$:deciliter
decillion:10^{33}; :10^{60}
decima:$\frac{1}{10}$; :10
decimal degree:$\frac{1}{240}$; :$\frac{1}{60}$; :$\frac{1}{10}$
decimal fraction:10
decimalism:10
decimalization:10
decimal minute:$\frac{1}{600}$; :$\frac{1}{100}$; :$\frac{3}{5}$
decimal system:10
decimal vigesimal system:20
decimate:$\frac{1}{10}$; :10
decime:$\frac{1}{10}$
decimestrial:10
decimeter:$\frac{1}{10}$
decimetre $\frac{1}{10}$:decimeter

decimo:$\frac{1}{10}$; :10
decimole:10
decimo-sexto   16:sexto-decimo
decinormal:$\frac{1}{10}$
Decisions, Fifty   50:Fifty Decisions
decisive battles of the world
  15:fifteen decisive battles
decistere:$\frac{1}{10}$
Decius   7:Seven Sleepers
deck   4:History of Four Kings;
  40:omber; 48:pinochle deck;
  52:playing cards
Declaration of Independence
  4:Independence day
Declaration of Indulgence   7:seven
  bishops
*Decline and Fall of the Roman
  Empire* $\frac{1}{10}$:Saladin's tenth
Decoration day   30:Memorial day
découplé:2
decuman:4; :10
decuple:10
decuplet   10:decimole
decuria:10
decurion:10
decury:10
decussis:10
Dedication, Feast of   7:menorah
deeds, king of the   5:Five Great
  Kings
deem:$\frac{1}{10}$
deeme $\frac{1}{10}$:deem
deep six:6
deer   14:Trisala
defective fifth:5
deferred pay   2:twopence, 2
deficient number:Misc.
De Gloria Martyrum   7:Seven
  Sleepers
degorder:2
deirbhfine   5:geilfine
deka-   10:deca-
dekadrachm:10
dekagram   10:decagram
dekaliter   10:decaliter
dekameter   10:decameter
dekass:5; :10
dekastere   10:decastere
Delaware   13:original colonies;
  50:United States of America;
  100:hundred, 2

DeLeon, Frank   6:San Quentin Six
Delian problem:2
delight of the eyes   5:Alkoremmi
Delilah   7:Samson
Dellinger, David T.   7:Chicago
  Seven
Deloit, Haskins, & Sells   8:Big
  Eight, 2
Delos, oracle at   2:Delian problem
Delphi   4:Pythian games
Delphian Sibyl   10:sibyl
delta:3; :4
deltoid:3
deltoideus   3:deltoid
demand deposits   1:M-1A
Demeter   3:Hecate; 12:Olympians
demi-:½
demi-bateau:2
demi-cadence:½
demicarlino:½
demi-column:½
demifarthing:½
demigalonier:½
demi-grevière   ½:demi-jambe
demi-jambe:½
demi-kindred:½
demi-landau:½
demi-metope:½
demi-pique:½
demi-pontoon   2:demi-bateau
demiquaver:¹⁄₁₆
demi-relief   ½:mezzo-rilievo
demisang:½
demi-season:2
demi-semi:¼
demisemiquaver:¹⁄₃₂
demisphere   ½:hemisphere
demitone   ½:semitone
demi-volt:½
demi-wolf:½
demobilization furlough:14
Demosthenes   3:Olynthiac
  orations, Third Philippic;
  4:Eloquence; 10:Attic Orators
demy:½
demy quarto   ¼:quarto
denarius:¹⁄₈₆; :10
denary:10

Denmark   6:Six Islands; 7:Seven
  Years' War of the North;
  12:European Economic
  Community
dennet:2
deoxyribonucleic   2:DNA
Department of Defense   5:Pentagon
depas amphikypellon:2
depravity of
  man   7:Fundamentalism
deprecations   5:litany
Depression   100:Hundred Days, 2
Derbyshire   15:Danelagh
de Rita, Joe   3:Three Stooges
derodidymus:2
Derry   6:Six Counties
descent from the cross   14:stations
  of the cross
desoxyribonucleic acid   2:DNA
dessert-spoonful:2
destruction of temples   5:Fast of
  the Fifth Month
desultor:2
Deucalion:9
deuces wild:2
deuteragonist:2
deuteranopia:2
deuterium:2
deutero-:2
deuterocanonical:2
deuterocanonical books:2
deuterogamist:2
deuterogenic:2
deuteron   2:deuterium
deutero-Nicene:2
Deuteronomy:2; :5
Deuteronomy   5:Pentateuch;
  6:Hexateuch; 7:Heptateuch;
  8:Octateuch
deuteropathia   2:deuteropathy
deuteropathy:2
deuteroscopy:2
deuto-:2
deutocerebrum:2
deutomala:2
deutopsyche:2
deutotergite:2
Deuxième Bureau:2
deux-temps:2
Devil   22:major arcana
devil on two sticks:2

devil's dozen:13
Devil's four-poster:4
Devil's Own:88
Dewing, Thomas W.   10:Ten
dexter base   ¼:grand quarter
dexter base point   9:point
dexter chief   ¼:grand quarter
dexter chief point   9:point
dharma   3:aum
dharma and Buddha, union
   of   3:aum
Dharmakaya   5:Buddha, 1
Dhyanibodhisattva:8
Dhyanibuddha:5
Dhyanibuddha   5:Buddha, 1,
   Herukabuddha
di-:2
diabolo   2:devil on two sticks
diactinal   2:diactine
diactine:2
diadelphic:2
diaderm:2
diagyíos:2
diamond:4; :4½
diamonds   78:tarot
diamonds, nine of   9:Curse of
   Scotland
diamonds, seven of   7:manille
diamonds, ten of   10:big casino
diamond wedding anniversary
   75:wedding
Diana   3:Hecate, Trivia
diandrous:2
dianome   10:decadianome
diapente:5
diapsid:2
diarchy:2
diaster   2:dyaster
diastyle:3
diatessaron:4
diatonic:2
diaulos,:2
diaxon:2
Díaz Ruiz, Angel   13:Macheteros
dibrach:2
dibrachys   2:dibrach
dibranchiate:2
dicast:6000
dicatalexis:2
dice   1:die; 2:snake-eyes;
   12:boxcars

dicellate:2
dicephalus:2
dicerion:2
dicerous:2
dichasium:2
dichlamydeous:2
dicho-:2
dichogamism:2
dichord:2
dichotic:2
dichotomy:½; :2
dichroism:2
dichromat:2
dichromate   2:dichromat
dichromatism:2
dichromic:2
dichromism:2
dichronous:2
dicker:10
dicoccous:2
dicoelous:2
dicolon:2
dicondylian:2
dicot   2:dicotyledon
dicotyledon,:2
dicrotic:2
dicrotous   2:dicrotic
dicycle:2
dicynodont:2
didactyl:2
didactyle   2:didactyl
didactylous   2:didactyl
didecahedral:10
didelphian:2
didelphic   2:didelphian
didelphoid   2:didelphian
di-diurnal:2
didodecahedral:12
didodecahedron   24:diploid
Didot point   0.0376:point
didrachm:2
didrachm   10:decalitron
didrachma   2:didrachm
didymium:2
didymous:2
didynamian   2:didynamous
didynamic   2:didynamous
didynamous:2
die:1; :7
die   2:double dies
diecian   2:dioecious

255

diecious   2: dioecious
dies fasti   3: triverbial
diesis   2: double dagger
diet: 3
dietheroscope: 2
Digambara   14: Trisala
digamma: 2
digamy: 2
digastric: 2
digenesis: 2
digenous: 2
digit   1: unidactyl; 2: didactyl;
  3: tridactyl; 4: quadridigitate,
  tetradactyl; 6: hexadactylous,
  sexdigitate
diglossia: 2
diglot: 2
diglottism: 2
diglottist   2: diglot
diglyph: 2
dignities   12: house
dignities, house of: 10
digoneutic: 2
digonous: 2
digram   2: digraph
digraph: 2
digraphic: 2
digyn: 2
dihedral angle: 2
dihedron: 2
dihexagonal: 12
dihexahedron: 12
diiamb: 2
diiambus   2: diiamb
dikast   6000: dicast
Dike   3: Horae
dilambdodont: 2
dilemma: 2
dilemma   3: trilemma;
  4: tetralemma; 22: catch-22
dilogy: 2
dimastigate: 2
dime: $\frac{1}{10}$; :10
"Dime"   10: Rockefeller
dime a dozen, a: 10
dime novel: 10
dimension: 4
dimerous: 2
dime store: 10
dimetallic: 2
dimeter: 2

dimidiate: $\frac{1}{2}$; :2
diminished triad: 3
dimorphism: 2
dimyarian: 2
dimyary   2: dimyarian
Din   10: Sephiroth
Dinarchus   10: Attic Orators
dinero: $\frac{1}{10}$
diningroom, Roman   3: triclinium
Dino   3: Graeae
diobely: 2
diobol: 2
dioctahedral: 16
diode: 2
diodont: 2
dioecian   2: dioecious
dioecious: 2
dioestrum: 2
Diomedan Mares   12: Dodekathlos
dionym: 2
Dionysius   7: Seven Sleepers
Dionysius the Areopagite   9: angels
Dionysus   4: tetralogy, 1;
  12: Olympians
diophthalmus   2: binoculus
Dioscuri   2: Twins
diota: 2
diothelism: 2
diotic: 2
dioxid   2: dioxide
dioxide: 2
Dipamkara   5: Buddha, 2
dipenthemimeres: 2
dipetalous: 2
Diphilus   13: Comic Poets
diphthong: 2
diphthongia: 2
diphyletic: 2
diphyllous: 2
diphyodont: 2
diphysitism: 2
diplacusis: 2
diplanar: 2
diplasiasmus: 2
diplasic: 2
diplegia: 2
dipleidoscope: 2
diplex: 2
diplex   2: duplex telegraphy
diploglossate: 2
diploid: 24

diploneural:2
diplophonia   2:diphthongia
diplopia:2
diplopy   2:diplopia
diplosome:2
diplospire:2
dipneumonous:2
dipnoous:2
dipode:2
dipody,:2
dipolar:2
dipole:2
diprotodont:2
dipteral:2
dipterous:2
dipterygian:2
diptych:2
Direct Action   5:Vancouver Five
dirhombohedron:12
Dirty Half-Hundred:50
Dirty Shirts:101
discalenohedron:24
disease, fifth   5:erythema
   infectiosum
disepalous:2
disomatous:2
'disordered'   6:at sixes and sevens
dispermatous   2:dispermous
dispermous:2
dispermy:2
dispireme:2
dispondee:2
disquiparance   2:disquiparancy
disquiparancy:2
disquiparant:2
dissyllable:2
dissymmetry:2
distich:2
distich   2:distichous
distichous:2
distigmatic:2
distomatous:2
distylous:2
ditesseral:2
ditetragonal:2
Dium   10:Decapolis
divalent:2
divide fifty-fifty   50:go fifty-fifty
Divina Commedia   3:terza rima
Divine Essence   4:tetratheism

division   2:dualism, 1, duplicate;
   4:four species, 1
dix:10
dizain:10
dizoic:2
dizygotic:2
Dmytryk, Edward   10:Hollywood
   Ten
DNA:2
do a 180° turn   180:one-eighty
dobhash:2
doblon:2
Doc   7:"Snow White and the Seven
   Dwarfs"
Dockwra, William   1:penny post
doctor:7; :9
Doctrine of the Two Swords:2
dodeca-:12
dodecafid:12
dodecagon:12
dodecagyn:12
dodecahedron:12
dodecameral:12
Dodecanese:12
dodecapartite   12:dodecafid
dodecapetalous:12
dodecapharmacum   12:apostle's
   ointment
dodecaphony:12
dodecarch:12
dodecasemic:12
dodecastyle:12
dodecasyllable:12
dodecatemory:12
Dodekathlos:12
dodrans:$3/4$
doe, fallow   2:pricket's sister
Doge   41:forty-one
dogger:2
Dog of the Seven Sleepers   7:Seven
   Sleepers; 10:animals
doit:$1/8$
dollar:8; :100
dollar   100:cent, centavo
dollar-a-year man:1
dom:3
Dominicans   4:four orders
dominions   9:angels
dominoes:28
Domitian   5:Adoptive Emperors
dom pedro   3:dom

Donder   8: Santa's reindeer
Don Pedro   9: Worthies of London
doorkeeper of heaven   2: Janus
Dopey   7: "Snow White and the
   Seven Dwarfs"
doppia   2: doppio
doppio: 2
Dorat   7: Pleiad, 3
Dorian   3: Ionian
Dorian tetrachord: 4
Doric   3: triglyph; 8: classic orders
Doric column: 20
Doris   50: Nereid
dos Hermanas, Sala de las
   2: Hermanas
dotage   2: second childhood
double: 2
'double'   2: bi-, di-
double   5: grandsire
double agent: 2
double ax: 2
double bar: 2
double-barreled: 2
double bed: 2
double-bitt: 2
double-blind: 2
double-bodied: 2
double bogey: 2
double boiler: 2
double bourdon: 32
double-breasted: 2
double-brooded: 2
double chin: 2
double consonant: 2
doublecross: 2
'double-cross'   2: two-time
   (someone)
double-cut file: 2
double dagger: 2
double date: 2
double-dealer: 2
double-dealing: 2
double-decker: 2
double demisemiquaver: ¹⁄₆₄
double dies: 2
double dip: 2
double-distilled: 2
double doubloon: 4
double down: 2
double dummy: 2
double Dutch: 2

double duty: 2
double-dyed: 2
double eagle: –3
double-eagle: 2
double-edged: 2
double-ender: 2
double entendre: 2
double-entry bookkeeping: 2
double exposure: 2
double feature: 2
double fever: 2
double figures: 2
double file   2: checkering file
double-first: 2
double fitché: 2
double flat: 2
double-gild: 2
double glazing: 2
double-half-round file: 2
doubleheader: 2
double helix   2: DNA
double-hung: 2
double indemnity: 2
double jeopardy: 2
double knit: 2
double life: 2
double line: 2
double-lunged: 2
double magnum   2: jereboam
double-man: 2
double-meaning: 2
double-minded: 2
double negative: 2
double-nostriled: 2
double obelisk   2: double dagger
double or nothing: 0; : 2
double or quits   0: double or
   nothing; 2: double or nothing
double pair royal: 4
double paragon: 40
double play: 2
double pneumonia: 2
double possessive: 2
double procession   2: Filioque
double quatrefoil: 8
double quatrefoil   8: eightfoil
double-quick   2: on the double
double quotes: 2
double quotidian fever: 2
double rime   2: female rime,
   feminine rime

double room:2
doubles 5:double, 9
double saucepan 2:double boiler
double scull:2
double solitaire:2
double space:1
double standard:2
double star 2:binary star
double-star 2:dyaster
double-stop:2
double-struck:2
doublet:2
doublet 2:dimorphism, 2
double take:2
double-talk:2
double-talk 2:double Dutch
double tertian fever:2
doubleton:2
double-tongue:2
double-topsail:2
doublets 2:doublet
double up:2
double-whammy:2
doubloon:2
doubloon 2:doblon
doucepere 12:douzepere
doupion 2:dupion
douzain:12
douzaine:12
douzepere:12
Dover 5:Cinque Ports
Down 6:Six Counties
dozen:12
Drachenfels 7:Siebengebirge
drachm 2:didrachm;
  4:tetradrachm; 6:hexadrachm;
  8:octadrachm; 10:dekadrachm
drachma:100
draw one:1
dread 8:Agni
dreibund:3
Dreikaiserbund 3:League of the
  Three Emperors
dreikanter:3
Drga-lha skyes-gcig-bu 5:Five Great
  Kings
droshky:4
Drudwas 12:Knights of the Round
  Table, 2
Drumgo, Fleeta 3:Soledad
  Brothers

'drunk' ½:half seas over; 3:three
  sheets to the wind
Druze, Johnny 7:seven blocks of
  granite
dry 4:elemental quality, secondary
Dryden 12:Knights of the Round
  Table
dry land 6:Creation
duad 2:dyad; 3:Pythagorean letter
dual:2
Dual Alliance:2
dual carriageway:2
dual citizenship:2
dualism:2
duality:2
dualize:2
Dual Monarchy:2
dual personality:2
duarchy 2:diarchy
Dubai 7:United Arab Emirates
duck's egg:0
due-cento:1200
due corde:2
duel:2
duel 2:monomachy
duello:2
duet:2
duettino:2
duettist:2
duetto:2
due volte:2
dufoil:2
Duglas 12:Arthur
Duija:2
Dumas, Alexandre 3:*Three
  Musketeers*
dummy:3
duo:2
duodecagon 12:dodecagon
duodecahedron 12:dodecahedron
duodecennial:12
duodecillion:10^{39}; :10^{72}
duodecimal:12
duodecimfid:12
duodecimo:12
duodecimo 6:sexto
duodecim scripta:12
duodecuple:12
duodenary:12
duodene:12
duodenum:12

duodrama:2
duogravure:2
duole:2
duoliteral   2:biliteral
duologue:2
duoparental:2
duoviri   2:duumviri
Dupérier   7:Pleiad, 4
dupion:2
duplation:2
duple ratio:2
duple rhythm:2
duplex:2
duplex apartment   2:duplex
duplex house:2
duplex telegraphy:2
duplicate:2
duplicate   2:duplicate bridge
duplicate bridge:2
duplicature:2
duplicity:2
duplo-:2
dupondius:2
Durey, Louis   6:Six
Durst, David   9:Catonsville Nine
Dutch 200:200
Dutch door:2
duumvir   2:duumviri; 10:decemvir;
   15:quindecemvir
duumviri:2
dwarf   10:avatar
dyad:2
dyad   2:two
dyadic:2
dyadic system   2:binary system
dyadic disyntheme:2
dyadic syntheme:1
dyakis-dodecahedron:24
dyakis-hexacontahedron:120
dyarchy   2:diarchy
dyaster:2
Dynamic Duo:2
dyocaetriacontahedron:32
dyokaitriakontahedron
   32:dyocaetriacontahedron
dyophysitic:2
Dyophysitism   2:diphysitism
dyotheism   2:dualism
Dyothelitism:2
dysis:7

E

E:2; :250
Ē:250,000
E, e:5
eagle:−2; :10
eagle   2:double-eagle; 4:Evangelists
eaglet:3
ear   2:dichotic, diotic
Earl, Sir Walter   5:Darnel's Case
earth:3
Earth   3:Golden Apples of the
   Hesperides; 9:planet
earth   4:elemental quality; 5:Five
   Agents; 9:Nine Heavens
earths:3
earthy trigon   1/3:trigon
east   4:cardinal point
East Anglia   7:heptarchy, 2
Easter   2:Passion Sunday; 7:church
   year; 8:octave, 1;   40:Lent;
   50:quinquagesima; 60:Sexagesima
   Sunday; 70:Septuagesima Sunday
Eastern Sea   4:four seas
Eastern Chin   6:Six Dynasties
East Jersey   24:Twenty-four
   Proprietors
East Riding   1/3:riding
easy death   5:Five Blessings
Ebisu   7:Shichi Fukujin
écarté:32
ecliptic, division of   10:decuman, 3
ectoderm   2:diaderm
Ector de Maris   12:Knights of the
   Round Table, 1
écu   1/5:cinquième
Ecuador   13:OPEC
Eddas   7:sacred books
Edinburgh   12:Arthur
Edward II   2:unlucky number
Edward III   9:Worthies of London;
   30:Battle of the Thirty
Edward, the Black Prince   3:Prince
   of Wales's feathers; 9:Worthies of
   London
EEC   12:European Economic
   Community
effort, right   8:Noble Eightfold Path

Egbert, King of
  Wessex   7: heptarchy, 2
Eginhard   7: Pleiad, 2
Egypt   7: Homer, Seven Cities,
  Wonders of the Ancient World;
  50: Danaides
Egyptian calendar   365: vague year
Egyptian code of laws   3: Hermes
  Trismegistus
Ehrlich, Paul   606: salvarsan;
  914: neosalvarsan
Ehyeh-Asher-Ehyeh   7: names of
  God
Eight: 8
'eight'   8: octa-
eightball, behind the   8: behind the
  eightball
eight bells: 8
eighteenmo: 18
"eighteenth of April, . . . ": 18
eighteen-wheeler: 18
eightfoil: 8
Eightfold Path   4: Noble Truths;
  8: Noble Eightfold Path
eight-foot law: 8
eighth: 8
eighth cranial nerve
  8: vestibulocochlear nerve
eighth-note: $\frac{1}{8}$
eight-hour law: 8
eighth wonder of the world: 8
Eight Immortals of the Wine
  Cup: 8
eightpenny nail: 8
Eights: 8
Eights   8: Eights Week
Eights Week: 8
eighty-eight: 88
Eighty Years' War: 80
Eirene   3: Horae
Eisenhower, Dwight D.   5: five-star
  general
Ekron   5: Philistines
El   7: names of God; 10: names of
  God
Elaphebolion: 9
Eleanor of Aquitaine   100: Hundred
  Years' War
Electra   7: Pleiades
elegiac: 5

Elegiac Poets: 4
elemental quality: 4
Elements, Euclid   47: Pythagorean
  proposition
Eleven: 11
eleven, rule of   11: rule of eleven
eleven-plus: 11
elevens, casting out   11: casting out
  elevens
elevenses: 11
eleventh chord: 11
eleventh commandment: 11
eleventh cranial
  nerve   11: accessory nerve
eleventh hour: 11
Eliot, Charles W.   5: Harvard
  Classics
Eliwlod   12: Knights of the Round
  Table, 2
Elizabeth of Aragon   9: Worthy
  Women
Elohim   4: Tetragrammaton;
  7: names of God
Eloquence, the Four Monarchs
  of: 4
Elye   10: names of God
Ember days: 4
embolismic year: 13
'emergency'   10: ten-thirty-three
Emperor   22: major arcana
Empress   22: major arcana
'end'   24: omega
endecagon   11: hendecagon
Ends, Four   4: Four Ends
enemies   12: house
enemies or captivity, house of: 12
England   3: Triple Alliance, 1;
  5: Cinque Ports; 7: Seven Years'
  War; 15: Danelagh; 100: Hundred
  Years' War
England, patron saint of
  7: Champions of Christendom
English: 14
English Channel   4: four seas
enlightened one   5: Buddha, 2
enlightener   5: Buddha, 2
ennea-: 9
enneacontahedron: 90
ennead: 9
enneadic system: 9
enneadic system of numeration: 9

Enneads 9: ennead
enneaeteric:9
enneagon:9
enneagonal number:9
enneahedria, 9: enneahedron
enneahedron:9
enneander:9
enneapetalous:9
enneaphyllous:9
enneasemic:9
enneasepalous:9
enneaspermous:9
enneastyle:9
enneasyllabic:9
enneatical days:9
enneatical years:9
Enneoctonus:9
enstyle:2¼
enthymeme 4: four species, 2
enthymeme of the first order:1
enthymeme of the second
  order:2
entoderm 2: diaderm
Envy 6: Lucifera
envy 7: sins
Enyo 3: Graeae
Eocene 3: Tertiary
eons 8: ogdoad
epanadiplosis:2
epanastrophe:2
Ephesus 7: Seven Sleepers,
  Wonders of the Ancient World
Ephraim 10: Lost Tribes of Israel
Epicharmus 13: Comic Poets
Epic Poets:5
epic song 9: Muses
Epigoni 7: Seven Against Thebes
epiparados:2
Epiphany:12
Epiphany 7: church year;
  12: Twelfth-cake
Epps, Reginald 10: Wilmington Ten
epsilon:5
equal 6: six of one
equivalent 6: six of one
Erato 9: Muses
Erinyes 3: Furies
erminé:4
Ernst & Whinney 8: Big Eight, 2
erotic poetry 9: Muses
Erymanthean Boar 12: Dodekathlos

erythema infectiosum 5: fifth
  disease
Erythrean Sibyl 10: sibyl
escudo 2: doblon
Esculapius 9: Sabines
Esquiline 7: Rome
essence 5: quintessence
Essex 7: Associated Counties,
  heptarchy, 2; 15: Danelagh
estimation 5: five wits
eta:7
été 4: quadrille, 2
Eteocles 7: Seven Against Thebes
Eteoclus 7: Seven Against Thebes
eternal banquet 5: Alkoremmi
Ethelred II 2: unlucky number
ethical dualism:2
Eton College 4: Speech Day
Etruscan gods:9
Ets-Haïm 10: Sephiroth
Eucharist 3: trisacramentarian;
  7: Holy Sacrament; 40: forty hours
euchre 7: seven-handed euchre
Euclid 47: Pythagorean proposition
Eumenides 3: Furies
Eunomia 3: Horae
Euphrates 4: rivers of Eden
Euphrosyne 3: Graces
Eupolis 13: Comic Poets
Euripides 13: Tragic Poets
Euromarket 12: European
  Economic Community
European Economic
  Community:12
Europeans, judge of 3: Judges of
  Hades
Eurystheus 12: Dodekathlos
Euterpe 9: Muses
Evangelists:4
evangelists 4: tetramorph
Eve 7: Seventh Heaven
even:2
Evening Prayer 7: Canonical Hours
evenly even:4
Evensong 7: Canonical Hours
'everywhere' 4: four corners of the
  earth
evil principle 2: two
Ewain 12: Knights of the Round
  Table, 1
ewe, two-year-old 2: gimmer

262

ewer   14:Trisala
Exaltation of the Cross   4:Ember
   Days; 14:Holy-Cross day
example   4:four species, 2
exanthem subitum   6:roseola
   infantum
excelsior:3
excitement-depression
   3:tridimensional theory
excrement   4:four-letter word
exempt   4:exon
'exhausted'   86:'all gone'
exhortation to the women of
   Jerusalem 14:stations of the cross
existence   8:Agni
Exodus   5:Pentateuch;
   6:Hexateuch; 7:Heptateuch;
   8:Octateuch; 50:Pentecost, 1
exon:4
Extreme Unction   7:Holy
   Sacrament
eye   1:uniocular; 2:binocular;
   6:senocular; 8:octonocular;
   100:centoculated
eyeglass   1:monocle; 2:bifocal

F

F:40
F̄:40,000
F, f:4; :6
fa:4
face:36
face   2:bifrons; 4:quadrifrons
facial nerve:7
faction:4
Fahrenheit 451°:451
fair, fat, and forty:40
fair rent   3:three Fs
faith   7:virtues
fall between two stools:2
famine   4:Four Horsemen of the
   Apocalypse; 7:seven years of
   plenty
fantasy   5:five wits
Farinacci García, Jorge
   13:Macheteros
farthing:¼
farthing, half   ½:demifarthing

"Fashion Avenue"   7:Seventh
   Avenue
fast   40:Lent
Fast of the Fifth Month:5
Fast of the Fourth Month:4
Fast of the Tenth Month:10
Fat, fair and forty:40
Fates:3
Fates   3:Norn
Father   3:Trinity, tritheism;
   4:tetratheism
father/son relationship   5:five
   social relations
fathom:6
Fatima:7
Fatima   4:perfect women; 10:wives
fear of the Lord   7:seven gifts of the
   Holy Ghost
Fearsome Foursome:4
Feast of Dedication   7:menorah
Feast of St. Lucy   4:Ember Days
Feast of the Exaltation of the Cross
   4:Ember Days
feast of weeks   50:Pentecost, 1
February:2
Federal Army of the Potomac
   7:Seven Days' Battle
feed the hungry   7:corporal works
   of mercy
feeling   3:tridimensional theory;
   7:senses
**feel like a million
   dollars:1,000,000**
feel like two cents:2
feet   14:tetradecapod
felt-post   144:post
female beauty   7:Shichi Fukujin
female rime:2
feminine number:2
feminine rime:2; :3
feminine sign:2
feriae:2
feriae privatae   2:feriae
feriae publicae   2:feriae
Fermanagh   6:Six Counties
Fernández Diamante, Hilton
   13:Macheteros
Ferumbras   12:Paladins of
   Charlemagne
fesse-point   9:point
Festival of Lights   8:Hanukkah

fever, intermittent 2:double fever,
double quotidian fever, double
tertian fever; 4:quartan fever;
5:quintan fever; 6:sextan fever;
7:septan fever; 8:octan fever
"fiddlers three,":3
fidelity 9:fruits of the Holy Spirit
Fides 'Faith' 9:Sabines
Field of the Forty Footsteps:40
Fierabras 12:Paladins of
Charlemagne
fiery trigon ⅓:trigon
FIFO 1:first in
Fifteen:15
fifteen decisive battles:15
Fifteen Decisive Battles of the World
15:fifteen decisive battles
**"Fifteen Men on a Dead Man's
Chest,":15**
Fifteen O's 15:O's of St. Bridget
fifteen-puzzle:15
Fifteen Years' War:15
Fifth:5
fifth:⅕
Fifth Amendment 2:double
jeopardy
Fifth-Amendment Communist:5
Fifth Avenue:5
fifth column:5
fifth cranial nerve 5:trigeminal
nerve
fifth day 6:Creation
Fifth-day:5
fifth disease:5
Fifth Monarchy men:5
Fifth Month, Fast of the 5:Fast of
the Fifth Month
fifth position:5
fifth quarter:5
Fifth Republic:5
fifth wheel:5
Fifty Decisions:50
fifty-fifty:50
fifty-first state:51
"Fifty-four Forty or Fight":54
**"Fifty million Frenchmen can't
be wrong":50,000,000**
fiftypenny nail:50
'fight in progress' 10:ten-ten
figure of eight:8
figure-of-four trap:4

figure of Lissajous:2
file 2:checkering file, double-cut
file, double half-round file;
3:three-square file
Filioque:2
Filioque controversy 2:Filioque
fin:5
fin 2:dipterygian
finale 4:quadrille, 2
finding of the cross by Helena
14:stations of the cross
finding of the lost Child 7:Joys of
Mary
Fine, Larry 3:Three Stooges
fine arts 7:Shichi Fukujin
fine as fivepence:5
finger 2:didactyl; 3:tridactyl;
4:quadridigitate, tetradactyl;
5:pentadactyl; 6:hexadactylous,
sexdigitate
fingers:5
fipenny bit:5
fire 3:Agni; 4:elemental quality;
5:Five Agents
fire, river of 5:rivers of Hell
'fire alarm' 10:ten-seventy
fire fueled with ghee 14:Trisala
firkin:¼
firmament 9:Nine Heavens
firmness 10:Sephiroth
first:1
first aid:1
first base:1
first base, get to 1:get to first base
first-born 1:primigenial,
primogeniture
first brass 2:dupondius
First Cause:1
first-class:1
first class mail:1
first come, first served:1
first covenant:1
first cranial nerve 1:olfactory
nerve
first day 6:Creation
First-day:1
First-day 1:Sabbath, Sunday
first-day cover:1
first-degree burn:1
first-degree murder 1:murder in
the first degree

first difference:1
first digit:1
First Empire:1
first family:1
first floor:1
first-foot:1
First French Pleiad   7:Pleiad, 3
first fruits:1
first gear   1:first, 2
first good:1
first-hand:1
first-in, first-out:1
first intention:1
first lady:1
first lance   1:first set
First Law of Thermodynamics:1
first lieutenant:1
first light:1
firstling:1
first mate:1
first mortgage:1
first name:1
first nighter:1
first offender:1
first papers:1
first person:1
first position:1
first quality   4:elemental quality
first quarter:1
first-rate:1
First Reich:1
First Republic:1
first-run:1
first sergeant:1
first set:1
first speed   1:first, 2
first-string:1
first Sunday in Lent   4:Ember Days
first things first:1
First Triumvirate:1
first watch:1
first water:1
fish   10:avatar
Fishes   12:sign
fishes   14:Trisala
fishing   7:Shichi Fukujin
five:5
'five'   5:penta-, quinque-
Five, the   5:St. Petersburg Five
Five Agents:5

five-and-dime store   5:five-and-ten-cent store
five-and-ten-cent store:5
Five Articles and the Five Points:5
Five Articles Oath   5:Charter Oath
Five Blessings:5
Five Bloods:5
Five Blossoms:5
five-boater:5
five by five:5
five-cant file:5
Five Civilized Nations:5
Five Civilized Tribes   5:Five Civilized Nations
five-day fever   5:trench fever
five-dollar word:5
Five Dynasties:5
Five Elements   5:Five Agents
"five farthings, . . .":5
five-finger:5
fivefold:5
five-foot shelf   5:Harvard Classics
Five Freedoms of the Air:5
five-gaited:5
Five Great Kings:5
Five Hundred, Council of   500:Council of Five Hundred
five hundred rummy:500
Five Kings   7:sacred books
Five Knights' Case   5:Darnel's Case
fiveleaf   5:cinquefoil
Five Mile Act:5
Five Nations:5
five o'clock shadow:5
Five Particulars:5
Five Pecks of Rice:5
fivepence:5
fivepence   5:fipenny bit
fivepenny morris:5
fivepenny morris   9:nine men's morris
fivepenny nail:5
five percenter:5
Five Points   5:Five Points Gang
Five Points Gang:5
Five-pound Act:5
"five-pound note":5
Five-Power Constitution:5
fiver:5
fives:5

fives 5:bat-fives
fives-court:5
five senses 5:five wits
five social relations:5
fivespot:5
five-star admiral:5
five-star general:5
five-twenty 5:five-twenty bond
five-twenty bond:5
five wits:5
five W's:5
Five-Year Plan:5
fixity of tenure 3:three Fs
flag:4
Flag day:14
Flag day 13:stripes
flagellum 2:dimastigate
flags, regimental 2:pair of colors
flanconade:9
flanconade 9:flanconade
flank point 9:point
flappers 20:Roaring Twenties
flat:½
Flavius Vopiscus 6:Augustae
  Historiae Scriptores
Fleet Admiral 5:five-star admiral;
  19:salute
flight-feather 10:decempennate
flight into Egypt 7:Sorrows of
  Mary
Floll 12:Knights of the Round
  Table, 1
flood 1:first covenant;
  9:Deucalion
Florence, plague in 10:Decameron
Florida 50:United States of
  America
florin 100:cent
Florismart 12:Paladins of
  Charlemagne
flower 3:trianthous; 4:four-
  o'clock, 2
flowers 14:Trisala
'focus' 0:zero in on
focus 2:bifocal
folio-post quarto ¼:quarto
Folz, Hans 12:Meistersingers
Fons 10:names of God
Fontenoy 2:unlucky number
Fool 22:major arcana

foot 1:uniped; 4:quadrumanous,
  quadruped; 6:hexapod;
  10:decempedal
football team:11
Footsteps, Field of Forty 40:Field
  of Forty Footsteps
Ford, John 8:Grand Jury Eight
Fordham University 7:seven blocks
  of granite
Foreign Ambassador 19:salute
foreign intelligence 1:M. O. 1
forestall 3:ugly man
forgive injuries 7:spiritual works
  of mercy
'forked' 2:bifurcate
fork-rest:2
Forseti 12:Asir
Fort Dix Five:5
Fort Douglas 7:Seven Oaks
  Massacre
fortescue:40
fortitude 7:seven gifts of the Holy
  Ghost, virtues
fortnight:14
Fortuna 9:Sabines
fortune 7:Shichi Fukujin
Fortune 500:500
forty:40
Forty:40
forty and eight:40
forty days and forty nights:40
Forty-Eight:48
forty-eighter:48
forty-eightmo:48
Forty-five:45
forty-five:.45
Forty Footsteps, Field of 40:Field
  of Forty Footsteps
forty-four:.44
forty hours:40
Forty Immortals 40:Forty, 3
forty-legs:40
Forty-Niners:49
Forty-one:41
fortypenny nail:40
forty-rod lightning:40
Forty-second Street:42
Forty-second Street thief:42
forty-skewer 40:fortescue
forty-spot:40
forty stripes:40

Forty stripes save one   39: Thirty-
    nine Articles
forty winks:40
foundation   10: Sephiroth
four:4
'four'   4: quadri-, tetra-
four-a-cat:4
"four and twenty blackbirds":24
"four-and-twenty sailors . . . ,"
    :24
four-bagger:4
four bells:4
four-boater:4
Four Branches of the Mabinogi:4
four-color map problem:4
four corners:4
four corners of the earth:4
four corners rule:4
Four Ends:4
fourer   4: four, 5
foureye butterfly fish:4
four-eyed fish:4
four-eyes:4
Four-F:4
fourflusher:4
Four Freedoms:4
four-handed:4
four-handed   4: quadrumanous
Four-H Club:4
four horsemen:4
Four Horsemen:4
Four Horsemen of the
    Apocalypse:4
Four Hundred:400
fourings:4
four-in-hand:4
Four Kings, History of
    the   4: History of the Four Kings
four-leaf clover:4
four-letter word:4
four-line brevier:32
four-line pica:48
four-line small pica:44
Four Lings:4
Four Marys:4
Four Masters of Anhwei:4
Four Masters of the Yüan
    Dynasty:4
four-minute mile:4

Four Monarchs of Eloquence, the
    4: Eloquence, The Four Monarchs
    of
Four Noble Truths:4
four o'clock   4: fourings
four-o'cat   4: four-a-cat
four-o'clock:4
four of a kind:4
four old cat   4: four-a-cat
four-on:4
four on the floor:4
four orders:4
four-part:4
fourpence:4
fourpence   4: flag
fourpence-halfpenny:4½
fourpenny bit   4: fourpence
fourpenny nail:4
fourpenny piece   4: fourpence
four-point average:4
four-poster:4
four-pounder:4
Four-Power Pacific Treaty:4
Four Principles:4
four questions:4
fours, on all   4: on all fours
fourscore:80
four score and seven:87
four seas:4
four sheets in the wind, be   3: three
    sheets in the wind
foursome:4
four species:4
foursquare:4
four-striper:4
four-tailed bandage:4
"Fourteen Hundred":1400
fourteen penn'orth of it:14
fourteen-point-one continuous
    pocket billiards:14
Fourteen Points:14
Fourteen Points   4: Four Ends, Four
    Principles; 5: Five Particulars
fourteenth:14
Fourteenth Amendment:14
fourth:4
Fourth, the   4: Independence day
fourth best:4
fourth class mail:4
fourth cranial nerve   4: trochlear
    nerve

Fourth-day:4
fourth day   6:Creation
Fourth Estate:4
Fourth Month, Fast of the   4:Fast of
the Fourth Month
fourth position:4
fourth R:4
Fourth Republic:4
four-wheel drive:4
Four Wheeled Hussars:4
Fragonard, Jean-Honore   16:Louis
Seize
France   1:First Republic, War of the
First Coalition; 2:Dual Alliance, 1,
War of the Second Coalition; 3:Big
Three, 1,   Triple Alliance, 1, 2, 3,
Triple Entente, War of the Third
Coalition, Third Republic; 4:Big
Four, 2,   Four-Power Pacific
Treaty, Fourth Republic; 5:Big
Five, 1; 6:Six Months' War;
7:Group of Seven, 1,   Seven
Years' War; 10:Big Ten, 1;
12:European Economic
Community; 16:Sixteen;
100:Hundred Years' War;
500:Council of Five Hundred
France, flag of   3:tricolor
France, patron saint
of   7:Champions of Christendom
Franciade:4
Francis I   3:Three Emperors
Franciscans   4:four orders
Francis Joseph I   3:League of the
Three Emperors
Franco, Ed   7:seven blocks of
granite
Franco-Prussian War   2:Louis
Napoleon; 3:Third Republic
Frankfurt Five:5
Frankfurt Group   5:Frankfurt Five
Frankie and Johnnie   .44:forty-
four
fraternal twin:2
fraud   8:Hell
Frayser's Farm   7:Seven Days'
Battles
Frederick II   1200:due-cento
Frederick Augustus II   3:Alliance of
the Three Kings

Frederick William I   3:Alliance of
the Three Kings
Frederick William IV   3:Alliance of
the Three Kings
free, white, and 21:21
freedom from fear   4:Four
Freedoms
freedom from want   4:Four
Freedoms
freedom of speech and expression
4:Four Freedoms
freedom of worship   4:Four
Freedoms
Freedoms of the Air   5:Five
Freedoms of the Air
freemason   3:mastermason
free sale   3:three Fs
French "75,":75
French empire   300:bees
French Pleiad, First   7:Pleiad, 3
French Pleiad, Second   7:Pleiad, 4
French star:3
Freya   12:Asir
friar-bird   4:four o'clock, 1
Friar Servants of St. Mary, Order of
7:Seven Holy Founders
friars, mendicant   4:four orders
Friday:6
Friday   6:Sixth-day; 13:unlucky
number
friend/friend relationship   5:five
social relations
friends   12:house
**friends and benefactors, house
of:11**
Frimaire:3
Froines, John   7:Chicago Seven
fruits of the Holy Spirit:9
Fukurokuju   7:Shichi Fukujin
full size   2:double bed
fundamental color   3:primary color
Fundamentalism, five points of:5
**Fundamentalism, seven points
of:7**
Furies:3
furrow   2:bisulcate;
4:quadrisulcate
Furry-day:8
fusa:$\frac{1}{8}$
fusella:$\frac{1}{16}$
future   3:Norn

fylfot   4:tetraskele

G

G:5; :400
Ḡ:400,000
G, g:7
Gabon   13:OPEC
Gabriel   7:menorah
Gad   10:Lost Tribes of Israel
Gadara   10:Decapolis
Gaea   3:Cyclopes, Golden Apples of
   the Hesperides; 12:Titan
Gaheris   12:Knights of the Round
   Table, 1
Gaines' Mill   7:Seven Days' Battles
Galahad   12:Knights of the Round
   Table, 1;   150:Round Table
Galento, Tony   2:"Two Ton" Tony
Galgalath   3:Magi
gailon, half   ½:demigalonier
gallows   2:two-legged mare;
   3:three-legged mare, three trees
Galohalt   12:Knights of the Round
   Table, 1
Gamelion:7
*Gamesmanship, The Theory and
   Practice of*   1:one-upmanship
gamma:3
Ganelon   12:Paladins of
   Charlemagne
Gardiner, Henry
   Balfour   5:Frankfurt Five
Gareth   12:Knights of the Round
   Table, 1
garment center   7:Seventh Avenue
Gaspar   3:Magi
Gassin, Ann   13:Macheteros
Gates, General   15:fifteen decisive
   battles
Gates, John Warne
   1,000,000: "Bet-a-million" Gates
gates of Hell:9
gates   6:Troy; 10:decuman
Gath   5:Philistines
Gautama, the enlightened one
   5:Buddha, 2
Gawain   12:Knights of the Round
   Table, 1

Gay Nineties:90
Gaza   5:Philistines
geilfine:5
Gem Blades   5:five o'clock shadow
gemel:2
geminate:2
Gemini:2
Gemini   ⅓:trigon; 2:double-bodied,
   Twins; 12:sign
general   17:salute
General of the Air Force   19:salute
General of the Army   5:five-star
   general; 19:salute
Genesis   5:Pentateuch;
   6:hexaëmeron, 2,   Hexateuch;
   7:Heptateuch; 8:Octateuch
Geneva Accords   17:Seventeenth
   Parallel
Genevan psalter   100:Old Hundred
Genius, Festival of   5:Sans-
   culottides
gentian   4:diatessaron
Geoffroy de Frises   12:Paladins of
   Charlemagne
geometry   4:quadrivium; 7:seven
   arts
George II   2:unlucky number
George III   4:Speech Day
George, David Lloyd   3:Big Three,
   1
Georgia   13:original colonies;
   50:United States of America
Gerasa   10:Decapolis
Gerena, Victor   13:Macheteros
German Ocean   4:four seas
German revolution   48:forty-
   eighter
Germany   1:First Reich, M. O. 1;
   2:Dual Alliance, 2,   Second Reich;
   3:Hekatist, League of the Three
   Emperors, Third Reich, Triple
   Alliance, 3;   30:Thirty Years'
   War, 2
Germany, peace settlement with
   4:Four Ends, Four Principles,
   Fourth Republic; 5:Five
   Particulars; 14:Fourteen Points
Germany, unification of   3:Alliance
   of the Three Kings
Germinal:7
Geryon   12:Dodekathlos

get to first base:1

Gettysburg, Battle of   87:four score and seven

ghostly strength   7:seven gifts of the Holy Ghost

gibberish   2:double Dutch, 1, double-talk

Gibbon, Edward $^{1}/_{10}$:Saladin's tenth; 5:Adoptive Emperors

Gibraltar, Rock of   2:Pillars of Hercules

gig   2:dennet

Gihon   4:rivers of Eden

gill   2:dibranchiate, dipnoous

gimel:3

gimel   22:major arcana

gimmer:2

Girdle of Hippolyte   12:Dodekathlos

give drink to the thirsty   7:corporal works of mercy

give (something) the deep six   6:deep six

give two cents for (something), 2:two cents, 2

Glackens, William J.   8:Eight

gleek:3

Glem   12:Arthur

Glencoe, massacre of   9:Curse of Scotland

Glendale   7:Seven Days' Battles

glossopharyngeal nerve:9

gluttons   8:Hell

Gluttony   6:Lucifera

gluttony   7:sins

Glyconic:4

go   5:gobang

goat   6:Lucifera

Goat   12:sign

gobang:5

Gobelins   14:Louis Quatorze

God, names of   7:names of God; 10:names of God

Godfrey of Bouillon   9:Nine Worthies

godliness   7:seven gifts of the Holy Ghost

Godolphin, Lord   3:Triumvirate, 1

God rested   7:day of rest

go fifty-fifty:50

gold   7:alchemy

Golden Apples of the Hesperides:3

Golden Apples of the Hesperides 12:Dodekathlos

golden wedding anniversary 50:wedding

"Goldilocks and the Three Bears":3

Gold Rush   49:Forty-Niners

golf   −3:double eagle; −2:eagle; −1:birdie; 1:bogey, hole in one; 3:threesome, triple bogey; 4:best-ball foursome, foursome, Scotch foursome; 19:nineteenth hole

go-moku-narabe   5:gobang

González Claudio, Orlando 13:Macheteros

González, Virgilio R.   7:Watergate Seven

Good Friday   3:three hours; 6:Holy Week

Goody Two-Shoes:2

googol:10^{100}

googolplex:$10^{10^{10}}$

goose-egg:0

Gospels   4:diatessaron, quadripartite, 2

Gotan   6:Cities of Refuge

Gotham, Three Wise Men of 3:"Three wise men of Gotham"

gothic influence   1200:due-cento

Gothland   6:Six Islands

Governor of a State   19:salute

Gower   3:Triumvirate, 2

grace of the Holy Ghost 4:Evangelists

Graces:3

graciousness   9:fruits of the Holy Spirit

Graeae:3

Graham, Gere   6:San Quentin Six

grain $^{1}/_{24}$:karob

Grainger, Percy A.   5:Frankfurt Five

grammar   2:double negative, double possessive; dual; 3:trivium; 7:seven arts

Granada   2:Hermanas

grand:1000

grand jury:23

Grand Jury Eight:8

grand quarter:1/4
grandsire:5
Grant, Ulysses S.   5:five-star
   general
Grant Jr., James Earl   3:Charlotte
   Three
Gratiae   3:Graces
Gray Friars   4:four orders
Great Britain   3:Big Three, 1, 2,
   Triple Alliance, 2,   Triple Entente;
   4:Big Four, 2,   Four-Power
   Pacific Treaty; 5:Big Five,
   1,   Intolerable Acts; 7:Group of
   Seven, 1;   10:Big Ten,
   1;   54:"Fifty-four Forty"
Great Britain, regents of   2:unlucky
   number
great casino   10:big casino
Great Compromiser   3:Great
   Triumvirate
great dodecahedron:12
great dodecahedron   4:Kepler-
   Poinsot solid
great god   8:Agni
great hundred:120
great icosahedron:20
great icosahedron   4:Kepler-
   Poinsot solid
great primer:18
**great
   rhombicosidodecahedron:**62
great rhombicosidodecahedron
   13:Archimedean solid
great rhombicuboctahedron:30
great rhombicuboctahedron
   13:Archimedean solid
great sanhedrim   70:great
   sanhedrin
great sanhedrin:70
great stellated dodecahedron:12
great stellated dodecahedron
   4:Kepler-Poinsot solid
Great Triumvirate:3
Great Wall   7:Seven Wonders of the
   Middle Ages
Greece   7:Ionian Islands;
   12:European Economic
   Community
green   3:secondary color,
   trichromatic; 7:perfect color

Green, Frank   10:"Ten Little
   Indians"
Gregory XIV   14:Henry IV
Gregory of Tours   7:Seven Sleepers
Gresford churchbells   7:Seven
   Wonders of Wales
Grhapati   3:Agni
grief, river of   5:rivers of Hell
Grier, Rooosevelt   4:Fearsome
   Foursome
Griggs, Gregory   27:"twenty-seven
   different wigs"
Grislet   12:Knights of the Round
   Table, 1
groat   4:flag, fourpence
groove   2:bisulcate;
   4:quadrisulcate
gross:144
Groucho   3:Marx Brothers
ground floor   1:first floor
ground zero:0
Group of Seven:7
grouse-shooting season   12:Twelfth
Grumpy   7:"Snow White and the
   Seven Dwarfs"
guard, fencing   1:prime; 2:seconde;
   3:tierce; 4:quarte; 5:quinte;
   6:sixte; 7:septime; 8:octave, 3
Guardian Angel   7:Seventh Heaven
Guerin, duc de
   Lorraine   12:Paladins of
   Charlemagne
Guillaume de l'Estoc   12:Paladins
   of Charlemagne
Guineas, 1000:1000
Guineas, 2000:2000
Guinevere   150:Round Table
Gulf Oil   7:Seven Sisters, 2
Gummo   3:Marx Brothers
gunpowder   6:Six Dynasties
Guy de Bourgogne   12:Paladins of
   Charlemagne
Gwalchmai   12:Knights of the
   Round Table, 2
Gwenion   12:Arthur
Gwevyl   12:Knights of the Round
   Table, 2
Gyes   100:Hecatoncheires
Gyges   100:Hecatoncheires

H

H: 200
H: 200,000
H, h: 6; : 8
Hadad   6: Magi
Hades   12: Olympians; 50: Danaides
Hades, Judges of   3: Judges of Hades
Hadrian   5: Adoptive Emperors;
   6: Augustae Historiae Scriptores
Haggadah   4: four questions
haiku: 17
Hakatist: 3
Hakmah   10: Sephiroth
'half'   ½: demi-
half-and-half: ½
half-arsed   ½: half-assed
half-assed: ½
half-baked: ½
half-blood   ½: demisang
half-breed: ½
half-dollar: ½
half-groat   2: twopence, 1
half lion, half man   10: avatar
half man, half lion   10: avatar
half model: ½
'half more'   1½: sesqui-
halfpenny: ½
half seas over: ½
Halicanarsus   7: Wonders of the
   Ancient World
'Hall of the Two
   Sisters'   2: Hermanas
Halsey, William F.   5: five-star
   admiral
hammer   3: Thor
Hampden, Sir Edmund   5: Darnel's
   Case
Hampshire   15: Danelagh
Han Chun   6: Six Idlers of the
   Bamboo Streams
hand: 4; : 5
hand   2: bimanous;
   4: quadrumanous
Han dynasty   5: Five Pecks of Rice;
   6: Six Dynasties
Hanged Man   22: major arcana
hanging gardens of Babylon
   7: Wonders of the Ancient World

hang ten: 10
Haniel   7: menorah
Hannah, Gerry   5: Vancouver Five
Hanover   3: Alliance of the Three
   Kings
Hansemann   3: Hakatist
Hansen, Ann   5: Vancouver Five
Hanukkah: 8
Hanukkah   7: menorah
ha'penny   ½: halfpenny
Happy   7: "Snow White and the
   Seven Dwarfs"
harbor the homeless   7: corporal
   works of mercy
Harder, Konrad   12: Meistersingers
harmony   3: Hermes Trismegistus
Harold II   2: unlucky number;
   15: fifteen decisive battles
harp   2: dichord
Harpies: 2; : 3
Harpo   3: Marx Brothers
harpoon   1: first set
Harris, Lawrence   7: Group of
   Seven, 2
Harrisburg Six: 6
Hartford Fourteen   13: Macheteros
Hartford Thirteen   13: Macheteros
Harvard   3: Big Three, 3
Harvard Classics: 5
Harvard University   5: Harvard
   Classics; 8: Ivy League
Hasdrubal   15: fifteen decisive
   battles
Hassam, Childe   10: Ten
Hastings   5: Cinque Ports; 15: fifteen
   decisive battles
hat   2: bicorn; 3: tricorn; 10: ten-
   gallon hat
hate, river of   5: rivers of Hell
hat trick: 3
(have) two left feet: 2
(have) two strings to one's bow: 2
Hawaii   50: United States of
   America
Hawkwood, Sir John   9: Worthies of
   London
Hayden, Thomas   7: Chicago Seven
he: 5
he   4: Tetragrammaton
heads   100: centipitous
healing, master of ·   5: Buddha, 2

healing by first intention:1
healing by second intention:2
health   7:Shichi Fukujin
hearing   2:diplacusis; 7:senses
heartbeat   2:dicrotic
hearts   78:tarot
hearts, seven of   7:manille
heathen philosophers   8:Hell
heaven:5
heaven   4:last things; 6:Creation
heavenly body   5:Buddha, 1
heavy hydrogen   2:deuterium;
    3:tritium
hebdomad:7
hebdomadal:7
hebdomadal council:7
Hebe:6
Hebron   6:Cities of Refuge
Hecate:3
Hecate   3:Trivia
hecatomb:100
Hecatombaeon:1
hecatompedon:100
Hecatoncheires:100
hecatonstylon:100
hecatontarchy:100
hectare:100
hectogram:100
hectogramme   100:hectogram
hectoliter:100
hectolitre   100:hectoliter
hectometer:100
hectometre   100:hectometer
Hector   9:Nine Worthies
hectostere:100
heh   22:major arcana
Heimdall   12:Asir
Heinz:57
Helen   10:Trojan War
Hell:7; :8
Hell   4:last things
Hell, gates of   9:Gates of Hell
Hell, rivers of   5:rivers of Hell
Hellenes   3:Ionians
Heller, Joseph   22:catch-22
Hellespontine Sibyl   10:sibyl
"Hell's Kitchen"   10:Tenth Avenue
helmet of Pluto   3:Cyclopes
hemidemisemiquaver   1/64:sixty-
    fourth note; 1/32:demisemiquaver
hemiobolion:1/2

hemisome:1/2
hemisphere:1/2
Hend   10:wives
hendecagon:11
hendecahedron:11
hendecasyllable:11
Hendeka   11:Eleven
Henri III   7:Pleiad, 3
Henri, Robert   8:Eight
Henry II   2:unlucky number;
    14:Henry IV; 100:Hundred Years'
    War
Henry IV:14
Henry IV   9:Worthies of London
Henry VI, Part 2   3:thrice
Henry VIII   6:Six Articles
Henry of Cornhill   9:Worthies of
    London
Hephaestus   12:Olympians
hepta-:7
heptace:7
heptachord:7
heptad:7
heptaglot:7
heptagon:7
heptahedron:7
heptal   7:hebdomadal
heptameride:7
heptameron:7
heptameter:7
heptandrian   7:heptandrous
heptandrious   7:heptandrous
heptandrous:7
heptane:7
Heptanesus:7
heptangular:7
heptapetalous:7
heptaphyllous:7
heptapody:7
heptapody   7:heptameter
heptarchy:7
heptasepalous:7
heptastich:7
heptasyllable   7:septisyllable
Heptateuch:7
heptathlon:7
heptavalent:7
Hera   3:Golden Apples of the
    Hesperides; 12:Olympians
Heracles   12:Dodekathlos
Hercules   9:Etruscan gods, Sabines

Hercules, Labors
of 12:Dodekathlos
Hercules, Pillars of 2:Pillars of
Hercules
heretics 8:Hell
Hermanas, Sala de las Dos:2
hermaphrodite 2:bisexual
hermaphroditism 2:dichogamism
Hermas 6:apostolic fathers
Hermes 12:Olympians
Hermes Trismegistus:3
Hermit 22:major arcana
Herod 7:Dance of the Seven Veils
Hertford 7:Associated Counties
Hertfordshire 15:Danelagh
Herukabuddha:5
Hesed 10:Sephiroth
Hesiod 2:Harpies; 3:Harpies;
5:Epic Poets; 9:Muses
Hesperides 3:Golden Apples of the
Hesperides; 12:Dodekathlos
Hestia 6:Vestal Virgins;
12:Olympians
heth 8:cheth
Heveningham, Sir John 5:Darnel's
Case
hexa-:6
hexace:6
hexachord:6
hexactinal:6
hexad:6
hexadactylous:6
hexadecimal:16
hexadecimal
system 16:hexadecimal
hexadrachm:6
hexaëmeron:6
hexafoil:6
hexagon:6
hexagon 6:sexangle, sexagon
hexagram:6
hexahedron:6
hexahemeron 6:hexaëmeron
hexakisoctahedron
48:hexoctahedron
hexameral 6:hexamerous
hexamerous:6
hexameter:6
hexandrian 6:hexandrous
hexandrous:6
hexane:6

hexangular:6
hexapartite 6:hexamerous
hexapetalous:6
hexaphyllous:6
Hexapla:6
hexaplar:6
hexaplaric 6:hexaplar
hexapod:6
hexapodan 6:hexapod
hexapodous 6:hexapod
hexapody:6
hexapsalmos 6:hexapsalmus
hexapsalmus:6
hexapterous:6
hexastemonous:6
hexaster:6
hexastich:6
hexastichon 6:hexastich
hexastichous:6
hexastylar 6:hexastyle
hexastyle:6
hexasyllabic:6
hexatetrahedron:24
Hexateuch:6
hexatomic:6
hexavalent:6
hexoctahedron:48
Heywood, Thomas 7:Homer
"Hierarchie of the Blessed Angels"
7:Homer
hieroglyphic writing 3:Hermes
Trismegistus
Hierophant 22:major arcana
Higgins, Eugene 8:Eight
High Court of Justice 3:vice-
chancellor
High Priestess 22:major arcana
Hill of Bath 12:Arthur
hills of Rome 7:Rome
Hindu Philosophy, Six Schools of
6:Six Schools of Hindu Philosophy
Hippolyte 12:Dodekathlos
Hippomedon 7:Seven Against
Thebes
Hippomenes 3:Golden Apples of
the Hesperides
Hipponax 3:Iambic Poets
Hippos 10:Decapolis
history 9:Muses; 10:tripos
History of Goody Two-Shoes
2:Goody Two-Shoes

*History of Prince
Arthur* 12: **Knights of the Round
Table**, 1
History of the Four Kings:4
'hit and run accident' 10: **ten-fifty-
seven**
Hitler, Adolf 3: **Third Reich**
Ho Chihchang 8: **Eight Immortals
of the Wine Cup**
Hod 10: **Sephiroth**
Hoël, comte de Nantes 12: **Paladins
of Charlemagne**
Hoffman, Abbie 7: **Chicago Seven**
Hogan, John 9: **Catonsville Nine**
hole in one:1
Hollywood Ten:10
Holy Communion 2: **duplicate**, 4;
7: **Holy Sacrament**
Holy-Cross day:14
Holy Ghost 3: **Trinity**;
7: **candelabrum**; 50: **Pentecost**
Holy Ghost, grace of 4: **Evangelists**
Holy Grail 150: **Round Table**
Holy Orders 7: **Holy Sacrament**
Holy Roman Empire 1: **First Reich**;
3: **War of the Third Coalition**
Holyrood day 14: **Holy-Cross day**
Holy Sacraments:7
Holy Spirit 3: **Trinity, tritheism**;
4: **tetratheism**
Holy Week:6
Holy Week 1: **Palm Sunday**
Homer:7
Homer 2: **Twins**; 3: **Harpies, Siren**;
5: **Epic Poets**; 10: **Trojan War**
"Homer" 7: **Pleiad**, 2
Homer the Younger 7: **Pleiad**, 1;
13: **Tragic Poets**
homosexuals 3: **third sex**
Honegger, Arthur 6: **Six**
honest dealing 7: **Shichi Fukujin**
honor point 9: **point**
honors, hundred:100
honors, hundred and fifty:150
**hoots (in hell), not care (or give)
two:2**
hope 7: **virtues**
Hopkins, Mark 4: **Big Four**
Hopper, Edward 8: **Eight**
Horae:3

horn 1: **monocerous**; 2: **dicerous**;
3: **tricerion**; 4: **quadricorn**
horns of the dilemma:2
hot 4: **elemental quality, secondary**
Hotei:7
Hotei 7: **Shichi Fukujin**
Hottelet, Richard C. 8: **Murrow's
Boys**
Houdon, Jean-Antoine 16: **Louis
Seize**
hound of hell 3: **Cerberus**
hour:24
Hours 3: **Horae**; 7: **Canonical Hours**
house:12
House of Commons 40: **quorum**;
100: **Chiltern Hundreds**
House of Keys:24
House of Lords 3: **vice-chancellor**
House Unamerican Activities
Committee 10: **Hollywood Ten**
Howard, Jerry 3: **Three Stooges**
Howard, Moe 3: **Three Stooges**
Howard, Shemp 3: **Three Stooges**
Hsiang Hsiu 7: **Seven Sages of the
Bamboo Grove**
Hsiao Yun-ts'ung 4: **Four Masters
of Anhwei**
Hsi K'ang 7: **Seven Sages of the
Bamboo Grove**
Huang Kung-wan 4: **Four Masters
of the Yüan Dynasty**
Hudson's Bay Company 7: **Seven
Oaks Massacre**
human genealogy of Christ
4: **Evangelists**
hundred:100
hundred-court:100
Hundred Days:100
hundred-legs:100
hundredweight:100
Hundred Years' War:100
Hungary 5: **Five-Year Plan**
Hung-jen 4: **Four Masters of
Anhwei**
hungry, feed the 7: **corporal works
of mercy**
Huns 16: **Sixteen Kingdoms**
Huntingdon 7: **Associated Counties**
Huntington, Collis P. 4: **Big Four**

I

infection,
  secondary 2: deuteropathy
Inferno 8: Hell
infinite life 5: Dhyanibuddha
infinite light 5: Dhyanibuddha
infinity:∞
informatory double:2
'inharmonious' 6: at sixes and
  sevens
"Injuns" 10: "Ten Little Indians"
injuries, forgive 7: spiritual works
  of mercy
innings:9
Innocent III 1/10: Saladin's tenth
'insignificant' 1: one-horse town;
  2: two-bit
'insincere' 2: two-faced
in sixteens:16
instruct the ignorant 7: spiritual
  works of mercy
instrument of ten strings:10
intelligence 10: Sephiroth
intelligence and soul 3: aum
intelligence quotient 100: I.Q.
interamnian:2
intercession 5: prayer
intercessions 3: Rogation days;
  5: litany
intercolumniation:2
intermittent fever 2: double fever,
  double quotidian fever, double
  tertian fever; 3: tertian; 4: quartan
  fever; 5: quintan, quintan fever;
  6: sextan fever; 7: septan fever;
  8: octan fever
interpreter 2: dobhash
interrogation 1: M. O. 1
intestine, small 12: duodenum
**in the country of the blind the
one-eyed is king:1**
Intolerable Acts:5
'intoxicated driver' 10: ten-fifty-
five
'in trouble' 8: behind the eightball
in two shakes 2: two shakes of a
  lamb's tail
invocations 5: litany
Ion 13: Tragic Poets
Ionian:3
Ionian Islands:7
Ionian Islands 7: Heptanesus

Ionic:4
Ionic 8: classic orders
Ionic dimeter 2: dimeter
Ios 7: Homer
iota:9
Iowa 50: United States of America
Iowa, University of 10: Big Ten, 2
Iowa State University 8: Big Eight,
  1
I. Q.:100
Iran 13: OPEC
Iraq 6: Six-Day War; 13: OPEC
Ireland 6: Six Islands; 12: European
  Economic Community
Ireland, patron saint
  of 7: Champions of Christendom;
  17: St. Patrick's day
Ireland, principal families of 5: Five
  Bloods
Irene:14
Irish Sea 4: four seas
Irish Land League 3: three Fs
iron 7: alchemy
iron gloves 3: Thor
Iroquois Confederation 5: Five
  Nations
irresistible grace 5: Five Articles
  and the Five Points
Irving, Washington 20: Rip Van
  Winkle
Isaac 3: patriarch; 10: animals in
  paradise
Isabella of Aragon 9: Nine Worthy
  Women
Isaeus 10: Attic Orators
Isana 8: Agni
"I saw three ships . . . ,":3
Ishmael 10: animals in paradise
Island of the Seven Cities:7
Isle of Man 3: triskelion; 24: House
  of Keys
Isle of Wight 3: Needles
Isocrates 10: Attic Orators
isolationist 1: America firster
Israel 2: second covenant;
  6: hexagram, 4, Six-Day War
Israel, Lost Tribes of 10: Lost
  Tribes of Israel
Israel, Tribes of 12: Ahijah
Israel, Twelve Tribes of 12: Twelve
  Tribes of Israel

Issachar 10:Lost Tribes of Israel
Isthmian games 5:quinquennalis
Italy 2:Sicilies; 3:Triple Alliance,
3; 7:Group of Seven, 1; 10:Big
Ten, 1; 12:European Economic
Community
Italy, frontiers of 14:Fourteen
Points
Italy, patron saint of 7:Champions
of Christendom
Ithaca 7:Homer, Ionian Islands;
20:Odysseus
It takes two to tango:2
"I've got sixpence . . .,":6
Ivry 14:Henry IV
Ivy League:8

J

J:9
J, j:1; 7; :10
jack:11
Jackson, George 3:Soledad
Brothers; 6:San Quentin Six
Jacob 3:patriarch
Jacobite rebellion 15:Fifteen;
45:Forty-five
Jacobs, Jerry 10:Wilmington Ten
"Jacob's well" 9:Worthies of
London
Jael 9:Nine Worthy Women
Jaina 14:Trisala
James I 1000:millenary petition
James II 2:unlucky number;
7:seven bishops; 15:Fifteen;
45:Forty-five; 88:Stuarts
James III 88:Stuarts
James VI 8:Octavian
James, Duke of York 24:Twenty-
four Proprietors
Jamyn, Amadis de 7:Pleiad, 3
Janias 6:Hyksos
January:1
Janus:2
Janus-faced:2
Japan 4:Four-Power Pacific Treaty;
5:Charter Oath; 7:Group of Seven
1, Pearl Harbor Day; 10:Big Ten,
1
Japanese 7:Pearl Harbor day

Jarvik, Robert 7:Jarvik-7
Jarvik-7:7
Jebel Musa 2:Pillars of Hercules
Jemlikha 7:Seven Sleepers
Jenkins, Roy 4:Limehouse Four
jereboam:2
Jerusalem $\frac{1}{10}$:Saladin's tenth;
7:Seven Cities; 10:animals, Fast of
the Tenth Month
Jesus 2:second covenant; 3:Magi;
6:apostle; 7:Sorrows of Mary;
10:animals in paradise
Jesus missed 7:Sorrows of Mary
Jewels, Three 3:Threefold Refuge
Jewish calendar 12:ordinary year;
13:embolismic year
Jews 7:Hell
Jinas 14:Trisala
Joachim du Bellay 7:Pleiad, 3
Joan of Arc 15:fifteen decisive
battles
Job 150:Psalms
Jodelle 7:Pleiad, 3
Johanna of Naples 9:Nine Worthy
Women
John 4:Evangelists; 7:Seven
Sleepers
John II 30:Battle of the Thirty
John of Montfort 30:Battle of the
Thirty
Johnston, Frank H. 7:Group of
Seven, 2
John the Baptist 7:Dance of the
Seven Veils
Jonah 10:animals in paradise
Jones, Deacon 4:Fearsome
Foursome
Jordan 6:Six-Day War
Jorojin 7:Shichi Fukujin
Joseph 7:seven years of plenty
Joshua 6:Cities of Refuge,
Hexateuch; 7:Heptateuch;
8:Octateuch; 9:Nine Worthies;
10:Lost Tribes of Israel
Jotunheim 3:Yggdrasil
joy 3:Graces; 9:fruits of the Holy
Spirit
Joys of Mary:7
Judah 12:Twelve Tribes of Israel
Judah, Kingdom of 12:Twelve
Tribes of Israel

Judas Iscariot   13:unlucky number;
   30:pieces of silver
Judas Maccabaeus   9:Nine Worthies
judge   9:possession;
   105:centumvir; 6000:dicast
Judges   7:Heptateuch; 8:Octateuch
Judges of Hades:3
judgment   4:last things
Judgment   22:major arcana
Judgment, Pillar of   10:Sephiroth
Judith   9:Nine Worthy Women
Juggler   22:major arcana
Julian calendar   707:year of
   confusion
Julian year:365¼
Julius Caesar   2:First Triumvirate;
   9:Nine Worthies
Julius Capitolinus   6:Augustae
   Historiae Scriptores
July:7
jumelle:2
June:3
Juno   9:Etruscan gods
Jupiter   7:alchemy, stars; 9:Nine
   Heavens, planet
Jurassic   3:Triassic
jury   9:possession; 12:petit jury;
   23:grand jury
justice   7:virtues; 10:Sephiroth
Justice   22:major arcana
Justinian   4:faction
Justinian, Code of   50:Fifty
   Decisions

K

K:10; :250; :1000
K̄:250,000
K   5:M. I. 5
K, k:8; :11
Kadesh   6:Cities of Refuge
Kadijah   10:wives
Kagyur   3:Tripitaka; 14:Tantras
Kai   12:Knights of the Round Table,
   2
kalends   1:calends
Kalimnos   12:Dodecanese
Kalki   10:avatar
Kamaheruka   5:Herukabuddha
Kane, Ronald   6:San Quentin Six

Kansas   50:United States of America
Kansas, University of   8:Big Eight, 1
Kansas State University   8:Big
   Eight, 1
kaph:11
kaph   22:major arcana
kappa:10
karman   14:Trisala
karob:¹⁄24
Karpathos   12:Dodecanese
karyokinesis   2:dispireme, dyaster
Kasos   12:Dodecanese
Kastellorizon   12:Dodecanese
Kasyapa   5:Buddha, 2
Katmir   7:Seven Sleepers
Kay   12:Knights of the Round
   Table, 1
Kelly, Yvette   8:New York Eight
Ken   7:seven bishops
Kenana   10:wives
Kennedy, John F.   1000:thousand
   days
Kennemann   3:Hakatist
Kent   7:heptarchy, 2
Kentucky   50:United States of
   America
Kepler-Poinsot solid:4
Kepler solid   4:Kepler-Poinsot solid
Keschetiouch   7:Seven Sleepers
Kethe   100:Old Hundred
Kether   10:Sephiroth
Keys, House of   24:House of Keys
Khadijah   4:perfect women
Khalki   12:Dodecanese
kill two birds with one stone:2
kilo-:1000
king:12
king   78:tarot; 100:honors;
   150:honors
King, Ernest J.   5:five-star admiral
King Arthur   9:Nine Worthies;
   12:Arthur, battles of King
King Chao   6:Six Idlers of the
   Bamboo Streams
kingdom:3
kingdom   10:Sephiroth
Kings:3; :12
Kings, History of the
   Four   4:History of the Four Kings
kissar:5

Kissinger, Henry A.   **6:Harrisburg Six**

kit:3

Kitanemuk   **4:Serrano**

kite   **6:Six Dynasties**

Klug, Terry   **5:Fort Dix Five**

knave   **11:jack; 78:tarot**

knickerbockers   **4:plus fours**

knight   **78:tarot**

Knights of the Round Table:12

Knight Templar:10

Knorr, Iwan   **5:Frankfurt Five**

know how many (blue) beans make five:5

knowledge   **7:seven gifts of the Holy Ghost; 14:Trisala**

knowledge, tree of   **3:Yggdrasil**

Knox, John   **2:Books of Discipline**

kootoo   **3:kowtow**

koph:19

koph   **22:major arcana**

koppa:20

Koran   **7:sacred books, Seven Sleepers**

Kos   **12:Dodecanese**

koto:13

kotoo   **3:kowtow**

kotou   **3:kowtow**

kotow   **3:kowtow**

kowtow:3

Krasnes, Paul   **6:San Quentin Six**

Kratim   **7:Seven Sleepers; 10:animals in paradise**

Kratimer   **7:Seven Sleepers**

kreutzer   **20:zwanziger**

Krishna   **10:avatar**

Kriya   **14:Tantras**

Kronos   **12:Titan**

Ksitigarbha   **8:Dhyanibodhisattva**

Kundä Ling   **4:Four Lings**

Kundeling   **4:Four Lings**

Kuo Hsiang   **7:Seven Sages of the Bamboo Grove**

Kuomintang   **5:Five-Power Constitution**

Kurma   **10:avatar**

Kuwait   **13:OPEC**

Kyme   **7:Homer**

Kynon   **12:Knights of the Round Table, 2**

L

L:11; :50

L̄:50,000

L, 1:9; :12

la:6

Labor, Festival of   **5:Sans-culottides**

Labors of Hercules   **12:Dodekathlos**

labrys   **2:double ax**

labyrinth   **7:seven youths**

Lachesis   **3:Fates**

Ladon   **3:Golden Apples of the Hesperides**

Lake   **7:seven bishops**

Laksmi   **14:Trisala**

lambda:11

Lambert, prince de Bruxelles   **12:Paladins of Charlemagne**

lamed:12; :30

lamed   **22:major arcana**

Lamoracke   **12:Knights of the Round Table, 1**

landaulet   **½:demi-landau**

language   **1:monoglot; 2:bilingual, diglot, diglottism, dobhash; 3:triglot, trilingual; 4:quadrilingual; 7:heptaglot; 8:octoglot**

lard   **4:tetrapharmacon**

Lardner, Jr., Ring   **10:Hollywood Ten**

lark   **5:alauda**

Lars Porsena   **9:Etruscan gods**

Larue   **7:Pleiad, 4**

La Salle College   **5:Big Five, 2**

last-in, first-out:1

Last Supper   **13:thirteen, unlucky number**

last things:4

latitude   **3:three L's**

lauds   **6:hexapsalmus; 7:Canonical Hours**

Launcelot   **12:Knights of the Round Table, 1, 2**

law   **10:tripos**

law, eight-foot   **8:eight-foot law**

Lawfeld   **2:unlucky number**

laws of the decemvirs   **12:Twelve Tables**

Lawson, Ernest   8:Eight
Lawson, John Howard
   10:Hollywood Ten
lawyer   9:possession
Layden, Elmer   4:Four Horsemen
layer   4:quadrilaminate
lead   3:three L's; 7:alchemy
leaf   2:diphyllous, dufoil; 3:trefoil,
   trifoliate; 4:quadrifoliate,
   quadriphyllous, tetraphyllous;
   5:quinquefoliate; 6:hexaphyllous;
   7:heptaphyllous; 10:decaphyllous
leaflet   4:quadrifoliate, quatrefoil,
   tetraphyllous; 5:quinquefoliate;
   9:enneaphyllous
league   7:seven-league boots
League of the Three Emperors:3
Leahy, William D.   5:five-star
   admiral
Leaning Tower   7:Seven Wonders
   of the Middle Ages
leap year:366
Lear, Edward   5:"five-pound note"
learning   12:house
lease:7
leatherhead   4:four o'clock, 1
Le Brun   14:Louis Quatorze
Lechery   6:Lucifera
Leda   2:Twins
Lee, Henry "Light-Horse Harry"
   1:Washington
Lee, Robert E.   7:Seven Days'
   Battles
leg   4:quadruped; 10:decapod
legislative bodies   2:bicameral
Legnago   4:quadrilateral, 3
Leicestershire   15:Danelagh
Leinster   5:Five Bloods
lemniscate   8:figure of eight
lemur   1:Primates
Le Nain, Antoine   13:Louis Treize
Le Nain, Louis   13:Louis Treize
Le Nain, Mathieu   13:Louis Treize
length   3:tridimensional;
   4:dimension
Le Nôtre, Andre   14:Louis Quatorze
Lent:40

Lent   3:Oculi Sunday; 4:Ember
   Days, Mid-Lent; 5:Passion Sunday,
   Passion week; 6:Holy Week, Palm
   Sunday; 7:church year;
   40:Quadragesima Sunday,
   quarantine; 60:quinquagesima,
   Sexagesima Sunday;
   70:Septuagesima Sunday
Leo   1/3:trigon; 12:sign
Leodegraunce, King   150:Round
   Table
Lepidus   2:Second Triumvirate
lepta   100:drachma
Lerna   9:Hydra
Lernean Hydra   9:hydra;
   12:Dodekathlos
Leros   12:Dodecanese
Lesbian Poets:4
lesser sanhedrim   23:lesser
   sanhedrin
lesser sanhedrin:23
Le Sueur, Larry   8:Murrow's Boys
Lethe   5:rivers of Hell
letter   2:biliteral, digraph;
   3:trigrammatic, triliteral;
   4:quadriliteral; 5:quinqueliteral
Leucas   7:Ionian Islands
Leucosia   3:Siren
lever of the first class:1
lever of the second class:2
lever of the third class:3
Leviticus   5:Pentateuch;
   6:Hexateuch; 7:Heptateuch;
   8:Octateuch
Lewis, Thomas P.   9:Catonsville
   Nine
lexiarchus:6
Lhasa, convents at   4:Four Lings
Liang   6:Six Dynasties
liberal arts   3:trivium
Libra:7
Libra   1/3:trigon; 12:sign
Libya   13:OPEC
Libyan Sibyl   10:sibyl
Li Chin   8:Eight Immortals of the
   Wine Cup
Liddy, G. Gordon   7:Watergate
   Seven
Life   7:*Seven Lamps of Architecture*
life   12:house
life, house of:1

life, infinite   5:Dhyanibuddha
life, thread of   3:Fates
life, tree of   3:Yggdrasil
Life begins at 40:40
LIFO   1:last-in
Ligea   3:Siren
light   6:Creation
light, brilliant   5:Dhyanibuddha
light, infinite   5:Dhyanibuddha
Light Brigade:600
"Light-Horse Harry"   1:Washington
lighthouse   7:Wonders of the
   Ancient World
lightning   3:; 8:Agni
like sixty:60
limb   4:quadrimembral
Limehouse Four:4
limited atonement   5:Five Articles
   and the Five Points
Lincoln   7:Associated Counties
Lincoln, Abraham   87:four score
   and seven
Lincolnshire   15:Danelagh
Lindau   4:Tetrapolitan
line:14
line   6:hexameter
line: See   verse (in Index)
Lings, Four   4:Four Lings
lion   4:Evangelists; 6:Lucifera
Lion   12:sign
Lionell   12:Knights of the Round
   Table, 1
Li Po   6:Six Idlers of the Bamboo
   Streams; 8:Eight Immortals of the
   Wine Cup
Lipsos   12:Dodecanese
liquid fire, river of   5:rivers of Hell
Li Shihchih   8:Eight Immortals of
   the Wine Cup
Lismer, Arthur   7:Group of Seven, 2
Lissajous   2:figure of Lissajous
litany:5
little casino:2
Little Three:3
Litton Industries   5:Vancouver Five
Liu Ling   7:Seven Sages of the
   Bamboo Grove
Liu-Sung   6:Six Dynasties
livelihood, right   8:Noble Eightfold
   Path
living creatures   6:Creation

Llangollen bridge   7:Seven
   Wonders of Wales
Llano y Sierro, Gonzalo Queipo de
   5:fifth column
Lloyd   7:seven bishops
Llywarch Hen   12:Knights of the
   Round Table, 2
lobe   2:deutocerebrum, dicoccous,
   digastric; 3:triformed, 1;
   6:hexafoil; 7:septfoil; 8:octofoil;
   10:decemfid
local degree:$\frac{1}{360}$
locust   17:seventeen-year locust
logic   1:enthymeme of the first
   order; 2:enthymeme of the second
   order; 3:trivium; 7:seven arts
lohan:18
Lohr-Berg   7:Siebengebirge
Lombardi, Vince   7:seven blocks of
   granite
London   7:Seven Cities; 9:Nine
   Worthies
loneliness   1:monophobia
Lone Star State:1
long-distance run   3:triathlon
longevity   5:Five Blessings;
   7:Shichi Fukujin
Longfellow, Henry Wadsworth
   1:"One if by land";
   18:"eighteenth of April . . . "
long hundred   120:great hundred
long primer:10
long six:6
look nine ways:9
lookout   3:three L's
lord chancellor   3:vice-chancellor
lord of cattle   8:Agni
Lord's Day   1:Sabbath
Lords Spiritual   3:Parliament
Lords Temporal   3:Parliament
Los Angeles   66:Route 66
Lost Sunday   70:Septuagesima
   Sunday
Lost Tribes of Israel:10
lotus lake   14:Trisala
Louisiana   50:United States of
   America
Louis IX:9
Louis XI   100:Hundred Years' War
Louis XIII   13:Louis Treize;
   14:Henry IV

Louis XIV:14
Louis XIV   14:Louis Quatorze
Louis XV   ⅕:cinquième; 15:Louis
    Quinze
Louis XVI   3:tricolor; 16:Louis
    Seize
Louis XVIII:18
Louis Napoleon:2
Louis Quatorze:14
Louis Quinze:15
Louis Seize:16
Louis Treize:13
love   9:fruits of the Holy Spirit;
    10:Sephiroth
Lovers   22:major arcana
love seat:2
love seat   2:tête-à-tête
loving one   5:Buddha, 2
low   1:first
Lowenburg   7:Siebengebirge
low gear   1:first
Low Sunday   8:octave, 1
Lucifera:6
Lucius Minucius   10:Wicked Ten
luck   7:Shichi Fukujin;
    9:possession
luck, good   4:four-leaf clover
Luck, Seven Gods of   7:Shichi
    Fukujin
Luke   4:Evangelists
Luks, George   8:Eight
luminous protector   5:Buddha, 2
Luna   3:Trivia
Lundy, Lamar   4:Fearsome
    Foursome
lung   2:dipneumonous, dipnoous
lurch:61
lust   7:sins
lustrum:5
lute   2:dichord; 3:Hermes
    Trismegistus
Luther, Martin   95:Theses
Luxembourg   12:European
    Economic Community
Lycophron   7:Pleiad, 1; 13:Tragic
    Poets
Lycurgus   10:Attic Orators
Lydgate   3:Triumvirate, 2
Lydian tetrachord:4
Lynceus   50:Danaides
Lynne, John   6:San Quentin Six

lyre   3:Hermes Trismegistus;
    5:kissar
Lyric Poets:9
lyric song   9:Muses
Lysander of Sparta   30:Thirty
    Tyrants
Lysias   10:Attic Orators

M

M:1000
M̄:1,000,000
M, m:10; :12; :13
M-1A:1
M-1B:1
M-2:2
M-3:3
Mabinogion   4:Four Branches of the
    Mabinogi
Mabinogion   12:Knights of the
    Round Table, 2
MacArthur, Douglas   5:five-star
    general
Macaulay, Thomas Babington
    10:Wicked Ten
Maccabees   7:menorah;
    8:Hanukkah
MacDonald, J. E. H.   7:Group of
    Seven, 2
Macheteros, los:13
Machiavelli   1500:Cinquecentesti
Madimial   7:menorah
Madison Square Garden   6:six-day
    bicycle race
Magalath   3:Magi
Magen David   6:hexagram, 4
Magi:3; :6
Magi, adoration of the   7:Joys of
    Mary
Magians   7:Hell
magic   3:Hermes Trismegistus
Magician   22:major arcana
Magna Carta   5:Darnel's Case
magnitude   0:zero-dimensional;
    1:one-dimensional; 2:two-
    dimensional; 3:three-dimensional;
    4:dimension
magpie:1
magpies:2; :3; :4; :5
Mahadeva   8:Agni

Mahasthamaprapta
8:Dhyanibodhisattva
Mahavira   14:Trisala
mahayugas:1000
Maia   7:Pleiades
Maid   12:sign
mail   1:first class mail; 2:second
class mail; 3:third class mail;
4:fourth class mail
Maimuna   10:wives
main body   5:cinquain
Maine   50:United States of America
Maitreya   5:Buddha, 2;
8:Dhyanibodhisattva
majesty   10:Sephiroth
major arcana:22
major arcana   78:tarot
make a double:2
make the pot with two ears:2
Malaysia   5:Five-Year Plan
Malchus   7:Seven Sleepers
Malebolge:10
Malebolge   8:Hell
Maleverer, Sir Henry   9:Worthies of
London
Malkuth   10:Sephiroth
Malory, Sir Thomas   12:Knights of
the Round Table, 1
Maltz, Albert   10:Hollywood Ten
Malvern Hill   7:Seven Days' Battles
Mama Bear   3:"Goldilocks and the
Three Bears"
man   1:Primates; 6:Creation;
10:avatar
man, depravity
of   7:Fundamentalism
Man, Isle of   3:triskelion; 24:House
of Keys
Manasseh   10:Lost Tribes of Israel
Manhattan:24
manille:2; :7
Manitoba   10:Ten Provinces
Manius Rabuleius   10:Wicked Ten
Manjusri   8:Dhyanibodhisattva
manque:Misc.
man's face   4:Evangelists
manslaughter   2:murder in the
second degree
Mantua   4:quadrilateral, 3
Manusibuddha:5
Manusibuddha   5:Buddha, 1

Manusibuddhas   8:Dhyanibodhi-
sattva
'man with gun'   10:ten-thirty-two
Marathon   15:fifteen decisive
battles
March:3
Marcus Aurelius   5:Adoptive
Emperors
Marcus Cornelius Maluginensis
10:Wicked Ten
Marcus Sergius   10:Wicked Ten
Margaret of Angoulème
7:heptameron
Marguerite Valois   14:Henry IV
Marhaus   12:Knights of the Round
Table, 1
Maries, Four   4:Four Marys
mariés de la tour Eiffel   6:Six
Marine Commandant   19:salute
Marini,
Giambattista   1600:Seicentisti
Mariyeh   15:concubines
Mark   4:Evangelists; 12:Knights of
the Round Table, 1
Mark Antony   2:Second
Triumvirate
marksmanship   2:biathlon
mark twain:2
Marlborough   15:fifteen decisive
battles
Marlborough, Duchess of
3:Triumvirate, 1
Marlborough, Duke of
3:Triumvirate, 1
marriage   1:monogamy; 2:bigamy,
digamy; 3:trigamist; 12:house
marriage, house of:7
Mars   7:alchemy, stars; 9:Etruscan
gods, Nine Heavens, planet
Marshall, George C.   5:five-star
general
Marshall, Thomas Riley   5:"What
this country needs . . . "
marsupial   2:diprotodont
Martel, Charles   15:fifteen decisive
battles
Martínez, Eugenio   7:Watergate
Seven
Martinian   7:Seven Sleepers
Marx Brothers:3
Mary   4:perfect women

Mary, Joys of   7: Joys of Mary
Mary, Queen   9: Worthies of London
Mary, Queen of Scots   4: Four Marys
Mary, Sorrows of   7: Sorrows of Mary
Maryland   13: original colonies; 50: United States of America; 100: hundred, 2
Mary Stuart   88: Stuarts
masculine number: Misc.
masculine rime: 1
masculine sign: Misc.
Masonry   10: Knight Templar; 32: thirty-second degree Mason; 33: thirty-third degree Mason
Mass   3: triennial, 4
Massachusetts   5: Intolerable Acts; 13: original colonies; 50: United States of America
masses   6: Six Articles
master mason: 3
master mason   3: third degree
matador: 3
match, three on a   3: three on a match
mathematical   10: tripos
matins   7: Canonical Hours
Matrimony   7: Holy Sacrament
Matsya   10: avatar
matter and body   3: aum
Matthew   4: Evangelists
Maugris   12: Paladins of Charlemagne
Maundy Thursday   6: Holy Week
Mausoleum   7: Wonders of the Ancient World
Maximian   7: Seven Sleepers
May: 5
May day: 1
McAlister, Elizabeth   6: Harrisburg Six
McAllister, Ward   400: Four Hundred, 2
McCarthy, Senator Joseph E.   5: Fifth-Amendment Communist
McClellan, George B.   7: Seven Days' Battles
McCord, James W.   7: Watergate Seven

McKoy, James   10: Wilmington Ten
McLaughlin, Neil   6: Harrisburg Six
McMurroughs   5: Five Bloods
Meath   5: Five Bloods
mechanical heart   7: Jarvik-7
Mechanicsville   7: Seven Days' Battles
Mechlima   7: Seven Sleepers
Medici   3: pawnbroker; 6: palle
medieval and modern languages   10: tripos
medio $\frac{1}{10}$: decimo, 2
meditation   3: Muses; 14: Trisala
meditative body   5: Buddha, 1
Mediterranean Sea   7: Seven Seas
medium quarto $\frac{1}{4}$: quarto
Medusa   3: Graeae
meekness   9: fruits of the Holy Spirit
meeting with His mother   14: stations of the cross
meg-   1,000,000: mega-
mega-: 1,000,000
Megaera   3: Furies
Mei Ch'ing   4: Four Masters of Anhwei
Meiji   5: Charter Oath
Meissen   16: Louis Seize
Meistersingers: 12
Mekchilinia   7: Seven Sleepers
Melchior   3: Magi
Melendez Carrión, Ivon   13: Macheteros
Melete   3: Muses
Melpomene   9: Muses
Melville, Marjorie B.   9: Catonsville Nine
Melville, Thomas   9: Catonsville Nine
mem: 13; :40
mem   22: major arcana
Memmingen   4: Tetrapolitan
Memorial day: 30
memory   5: five wits
Memory   7: *Seven Lamps of Architecture*
Ménage   7: Pleiad, 4
ménage à trois: 3
Menander   13: Comic Poets
Menelaus   10: Trojan War
menorah: 7; :9

mental age   100:I. Q.
Mercia   7:heptarchy, 2
Mercury   7:alchemy, stars; 9:Nine
   Heavens, planet
mercy   10:Sephiroth
mercy, corporal works
   of   7:corporal works of mercy
mercy, spiritual works
   of   7:spiritual works of mercy
Merlima   7:Seven Sleepers
Merlin   150:Round Table
Merope   7:Pleiades
Meshach   3:Shadrach
Mesopotamia   2:interamnian
Mesozoic   3:Tertiary, Triassic
Messiah, the   6:Magi
Messias   10:names of God
Messidor:10
metal   2:bimetallism; 5:Five Agents
Metaurus   15:fifteen decisive
   battles
Metcalf, Willard Leroy   10:Ten
meter   2:dyadic; 4:Glyconic
Methodist Episcopal Church
   25:Twenty-five Articles
metope, half $1/2$:demi-metope
Mexico   3,000,000:three-million
   bill
mezzo-rilievo:$1/2$
M. I. 1   1:M. O. 1; 6:M. I. 6
M. I. 3   1:M. O. 1
M. I. 5:5
M. I. 6:6
M. I. 9   1:M. O. 1
M. I. 19   1:M. O. 1
mice, three blind   3:"Three blind
   mice"
Michael   7:menorah
Michelangelo   1500:Cinquecentesti
Michigan   50:United States of
   America
Michigan, University of   10:Big
   Ten, 2
Michigan State University   10:Big
   Ten, 2
midangle:45
Middle Comedy   13:Comic Poets
middle path   4:Noble Truths
Middlesex   15:Danelagh
Mid-Lent:4
mid-morn:9

midnight:12
midsummer:21
midwinter:21
"Mighty Handful"   5:St. Petersburg
   Five
mil:$1/1000$
Milhaud, Darius   6:Six
Military Operations Directorate
   1:M. O. 1
mille:10
millenarian:1000
millenary:1000
millenary petition:1000
milleniad:1000
millennium:1000; :360,000
mille passus   1:milliarium
Miller, Don   4:Four Horsemen
millesimal:$1/1000$
milli-:$1/1000$; :1000
milliad   1000:milleniad
milliard:10^9
milliare:$1/1000$
milliarium:1
milliary:1000
millier:1000
million:1,000,000
'million'   1,000,000:mega-
millionaire:1,000,000
mill-ream:480
Mills, Donald   4:Mills Brothers
Mills, Harry   4:Mills Brothers
Mills, Herbert   4:Mills Brothers
Mills, John Jr.   4:Mills Brothers
Mills Brothers:4
milreis:1000
Miltiades   15:fifteen decisive
   battles
Milton   9:Gates of Hell
Mimamsa   6:Six Schools of Hindu
   Philosophy
Mimnermus   4:Elegiac Poets
mind, king of the   5:Five Great
   Kings
mindfulness, right   8:Noble
   Eightfold Path
mineral kingdom   3:kingdom
Minerva   9:Etruscan gods, Nine
   Worthy Women
Ming   3:three-color Ming ware
Ming dynasty   4:Four Masters of the
   Yüan Dynasty

minion: 7
Minnesota   50: United States of
   America
Minnesota, University of   10: Big
   Ten, 2
minor arcana   78: tarot
Minos   3: Judges of Hades; 7: seven
   youths; 8: Hell
Minotaur   7: seven youths;
   12: Dodekathlos
Minuit, Peter   24: Manhattan
Miocene   3: Tertiary
Miracles, Christ to work
   5: Fundamentalism
mirage   2: dietheroscope
Mirja   6: Magi
Mische, George   9: Catonsville Nine
Mississippi   50: United States of
   America
Missouri   50: United States of
   America
Missouri, University of   8: Big Eight,
   1
Mitra   3: aum
mittelhand: 2
Mneme   3: Muses
Mnemosyne   9: Muses; 12: Titan
M. O. 1: 1
M. O. 2   1: M. O. 1
M. O. 3   1: M. O. 1
M. O. 4   1: M. O. 1
M. O. 5   1: M. O. 1
Mobil Oil   7: Seven Sisters, 2
modern pentathlon   5: pentathlon
Mohawk   5: Five Nations
Moirae   3: Fates
moist   4: elemental quality
mollusk   1: univalve; 2: bivalve
monad   3: Pythagorean letter
Mon-bu-pu-tra   5: Five Great Kings
Monday: 2
Monday   2: Second-day
Monday-morning quarterbacking
   20: twenty-twenty hindsight
money   9: possession
Mongolians   16: Sixteen Kingdoms
monkey   1: Primates
Monmouth rebellion   40: Field of
   the Forty Footsteps
mono-: 1
monoceros   1: unicorn

monocerous: 1
monochroic   1: monochromatic
monochromatic: 1
monochrome: 1
monocle: 1
monocracy: 1
monocular   1: uniocular
monodactyl   1: unidactyl
monogamous   1: monogynous
monogamy: 1
monoglot: 1
monograph: 1
monogynous: 1
monolatry: 1
monolith: 1
monomachy: 2
mononym: 1
monophobia: 1
monophthong: 1
monorhine: 1
monosyllable: 1
Monotheletism   1: Monothelitism
Monothelism   1: Monothelitism
Monothelitism: 1
monoxylon: 1
monozygotic: 1
Montana   50: United States of
   America
Montauban, Guillaume de   30: Battle
   of the Thirty
Montenegro, restoration of
   14: Fourteen Points
Montfort, John of   30: Battle of the
   Thirty
moon   6: Creation; 7: alchemy,
   stars; 9: Nine Heavens
Moon   22: major arcana
Moor, Wayne   10: Wilmington Ten
Moore, Clement   8: Santa's reindeer
mora   2: dichronous; 9: enneasemic;
   10: decasemic; 12: dodecasemic
moral sciences   10: tripos
Mordred   12: Knights of the Round
   Table, 1
Morgan, J. P.   1,000,000: "Bet-a-
   million" Gates
Morning Prayer   7: Canonical Hours
Morris   5: Big Five Packers
mortal body   5: Buddha, 1
mortal sins   7: sins, seven deadly

287

Moses   2:Deuteronomy; 4:Fast of
the Fourth Month, perfect women;
5:Deuteronomy; 6:apostle, Cities
of Refuge; 7:Seventh Heaven;
10:animals in paradise,
Decalogue, Lost Tribes of Israel;
50:Pentecost, 1
Moses's law:39
Mosque of St. Sophia   7:Seven
Wonders of the Middle Ages
Motassem   5:Alkoremmi
Motion, Newton's Laws of
1:Newton's First Law of Motion;
2:Newton's Second Law of Motion;
3:Newton's Third Law of Motion
Mount Holyoke   7:Seven Sisters, 1
mouth   2:distomatous
Moylan, Mary   9:Catonsville Nine
mu:12
Mueglen, Heinrich von
12:Meistersingers
Mugello   3:pawnbroker
Muhammad   6:apostle; 7:Seventh
Heaven; 10:animals in paradise,
wives; 15:concubines
Muhammad, daughter of   4:perfect
women
Muhammad, first wife of   4:perfect
women
Muhammad's concubines
15:concubines
Muhammad's wives   10:wives
multiplication   2:duplation; 4:four
species, 1
muni:7
Murder Incorporated   20:Roaring
Twenties
murder in the first degree:1
murder in the second degree:2
murder one   1:murder in the first
degree
murder two   2:murder in the
second degree
Murner, Thomas 12:Meistersingers
Murray, Robert   1:penny post
Murrow, Edward R.   8:Murrow's
Boys
Murrow's Boys:8
Muscablüt, Master 12:Meistersingers
Muses:3; :9

music   3:Muses; 4:quadrivium;
7:seven arts; 9:Muses;
12:Meistersingers
Muslims   7:Hell
Mussorgsky, Modest P.   5:St.
Petersburg Five
Mutsuhito   5:Charter Oath
Myers, Jerome   8:Eight
myriad:10,000
myriarch:10,000
myrrh   4:diatessaron
Myson of Chen   7:Seven Wise Men
mystery   3:three, 2; 150:psalter, 1

N

N:90
N̄:90,000
N, n:11; :13; :14
name   1:mononym; 2:dionym;
3:trionym
name, Christian   1:first name
name, given   1:first name
name, scientific   1:mononym,
uninominal; 2:binomial, dionym;
3:trionym
name relation   2:disquiparancy
names of God:7; :10
Namo   12:Paladins of Charlemagne
Nanking   7:Seven Wonders of the
Middle Ages
Naphtali   10:Lost Tribes of Israel
Naples, kingdom of   2:Sicilies
Napoleon:2
Napoleon   1:First Empire, War of
the First Coalition; 2:War of the
Second Coalition; 3:Three
Emperors, War of the Third
Coalition; 15:fifteen decisive
battles; 100:Hundred Days, 1
Napoleon III   3:Third Republic
Narasinha   10:avatar
nastyman   3:ugly man
Nativity   7:Joys of Mary
natural:7; :11
natural inability   5:Five Articles and
the Five Points
natural sciences   10:tripos
Nautilus   20,000: *Twenty Thousand
Leagues Under the Sea*

Navarre   14:Henry IV
navigation, freedom of 14:Fourteen
   Points
Nayme de Bavière   12:Paladins of
   Charlemagne
Nazi   3:Third Reich
Nebraska   50:United States of
   America
Nebraska, University of   8:Big
   Eight, 1
Nebuchadnezzar   3:Shadrach;
   10:Fast of the Tenth Month
nectar of the soul   5:Alkoremmi
Needles:3
Nemean Lion   12:Dodekathlos
Nemo, Captain   20,000:*Twenty
   Thousand Leagues Under the Sea*
Neoclassicism   16:Louis Seize
neodymium   2:didymium
Neo-Platonism   9:ennead
neosalvarsan:914
Neptune   9:planet
Nereid:50
Nereus   50:Nereid
Nerva   5:Adoptive Emperors
nerve   2:diploneural
Netherlands   3:Triple Alliance, 1, 2;
   12:European Economic
   Community; 18:Treaty of the
   Eighteen Articles; 24:Treaty of the
   Twenty-four Articles; 80:Eighty
   Years' War
Netsah   10:Sephiroth
neuron   2:diaxon, 3
neuter:3
Nevada   50:United States of
   America
new birth for salvation
   7:Fundamentalism
New Brunswick   10:Ten Provinces
New Comedy   13:Comic Poets
Newfoundland   10:Ten Provinces
New Hampshire   13:original
   colonies; 50:United States of
   America
New Jersey   5:Fort Dix Five;
   13:original colonies; 24:Twenty-
   four Proprietors; 50:United States
   of America
Newmarket   1000:Guineas;
   2000:Guineas

New Mexico   50:United States of
   America
New Netherlands   24:Manhattan
new threes:3
Newton's First Law of Motion:1
Newton's Second Law of
   Motion:2
Newton's Third Law of Motion:3
New-Year's day:1
New York   13:original colonies;
   50:United States of America
New York City   24:Manhattan
New York Eight:8
New York Eight   8:Grand Jury
   Eight, New York Eight Plus
New York Eight Plus:8
New York Nine   8:New York Eight
   Plus
Ney   100:Hundred Days, 1
Niagara Conference
   7:Fundamentalism
Niagara University   3:Little Three
nibble:4
Nicander   7:Pleiad, 1
Nicanor   7:seven men of good
   repute
Nicene council   2:deutero-Nicene
Nicene Creed   2:Filioque;
   3:Christian creeds
Nicholas I   3:Third Department
nickel:$\frac{1}{20}$
nickel-and-dime (someone) to
   death:5
Nicolaus   7:seven men of good
   repute
nictitating membrane:3
Nidhug   3:Yggdrasil
Niflheim:12
Niflheim   3:Yggdrasil
Nigeria   13:OPEC
nil:0
Nimitz, Chester W.   5:five-star
   admiral
Niña   3:Columbus
nine:9
'nine'   9:ennea-
nine-banded armadillo:9
"nine days old":9
Nine Days' Wonder:9
Nine Elms:9
nine-eyed:9

ninefold:9
Nine Heavens:9
nine-holes:9
nine-killer:9
nine-lived:9
nine men's merels   9:nine men's morris
nine men's morris:9
nine men's morris   5:fivepenny morris
nine-murder   9:nine-killer
nine old men   9:Supreme Court
ninepegs   9:ninepins
nine-penny morris   9:nine men's morris
ninepins:9
nines, casting out   9:casting out nines
nines, rule of   9:rule of nines
nineteenth hole:19
nineteen to the dozen:19
Ninety Days' Wonder:90
Nine Worthies:9
nineworthiness:9
Nine Worthy Women:9
ninth cranial nerve
   9:glossopharyngeal nerve
ninth part of a man:9
Niord   12:Asir
Nirmanakaya   5:Buddha, 1
Nisiros   12:Dodecanese
Nissan   14:Quartodeciman;
   50:Pentecost, 1
Ni Tsan   4:Four Masters of the Yüan Dynasty
Nivôse:14
Niza, Fray Marcos de   7:Cibola, Seven Golden Cities of
Noah   1:first covenant; 6:apostle;
   7:Seventh Heaven
nobility   4:Fourth Estate
Noble Eightfold Path:8
Noble Eightfold Path   4:Noble Truths
Noble Truths:4
nombril   9:point
nonage:⅑
nonagenarian:90
nonagon:9
'noncommittal'   ½:half-baked, 2
None   7:Canonical Hours

nones:9
nonillion:10³⁰; :10⁵⁴
Nonnen-
   Stromberg   7:Siebengebirge
nonpareil:6
nonuple:9
noon:9; :12
noon   6:sixth hour
Norfolk   7:Associated Counties;
   15:Danelagh
Norn:3
north   4:cardinal point
Northamptonshire   15:Danelagh
North Carolina   13:original colonies; 50:United States of America
North Church   1:"One if by land, and two if by sea, . . ."
North Dakota   50:United States of America
'northern'   7:septentrional
Northern Ireland   6:Six Counties
Northern Sea   4:four seas
North Riding   ⅓:riding
North Sea   4:four seas; 7:Seven Seas
Northumbria   7:heptarchy, 2;
   15:Danelagh
North West Company   7:Seven Oaks Massacre
Northwestern University   10:Big Ten, 2
Northwest Territories   10:Ten Provinces
Norway   6:Six Islands
nothing:0
Notre Dame backfield   4:Four Horsemen
Nottinghamshire   15:Danelagh
not touch (something *or* someone) with a ten-foot pole:10
no two ways about it:2
nouveaux jeunes   6:Six
Nova Scotia   10:Ten Provinces
November:9; :11
novemdecillion:10⁶⁰; :10¹¹⁴
novena:9
novenary:9
novene:9
novennial:9

oculomotor nerve:3
Ocypete   2:Harpies; 3:Harpies
'odd'   3:queer as a three-dollar bill
odd trick:7
Odin   3:Thor; 9:Valkyr; 12:Asir
Odysseus:20
Odysseus   1:Polyphemus
Odyssey   3:Cyclopes
Oedipus   7:Seven Against Thebes
Oeneus   7:Seven Against Thebes
'officer needs assistance'   10:ten-
   seventy-eight
ogdoad:8
Ogier, the Dane:7
Ogier the Dane   12:Paladins of
   Charlemagne
Ohio   50:United States of America
Ohio State University   10:Big Ten, 2
Ojeda Ríos,
   Filiberto   13:Macheteros
'okay'   10:ten-four
O'Keefe, John   40:Fat, fair and
   forty
Oklahoma   50:United States of
   America
Oklahoma, University of   8:Big
   Eight, 1
Oklahoma State University   8:Big
   Eight, 1
O'Lachlans   5:Five Bloods
Olaf Redbeard:7
Ölberg   7:Siebengebirge
Old Bailey   5:"five farthings"
Old Comedy   13:Comic Poets
Old Hundred:100
Old Hundredth   100:Old Hundred
"Old King Cole"   3:"fiddlers
   three"
Old Man of the Sea   7:"Sindbad the
   Sailor, Seven Voyages of"
"Old Pretender"   15:fifteen
Old Testament   5:Pentateuch;
   6:Hexapla, Hexateuch;
   50:Pentecost, 1;   150:Psalms
olfactory nerve:1
Oligocene   3:Tertiary
Oliver   12:Paladins of Charlemagne
Olsen, Merlin   4:Fearsome
   Foursome
Olympia   7:Wonders of the Ancient
   World

Olympiad:4
Olympian games   5:quinquennalis
Olympians:12
Olympus, Mt.   12:Olympians
Olynthiac orations:3
Olynthus   3:Olynthiac orations
omber   40:ombre
ombre:40
omega:24
Omega   10:names of God
omicron:15
on all fours:4
once:1
once bitten, twice shy:2
once in a blue moon:1
oncia $1/10$:decimo
one:1
'one'   1:mono-
"One, two, Buckle my
   shoe . . .,":1
one, two, three a-leery:1
one-a-cat:1
one-alarm fire:1
one-armed bandit:1
one-cent sale:1
one-dimensional:1
one-eighty, do a:180
one for the road:1
one-horse town:1
one hundred percent:100
Oneida   5:Five Nations
"One if by land, and two if by
   sea, . . .":1
O'Neill, Norman   5:Frankfurt Five
"One I love . . .,":1
O'Neils   5:Five Bloods
one-man dog:1
oneness:1
one-night stand:1
one-o'-cat   1:one-a-cat
one old cat   1:one-a-cat
oneself   1:number one
one-twenty in the water bag:120
one-upmanship:1
"On Homer"   7:Homer
only:1
Onondaga   5:Five Nations
Ontario   10:Ten Provinces
on the double:2
OPEC:13
operaglasses   2:jumelle

operational intelligence   1:M. O. 1
Opinion, Festival of   5:Sans-
culottides
optic nerve:2
orange   3:secondary color;
7:perfect color
oranges and lemons . . .   5:"five
farthings"
ordinary regular
dodecahedron:12
ordinary year:12
Oregon   50:United States of
America
Oregon, University of   10:Pacific
Ten
Oregon State University   10:Pacific
Ten
Oregon Territory   54:"Fifty-four
Forty or Fight"
Organization of Petroleum Exporting
Countries 13:OPEC
Oriental Tales   7:Seven Sleepers
Origen   6:Hexapla
original colonies:13
Orkneys   6:Six Islands
Orlando   12:Paladins of
Charlemagne
Ornitz, Samuel   10:Hollywood Ten
orphans, visit the   7:corporal
works of mercy
orthron   6:hexapsalmus
O's of St. Bridget:15
other:2
Ottoman Empire, autonomy in
14:Fourteen Points
Ottoman Turks   15:Fifteen Years'
War
Otuel   12:Paladins of Charlemagne
ounce   ¹/₁₀:decimo
Overton churchyard   7:Seven
Wonders of Wales
Owain   12:Knights of the Round
Table, 2
Owen, David   4:Limehouse Four
"Owl and the Pussycat"   5:"five-
pound note"
ox   4:Evangelists; 10:animals in
paradise
oyster   2:bivalve

P

P:400
P̄:400,000
P, p:12; :15; :16
Pacific Eight   10:Pacific Ten
Pacific Ocean   7:Seven Seas
Pacific Ten:10
pack   52:playing cards
Padmaheruka   5:Herukabuddha
paeon diagyíos:2
paeon epibatus:2
paeon(ic foot)   2:diagyíos, paeon
diagyós, paeon epibatus
Paginet   12:Knights of the Round
Table, 1
Pahad Geburah   10:Sephiroth
Pain, Noble Truth of   4:Noble
Truths
Pain, Noble Truth of the Cause of
4:Noble Truths
Pain, Noble Truth of the Cessation of
4:Noble Truths
pair:2; :3
pairial   3:pair royal
pair of colors:2
pair royal:3
'pairs, in'   2:dicho-
pair-toed   2:zygodactyl
palace of perfumes   5:Alkoremmi
Paladins of Charlemagne:12
Palais de Justice   13:Louis Treize
Palais du Luxembourg   13:Louis
Treize
Palatine   7:Rome
palle:6
Palm Sunday:1; :6
Palomides   12:Knights of the
Round Table, 1
pane   4:quarrel
pantalon   4:quadrille, 1
Panyasis   5:Epic Poets
Papa Bear   3:"Goldilocks and the
Three Bears"
paper   24:quire
paper, handmade   144:post, white
post; 480:mill-ream
Papias   6:apostolic fathers
pappataci   3:phlebotomous fever

Paquin, Leo   7: seven blocks of granite
paragon:20
parakeet, Australian   28: twenty-eight
paralysis   2: diplegia
Parashurama   10: avatar
Parcae   3: Fates
parcel post   4: fourth-class mail
parents   2: duoparental; 12: house
parents, house of:4
parial   3: pair royal
Paris   3: tricolor; 7: Seven Cities; 10: Trojan War
Parish, Milton   8: Grand Jury Eight
Paris Peace Conference   3: Big Three, 1;   10: Big Ten, 1
Parker, Charles   3: Charlotte Three
Parliament:3
Parliament   3: Triennial Act; 5: Intolerable Acts; 7: Septennial Act
Parmenas   7: seven men of good repute
Parnassus, Mount   9: Deucalion
Parousia   2: Second Coming
part   2: bifarious, bifid, dimerous; 3: trifid, triformed, 1; 4: quadrigeminous, quadripartite, 1,   quadrisulcate, tetractomy; 5: pentamerous, quinary, quinquepartite; 6: hexamerous; 7: heptameride, septempartite; 8: octamerous, octoad, 1
parthenogenesis   2: digenesis
Parthenon   100: hecatompedon
Parthenopaeus   7: Seven Against Thebes
Parthenope   3: Siren
particular predestination   5: Five Articles and the Five Points
Particulars, Five   5: Five Particulars
partition   4: quadriseptate; 5: quinqueseptate
passe:Misc
Passion of Christ   14: stations of the cross; 15: O's of St. Bridget
Passion Sunday:2; :5
Passion Week:5

Passover   4: four questions; 14: Quartodeciman; 50: Pentecost, 1
past   3: Norn
Pastor   10: names of God
pastourelle   4: quadrille, 2
Pasupati   8: Agni
pater   10: decad ring
Path that Leads to the Cessation of Pain, Noble Truth of the   4: Noble Truths; 8: Noble Eightfold Path
patience   9: fruits of the Holy Spirit, possession
Patmos   12: Dodecanese
patriarch:3
Patrick, Marvin   10: Wilmington Ten
"Paul Revere's Ride"   18: "eighteenth of April,.. ."
paving stone   4: quarrel
pawnbroker:3
Paxos   7: Ionian Islands
pe:17; :80
pe   22: major arcana
peace   9: fruits of the Holy Spirit
peace of Nystad   21: Twenty-one Years' War
pearl:5
Pearl Harbor day:7
Pease porridge hot   9: "nine days old"
Peat, Marwick, Mitchell   8: Big Eight, 2
'peculiar'   3: queer as a three-dollar bill
Pe-har   5: Five Great Kings
Pei Chêng   6: Six Idlers of the Bamboo Streams
Pella   10: Decapolis
Pelleas   12: Knights of the Round Table, 1
Peloponnesian War   30: Thirty Years' Truce, Thirty Years' War, 1
Penance   7: Holy Sacrament
pendactyl   5: pentadactyl
pendecagon   15: quindecagon
Peninsular Campaign   7: Seven Days' Battles
penitence   5: prayer

Pennsylvania   13:original colonies;
   50:United States of America;
   100:hundred, 2
Pennsylvania, University of   5:Big
   Five, 2;   8:Ivy League
penny:$^1/_{100}$
penny $^1/_4$:farthing
penny post:1
penta-:5
pentachord:5
pentacle:5
pentact:5
pentactinal   5:pentact
pentad:5
pentadactyl:5
pentadactylous   5:pentadactyl
pentagon:5
Pentagon:5
pentagonal dodecahedron:12
pentagram   5:pentacle
pentahedron:5
pentalpha   5:pentagon
pentamerous:5
pentameter:5
pentangle   5:pentacle
Pentapolis:5
pentarchy:5
pentastichous   5:quinquefarious
pentasyllabic:5
Pentateuch:5
Pentateuch   2:Deuteronomy;
   5:Deuteronomy
pentathlon:5
penteconter:50
Pentecost:50
Pentecost   4:Ember Days;   7:church
   year, seven men of good repute
pentecoster:50
pentecostys:50
penteteric:5
penuchle   52:playing cards
Pephredo   3:Graeae
Peratoras   3:Magi
per cent:100
per centum   100:per cent
Percival   12:Knights of the Round
   Table, 1
'perfect'   100:one hundred percent
perfect "10," a:10
perfect color:7

Perfection, Tantra of   14:Tantras
Perfection, Tantra of Supreme
   14:Tantras
perfect knowledge   14:Trisala
perfect women:4
Periander   7:Seven Sages of Ancient
   Greece
perianth   2:dichlamydeous
Pericles   30:Thirty Years' Truce
periodical cicada   17:seventeen-
   year locust
Permian   3:Triassic
permillage:1000
Persephone   3:Hecate
Perseus   3:Graeae
perseverance of saints   5:Five
   Articles and the Five Points
Persian Sibyl   10:sibyl
'personally'   1:at first hand
Peschiera   4:quadrilateral, 3
peso   100:centavo
peso duro   8:piece of eight
pestilence   4:Four Horsemen of the
   Apocalypse
petal   2:dipetalous;
   6:hexapetalous; 7:heptapetalous;
   9:enneapetalous; 10:decapetalous;
   12:dodecapetalous
Petersberg   7:Siebengebirge
Peter the Great   15:fifteen decisive
   battles; 21:Twenty-one Years' War
Petit   7:Pleiad, 4
petition   5:prayer, litany
petit jury:12
Petrarch   3:Triumvirate, 3;
   6:sestina
Petrarchan sonnet   14:sonnet
Pharaoh   7:seven years of plenty
Pharaoh, wife of   4:perfect women
Pharos   7:Wonders of the Ancient
   World
Pherecrates   13:Comic Poets
phi:21
Phidias   7:Wonders of the Ancient
   World
Philadelphia   10:Decapolis
Philemon   13:Comic Poets
Philetas   4:Elegiac Poets
Philip   7:seven men of good repute
Philip Augustus $^1/_{10}$:Saladin's tenth

Philip of Macedon   3:Third
  Philippic
Philippic   3:Third Philippic
Philippics   3:Olynthiac orations
Philippides   13:Comic Poets
Philiscus of Corcyra   13:Tragic
  Poets
Philistines:5
Philosophers:5
phlebotomous fever:3
Phlegethon   5:rivers of Hell
phlogiston   3:earths
Phoebe   12:Titan
Phorcydes   3:Graeae
Phorcys   3:Graeae
Phrygian Sibyl   10:sibyl
pi:3.1415927; :16
pica:12
pice:¼
pickpocket   42:Forty-Second Street
  thief
picnostyle:1½
piece of eight:8
pieces of silver:30
Pierce, Nat   7:seven blocks of
  granite
Pillar of Judgment   10:Sephiroth
Pillars of Hercules:2
pillow lace   8:octagon loop
Pindar   9:Lyric Poets
pinochle deck:48
pinochle deck   52:playing cards
Pinta   3:Columbus
piquet:32
Pisa   7:Seven Wonders of the
  Middle Ages
Pisa, Leonardo da   10:decimal
  system
Pisander   5:Epic Poets
Pisano, Niccola   1200:due-cento
Pisces   ⅓:trigon; 2:double-bodied;
  12:sign
Pishon   4:rivers of Eden
pistole   2:doubloon; 4:double
  doublon
pitch   4:tetrapharmacon
Pittacus   7:Seven Sages of Ancient
  Greece, Seven Wise Men
planet:9
planet   10:decanate

planetoid   3:June; 4:Vesta;
  10:Hygeia; 16:Psyche; 23:Thalia
Plantagenet   100:Hundred Years'
  War
plate   4:quadrilaminate
Plato   5:Philosophers; 7:Seven
  Wise Men; 13:Comic Poets
Platonic solid:5
playing cards:52
play second fiddle:2
plead the Fifth:5
pleasantness-unpleasantness
  3:tridimensional theory
Pleiad:7
Pleiades:7
Pleiad of Alexandria   7:Pleiad, 1
Pleiad of Charlemagne   7:Pleiad, 2
Pleiads, the   7:Pleiades, the
Pliny   9:Amalthea
Pliocene   3:Tertiary
Ploërmel   30:Battle of the Thirty
Plotinus   9:ennead, 3
Plummer, Viola   8:New York Eight
plus fours:4
Pluto   3:Cyclopes; 9:planet
Pluto, daughter of   6:Lucifera
Plutus   8:Hell
Pluviôse:5
pneuma   3:trichotomy
Pnyx   6:lexiarchus
Poetic Pleiades   7:Pleiad, 1
point:1/72; :0.0376; :9; :11¼
point   10:decemdentate
point of the compass:32
Points, Fourteen   14:Fourteen
  Points
points of a good horse:15
Poland   9:Nine Worthies;
  13:Thirteen Years' War
Poland, independence of
  14:Fourteen Points
Polk, James K.   54:"Fifty-four Forty
  or Fight"
poll tax:24
Pollux   2:Gemini, Twins
Polycarp   6:apostolic fathers
Polydeuces   2:Twins
Polyhymnia   9:Muses
Polymnia   9:Muses
Polyneices   7:Seven Against Thebes
Polyphemus:1

Polyphemus   3:**Cyclopes**
Pompey   1:**First Triumvirate;**
   10:**Decapolis**
Ponthus de Thiard   7:**Pleiad, 3**
Pontine marshes   19:**Decennovium**
pontoon-boat   2:**demi-bateau**
pony:25
Poor Clares   2:**Second Order of St.**
   Francis
Poor Ladies   2:**Second Order of St.**
   Francis
Pope   22:**major arcana**
Porcelain Tower   7:**Seven Wonders**
   of the Middle Ages
porta principalis dextra   4:**decuman**
porta principalis
   sinistra   4:**decuman**
Porthos   3:*Three Musketeers*
portico   4:**tetrastyle;** 8:**octastyle;**
   10:**decastyle;** 12:**dodecastyle**
portrait   ¾:**three-quarter**
Portugal   12:**European Economic**
   Community
Poseideon:6
Poseidon   3:**Cyclopes;**
   12:**Olympians**
possession:9
post:144
postal service   1:**penny post**
Potomac, Army of the   7:**Seven**
   Days' Battles
Potter, Stephen   1:**one-upmanship**
poule   4:**quadrille, 2**
Poulenc, Francis   6:**Six**
pound/shilling/penny system
   10:**decimalization**
Poussin, Nicolas   14:**Louis Quatorze**
Power   7:*Seven Lamps of Architecture*
power, infallible   5:**Dhyanibuddha**
powers   9:**angels**
poyou:6
Prairial:9
praise and adoration   5:**prayer**
Prancer   8:**Santa's reindeer**
praseodymium   2:**didymium**
prayer:5
Prayer, Apostolic Hours of
   3:**Apostolic Hours of Prayer**
prayers   5:**litany**
pray for salvation of neighbor
   7:**spiritual works of mercy**

precious birth   5:**Dhyanibuddha**
Precious Stones, Nine:9
pregnancy, first   1:**primigravida**
Premier or Prime Minister of a
   Foreign Country   19:**salute**
Prendergast, Maurice   8:**Eight**
present   3:**Norn**
presentation in the temple   7:**Joys**
   of Mary
preservation of the Bible
   5:**Fundamentalism;**
   7:**Fundamentalism**
President, ex-President, President-
   elect   21:**salute**
President of the Senate pro tempore
   19:**salute**
press   4:**Fourth Estate**
'pretender'   4:**fourflusher**
pretorian   4:**decuman**
prial   3:**pair royal**
Priam   10:**Trojan War**
Price, Waterhouse   8:**Big Eight, 2**
pricket's sister:2
pride   7:**sins**
prima ballerina:1
prima buffa:1
primacy:1
prima donna:1
primal:1
Primalia:3
Primalia   3:**Animalia, Vegetabilia**
primal scream:1
primal urge:1
primary:1
primary   2:**secondary education;**
   10:**decempennate**
primary color:3
primary deviation:1
primary education:1
primary planet:9
Primates:1
prime:1
prime   7:**Canonical Hours**
primigenial:1
primigravida:1
primipara:1
primo:1
primogenial   1:**primigenial**
primogeniture:1
prince   3:**diet**

Q

quadric:2
quadric   2:quantic
quadricentennial:400
quadriceps:4
quadricorn:4
quadricostate:4
quadricrescentic:4
quadricrescentoid
    4:quadricrescentic
quadricuspidate:4
quadricycle:4
quadridigitate:4
quadrifarious:4
quadrifid:4
quadrifoliate:4
quadrifrons:4
quadriga:4
quadrigemina:4
quadrigeminous:4
quadrigenarious:400
quadrilaminar   4:quadrilaminate
quadrilaminate:4
quadrilateral:4
quadrilingual:4
quadriliteral:4
quadrille:4
quadrillion:10^{15}; :10^{24}
quadrimanous   4:quadrumanous
quadrimembral:4
quadrinomial:4
quadrinomical   4:quadrinomial
quadrinominal   4:quadrinomial
quadripara:4
quadriparous:4
quadripartite:4
Quadripartite of Ptolemy
    4:quadripartite, 2
quadripennate:4
quadriphyllous:4
quadrireme:4
quadrisacramentalist
    4:quadrisacramentarian
quadrisacramentarian:4
quadriseptate:4
quadriserial:4
quadrisulcate:4
quadrisyllabic:4
quadrivalent   4:tetravalent
quadrivium:4
quadrivium   7:seven arts
quadroon:¼

quadrumane:4
quadrumanous:4
quadruped:4
quadruple:4
quadruplet:4
quadruplex:4
quadruplicate:4
Quakers   24:Twenty-four
    Proprietors
quantic:2
quarantine:40
quarrel:4
quart:¼; :4
quartan:4
quartan fever:4
quartan fever   4:quartan
quarte:4
quarter:¼
quarter-horse:¼
quarterly:4
quarterly waiter   ¼:quarter-waiter
quarter-point:¼
quarter-section:160
quarter-waiter:¼
quarter-watch:½
quartet:4
quartette   4:quartet
quartic:4
quartic   2:quantic
quartic equation:4
quartile:90
quartile aspect   4:tetragon, 2
quarto:¼; :4
Quartodeciman:14
quater-centenary:400
Quaternary   3:Tertiary
quaternary number:10
quaternary number   10:decade, 2
quaternion:4
Quaternity:4
quatorzain:14
quatorze:14
quatrain:4
quatre:4
quatrefoil:4
quattrocentist:1400
quattrocento:1400
quattuordecillion:10^{45}; :10^{84}
quaver   ⅛:fusa, eighth-note
Quebec   10:Ten Provinces
Quebec Act   5:Intolerable Acts

queen   78:tarot; 100:honors;
   150:honors
Queen Anne's bounty:10
queer as a three-dollar bill:3
quicksilver   7:alchemy
Quilter, Roger   5:Frankfurt Five
quinarius:5
quinary:5
quince:15
quincentenary:500
quincunx:5
quindecad:15
quindecagon:15
quindecemvir:15
quindecemvirs   2:duumviri;
   10:decemvir
quindecennial:15
quindecillion:10^{48}; :10^{90}
quinquagenarian:50
quinquagesima:50
quinquagesimal:50
Quinquagesima Sunday
   50:quinquagesima
quinquangular:5
Quinquarticular controversy   5:Five
   Articles and the Five Points
quinque-:5
quinquedigitate   5:pentadactyl
quinquefarious:5
quinquefid:5
quinquefoliate:5
quinquefoliated   5:quinquefoliate
quinquelateral:5
quinqueliteral:5
quinquennalis:5
quinquenniad:5
quinquennial:5
quinquennium,   5:quinquenniad
quinquepartite:5
quinqueradial   5:quinqueradiate
quinqueradiate:5
quinquereme:5
quinquesect:5
quinqueseptate:5
quinqueserial:5
quinquesyllabic:5
quinquevalent:5
quinquevir:5
quint:5
quintain:5
quintal:100

quintan:5
quintana fever   5:trench fever
quintan fever:5
quinte:5
quintessence:5
quintet:5
quintette   5:quintet
quintic:5
quintic   2:quantic
quintic equation:5
quintile:$\frac{1}{5}$
quintillion:10^{18}; :10^{30}
quintroon:$\frac{1}{8}$
quintuple:5
quintuplet:5
quintuplicate:5
Quintus Fabius
   Vibulanus   10:Wicked Ten
Quintus Poetelius   10:Wicked Ten
quinzain:15
quinzaine   15:quinzain
quire:24
Quirinal   7:Rome
quizzing-glass   1:monocle
quorum:40
quorum of Twelve:12

R

R:80
R̄:80,000
R, r:14; :17; :18
racial segregation   14:Fourteenth
   Amendment . . .
rack   5:five-gaited
Radcliffe   7:Seven Sisters, 1
radiocarbon dating   14:carbon-14
Radowitz, General Joseph Maria von
   3:Alliance of the Three Kings
rain   40:St. Swithin's day
rainbow   1:first covenant
ram   10:animals in paradise
Ram   12:sign
Rama   10:avatar
Ramachandra   10:avatar
Rama with an ax   10:avatar
Ramírez Talavera, Norman
   13:Macheteros
Ramoth   6:Cities of Refuge

Raphael   7:menorah;
   1500:Cinquecentesti
'rapid'   3:Harpies
Rapin   7:Pleiad, 4
'rarely'   1:once in a blue moon
Ras al-Khaimah   7:United Arab
   Emirates
Ratnaheruka   5:Herukabuddha
Ratnapani   5:Manusibuddha
Ratnasambhava   5:Dhyanibuddha,
   Herukabuddha
ray   3:triactinal; 5:pentact,
   quinqueradiate; 6:hexactinal,
   hexaster; 10:decapod
real   8:dollar, piece of eight
real vellon   ¹/₁₀:decima
ream:500
ream of insides   480:mill-ream
reception of the cross   14:stations
   of the cross
rectangle:4
red   3:primary color, trichromatic;
   7:perfect color
Reddy, Thomas J.   3:Charlotte
   Three
Red Hot Video   5:Vancouver Five
reduced threes   3:new threes
Refuge, Threefold   3:Threefold
   Refuge
Regenbogen, Master Barthel
   12:Meistersingers
Rehana   10:wives
rei   1000:milreis
Reich   1:First Reich; 2:Second
   Reich; 3:Third Reich
Reid, Robert   10:Ten
relapsing fever:7
religion   7:sacred books; 12:house
religion and learning, house of:9
remembrance   3:Muses
Remi-Belleau   7:Pleiad, 3
Repeal:21
repetition   2:dilogy, diplasiasmus
reserve   5:cinquain
resh:20
resh   22:major arcana
resin   4:tetrapharmacon
Resurrection   5:Fundamentalism
retreat of joy   5:Alkoremmi
Reuben   10:Lost Tribes of Israel

Revere, Paul   1:"One if by land,
   and two if by sea, . . .";
   18:"eighteenth of April, . . ."
Rewards, Festival of   5:Sans-
   culottides
Rhadamanthus   3:Judges of Hades
Rhea   3:Hecate; 12:Titan
rhetoric   2:epanadiplosis;
   3:trivium; 7:seven arts
rho:17
Rhode Island   13:original colonies;
   50:United States of America
Rhodes   12:Dodecanese
Rhodes, Colossus of   7:Wonders of
   the Ancient World
Rhodos   7:Homer
rhombic dodecahedron:12
rhombicosidodecahedron:62
rhombicuboctahedron:26
rhombohedron:6
rib   4:quadricostate;
   10:decemcostate
Rice, Grantland   4:Four Horsemen
Richard I   ¹/₁₀:Saladin's tenth
Richard II   2:unlucky number
Richard, duc de Normandy
   12:Paladins of Charlemagne
riches   12:house
riches, house of:2
Richmond   7:Seven Days' Battles
Riculfe   7:Pleiad, 2
riddle of claret:13
riding:¹/₃
rig   2:double-topsail
right action   8:Noble Eightfold Path
right concentration   8:Noble
   Eightfold Path
right conduct   14:Trisala
right effort   8:Noble Eightfold Path
right faith   14:Trisala
right knowledge   14:Trisala
right livelihood   8:Noble Eightfold
   Path
right mindfulness   8:Noble
   Eightfold Path
Rights, Bill of   10:Bill of Rights
right speech   8:Noble Eightfold
   Path
right thought   8:Noble Eightfold
   Path
right view   8:Noble Eightfold Path

rime-royal:7
Rimsky-Korsakov, Nikolay A. 5:St.
  Petersburg Five
Rinaldo 12:Paladins of
  Charlemagne
Riol du Mans 12:Paladins of
  Charlemagne
Ríos, Pepe 8:New York Eight Plus
'riot' 10:ten-thirty four
Rip Van Winkle:20
rishi 7:muni
Ritz, Al 3:Ritz Brothers
Ritz, Harry 3:Ritz Brothers
Ritz, Jim 3:Ritz Brothers
Ritz Brothers:3
river-lamprey 7:seven holes
rivers of Eden:4
rivers of Hell:5
roarer 8:Agni
Roaring Forties:40
Roaring Twenties:20
Robin 2:Dynamic Duo
roc 7:"Sindbad the Sailor, Seven
  Voyages of"
Rockefeller, John "Dime":10
Rockne, Knute 4:Four Horsemen
Rock of Gibraltar 2:Pillars of
  Hercules
Rococo 15:Louis Quinze; 16:Louis
  Seize; 1600:Seicentisti
Rodgers, Richard 10:Tenth Avenue
Rodgers, William 4:Limehouse
  Four
Rodrigues-Torres, Carlos 5:Fort
  Dix Five
Rogation days:3
Roland 12:Paladins of
  Charlemagne
romanesque influence 1200:due-
  cento
Romania, restoration
  of 14:Fourteen Points
Rome:7
Rome 7:Seven Cities, Seven
  Wonders of the Middle Ages
Romney 5:Cinque Ports
Romulus 9:Sabines; 100:centuria
Ronsard 7:Pleiad, 3
Roosevelt, Franklin Delano 3:Big
  Three, 2; 4:Four Freedoms;
  100:hundred days, 2

rope-jumping 2:double Dutch
rosary 150:psalter, 2
rosary ring 10:decad ring
rose 7:perfect color
Rosenblüt, Hans 12:Meistersingers
roseola infantum:6
rouge-et-noir:30
round o:0
Round Table:150
Round Table, Knights of the
  12:Knights of the Round Table
Route 66:66
row 2:distichous; 4:quadriserial;
  5:quinquefarious
Royal Bengal Fusileers 101:Dirty
  Shirts
royal bodyguard, yeomen
  of 4:exon
royal colonies 13:original
  colonies; 24:Twenty-four
  Proprietors
Royal Horse Artillery 4:Four
  Wheeled Hussars
royal quarto ¼:quarto
royal stag:12
royalty of Christ 4:Evangelists
Rubens, Peter Paul 13:Louis Treize
Rubin, Jerry C. 7:Chicago Seven
Rucellai 1500:Cinquecentesti
Rudra 8:Agni
Rue de la Ferronnerie 14:Henry IV
rugby:15
rule of 78:78
rule of eleven:11
rule of nines:9
rule of three:3
rule of two:2
ruler 8:Agni
rummy 500:five hundred rummy
"Rumpelstiltskin":3
rupee 100:cent
Ruskin, John 7:*Seven Lamps of
  Architecture*
Russell, Jeffrey 5:Fort Dix Five
Russia 2:Dual Alliance, 1, 2;
  3:League of the Three Emperors,
  Triple Alliance, Triple Entente;
  7:Seven Years' War; 13:Thirteen
  Years' War; 21:Twenty-one Years'
  War; 54:"Fifty-four Forty";
  600:Light Brigade

Russian Five   5: St. Petersburg Five
"Russian Five, The French Six, and
   Erik Satie" 6: Six
Russia's right of self-determination
   14: Fourteen Points
Ruth   8: Octateuch
Rye   5: Cinque Ports

S

S̲:7; :70
S̅:7000
S, s:15; :18; :19
Sabaoth   10: names of God
Sabbath:1; :7
Sabians   7: Hell
Sabines:9
Sabines   9: Sabines
sacerdotal office of Christ
   4: Evangelists
Sachs, Hans   12: Meistersingers
Sacrament, Holy   7: Holy Sacrament
sacred books:7
Sacred Order   9: Precious Stones
sacred songs   9: Muses
sacred thread   2: Duija
Sacrifice   7: *Seven Lamps of
   Architecture*
sacrifice double:2
sacrifice of Christ   4: Evangelists
saddle   ½: demi-pique
sadi:18
Sadi   4: Eloquence
Safiya   10: wives
sage   7: muni
Sagittarius   ⅓: trigon; 2: double-
   bodied; 12: sign
Sagris   12: Knights of the Round
   Table, 1
St. Alexis Falconieri   7: Seven Holy
   Founders
St. Andrew   7: Champions of
   Christendom
St. Anthony   7: Champions of
   Christendom
St. Bartholomew Amidei   7: Seven
   Holy Founders
St. Bartholomew massacre
   24: Bartholomew day

St. Benedict dell'Antella   7: Seven
   Holy Founders
St. Bonaventure College   3: Little
   Three
St. Bonfilius   7: Seven Holy
   Founders
St. Bridget   15: O's of St. Bridget
St. Clare   2: Second Order of St.
   Francis
St. Clement's   5: "five farthings"
St. David   7: Champions of
   Christendom
St. David's day:1
St. Denys   7: Champions of
   Christendom
St. Francis   2: Second Order of St.
   Francis; 3: Third Order of St.
   Francis
St. George   7: Champions of
   Christendom
St. Gerard Sostegni   7: Seven Holy
   Founders
St. James   7: Champions of
   Christendom
St. John Bonagiunta   7: Seven Holy
   Founders
St. Joseph's College   5: Big Five, 2
St. Lucy, Feast of   4: Ember Days
St. Martin's   5: "five farthings"
St. Nicholas day:6
St. Patrick   7: Champions of
   Christendom
St. Patrick's day:17
St. Petersburg Five:5
St. Ricoverus Uguccione   7: Seven
   Holy Founders
St. Swithin's day:15; :40
St. Valentine's day:14
St. Winifred's well   7: Seven
   Wonders of Wales
Saites   6: Hyksos
Sakyamuni   5: Buddha, 2
Saladin   1/10: Saladin's tenth
Saladin's tenth:1/10
Salamis   7: Homer
Salates   6: Hyksos
Saleh   10: animals in paradise
Salic legal system   100: centenarius
Salome   7: Dance of the Seven Veils
salute:17; :19; :21
salvarsan:606

salvarsan   914:neosalvarsan
salvation, new birth for
   7:Fundamentalism
salvation of neighbor, pray for
   7:spiritual works of mercy
salvation to the believer
   7:Fundamentalism
Samantabhadra   5:Manusibuddha;
   8:Dhyanibodhisattva
Sambhogakaya   5:Buddha, 1
samekh:15
samekh   22:major arcana
Samian Sibyl   10:sibyl
Samkhya   6:Six Schools of Hindu
   Philosophy
Samson:7
Samson, duc de Bourgogne
   12:Paladins of Charlemagne
Samuel   3:Kings
sancho pedro   3:dom
Sancroft   7:seven bishops
sandfly fever   3:phlebotomous
   fever
Sandras, Giles de Courtilz
   de   3:*Three Musketeers*
sandwich   2:double-decker;
   3:triple-decker
Sandwich   5:Cinque Ports
sanhedrin   23:lesser sanhedrin;
   70:great sanhedrin
San Quentin Six:6
Sans-culottides:5
Santa   9:Sabines
Santa Claus   6:St. Nicholas day;
   8:Santa's reindeer
Santa Maria   3:Columbus
Santa's reindeer:8
Santeuil   7:Pleiad, 4
san ts'ai   3:three-color Ming ware
Sappho   4:Lesbian Poets; 9:Lyric
   Poets
Sarasin   3:Magi
Saratoga   15:fifteen decisive battles
Sarva   8:Agni
Sarvanivaranaveskambhin
   8:Dhyanibodhisattva
Saskatchewan   10:Ten Provinces
Satan   6:Lucifera
satellites   2:secondary planet
Sator   3:Magi
Saturday:7

Saturday   7:Sabbath, Seventh-Day,
   Seventh-Day Adventist
Saturn   7:alchemy, stars;
   9:Etruscan gods, Nine Heavens,
   planet
Saudi Arabia   13:OPEC
Savage's Station   7:Seven Days'
   Battles
sawbuck:10
Saxony   3:Alliance of the Three
   Kings; 7:Seven Years' War
scabies   7:seven-year itch, 1
Scaea   6:Troy
scalenohedron:12
scales   7:Libra
Scheherezade   1001:Arabian Nights
Scoblick, Anthony   6:Harrisburg
   Six
score:20
Scorpio   1/3:trigon; 12:sign
Scorpion   12:sign
Scotch Sea   4:four seas
Scotch foursome:4
Scotland, Church of   2:Books of
   Discipline
Scotland, Curse of   9:Curse of
   Scotland
Scotland, patron saint of
   7:Champions of Christendom
Scott, Adrian   10:Hollywood Ten
Scott, Cyril   5:Frankfurt Five
Scottish Rite of Masonry   32:thirty-
   second degree Mason; 33:thirty-
   third degree Mason
Scythopolis   10:Decapolis
Seale, Bobby   7:Chicago Seven
sea nymphs   50:Nereid
secentismo:1600
second:1/60; :2
'second'   2:deutero-
secondary:4
Secondary   3:Triassic
secondary color:3
secondary deviation:2
secondary education:2
secondary planet:2
secondary school   2:secondary
   education
second brass   2:dupondius
second childhood:2
second class mail:2

Second Coming:2
Second Coming of Christ
  7:**Fundamentalism**
second covenant:2
second cranial nerve   2:**optic nerve**
second day   6:**Creation**
second death:2
second-degree burn:2
second-degree murder   2:**murder in
  the second degree**
seconde:2
Second French Pleiad   7:**Pleiad, 4**
second-hand:2
second intention:2
**Second Law of
  Thermodynamics:2**
Second Order of St. Francis:2
second person:2
second position:2
Second Reich:2
seconds:2
second sight:2
second string:2
Second Triumvirate:2
second wind:2
Secretary of Air Force, Army,
  Defense, Navy 19:**salute**
Secret Council   8:**Octavian**
secret diplomacy   14:**Fourteen
  Points**
Secret Intelligence Service   6:**M. I. 6**
secret service   6:**M. I. 6**
section   160:**quarter-section**
section eight:8
secular sword   2:**Doctrine of the
  Two Swords**
secundigravida:2
secundipara:2
secundogeniture:2
security   6:**M. I. 6**
Security Service   5:**M. I. 5**
Seder   4:**four questions**
seed   9:**enneaspermous**
see double:2
seeing   7:**senses**
Segarra III Palmer, Juan Enrique
  13:**Macheteros**
segment   5:**quinquefid**
segregation, racial   14:**Fourteenth
  Amendment . . .**
seicentismo   1600:**secentismo**

Seicentista   1600:**Seicentisti**
Seicentisti:1600
Selima   6:**Magi**
semi-brevier:4
semicentennial:50
semidiurnal:6; :12
Seminole   5:**Five Civilized Tribes**
semiquaver $1/16$:**demiquaver,
  sixteenth-note**
Semiramis   9:**Nine Worthy Women**
Semitic languages   10:**tripos**
semitone:$1/2$
semivowel   20:**Y**
Seneca   5:**Five Nations**
senility   2:**second childhood**
sennight:7
senocular:6
senses:7
sepal   2:**disepalous;**
  7:**heptasepalous;**
  9:**enneasepalous;** 10:**decasepalous**
Sephiroth:10
septa-:7
septan fever:7
septangle:7
September:7; :9
septempartite:7
septemvir,:7
septemviri epulones   7:**septemvir**
septenary:7
septendecillion:10^{54}; :10^{102}
septennial:7
Septennial Act:7
Septennial Act   3:**Triennial Act**
septennium:7
septentrional:7
septet:7
septette   7:**septet**
septfoil:7
septicentennial:700
septiform:7
Sept-Îles   7:**Ionian Islands**
septillion:10^{24}; :10^{42}
septimal:7
septime:7
Septinsular Republic   7:**Ionian
  Islands**
septisyllable:7
septivalent:7
septuagenarian:70
septuagenary:70

septuagesima:70
Septuagesima Sunday:70
Septuagint:70
Septuagint   3:Kings; 6:Hexapla;
   70:Seventy
septuplet:7
seraphim   9:angels
Serapion   7:Seven Sleepers
Serbia, restoration of   14:Fourteen
   Points
serenity   5:Five Blessings
Serrano:4
serve two masters:2
Servites   7:Seven Holy Founders
sesqui-:1½
sesquialteral:1½
sesquialterate   1½:sesquialteral
sesquialterous   1½:sesquialteral
sesquicentennial:150
sesquiduplicate:2½
sesquioctaval:1⅛
sesquiquartal:1¼
sesquiquintal:1⅕
sesquiseptimal:1⅐
sesquisextal:1⅙
sesquitertial   1⅓:sesquitertian
sesquitertian:1⅓
sesquitertianal   1⅓:sesquitertian
sestertius:2½
sestertius   2:dupondius
sestet:6
sestetto   6:sextet
sestiad:6
sestina:6
sestine   6:sestina
set   2:diadelphic; 5:quint;
   15:quindecad
Seton, Mary   4:Four Marys
Sevareid, Eric   8:Murrow's Boys
seven:7
'seven'   7:septa-
seven   7:septenary
Seven Against Thebes:7
seven arts:7
seven bishops:7
seven blocks of granite:7
seven-branched candlestick:7
Seven Cities:7
seven-day fever   7:relapsing fever,
   septan fever
Seven Days' Battles:7

Seven Dials:7
seven eyes   7:seven holes
seven gifts of the Holy Ghost:7
sevengills:7
Seven Gods of Luck   7:Shichi
   Fukujin
seven-handed euchre:7
seven holes:7
Seven Holy Founders:7
Seven Islands   7:Ionian Islands
Seven Lamps of Architecture:7
seven-league boots:7
seven maidens   7:seven youths
seven men of good repute:7
seven mortal sins   7:sins, seven
   deadly
sevennight   7:sennight
Seven Oaks Massacre:7
Seven Odes:7
Sevenoke, Sir William   9:Worthies
   of London
Seven Sages of Ancient Greece:7
Seven Sages of the Bamboo
   Grove:7
seven sciences   7:seven arts
Seven Seas:7
Seven Sisters:7
Seven Sisters   7:stars
Seven Sleepers:7
Seven Sleepers   10:animals in
   paradise
seventeen:17
seventeen-day fever   7:relapsing
   fever
Seventeenth Amendment:17
Seventeenth Parallel:17
seventeen-year locust:17
seventh, commit the   7:commit the
   seventh
Seventh Amendment:7
Seventh Avenue:7
seventh cranial nerve   7:facial
   nerve
Seventh-day:7
seventh day   6:Creation; 7:day of
   rest
Seventh-Day Adventist:7
Seventh Heaven:7
seventh heaven, to be in:7
seventh-inning stretch:7
seventy   70:septuagenary

Seventy, the:70
Seventy, the   70:great sanhedrin, Septuagint
seventy disciples:70
seventy-four:74
seventy times seven:490
Seven Weeks' War:7
Seven Wise Men:7
Seven Wonders of the Middle Ages:7
Seven Wonders of Wales:7
Seven Worthies of the Bamboo Grove   7:Seven Sages of the Bamboo Grove
seven-year itch:7
seven years of famine   7:seven years of plenty
seven years of plenty:7
Seven Years' War:7
Seven Years' War of the North:7
seven youths, seven maidens:7
Sèvres   16:Louis Seize
Seward, Thomas   7:Homer
sex   4:four-letter word
sex-:6
sexadecimal:16
sexagenal   60:sexagenary
sexagenarian:60
sexagenary:60
Sexagesima Sunday:60
sexangle:6
sexcentenary:600
sexdecillion:10^{51}; :10^{96}
sexdigitate:6
sexennial:6
sexi-:6
sexisyllabic   6:hexasyllabic
sexivalent:6
sext:6
sext   6:sixth hour; 7:Canonical Hours
sextain:6
sextan:6
sextan fever:6
sextans:$\frac{1}{6}$
sextant:$\frac{1}{6}$
sexte   6:sext
sextennial   6:sexennial
sextet:6
sextette   6:sextet
sextile:60

sextillion:10^{21}; :10^{36}
sexto:6
sexto-decimo:16
sextumvirate:6
sextuple:6
sextuplet:6
sextuplicate:6
sexual encounter   1:one-night stand, 2
sexual urge   1:primal urge
Shabuoth   50:Pentecost, 1
Shaddai   7:names of God
Shadrach, Meshach, and Abednego:3
Shaftesbury, Earl of   10:Ten Hours Act
Shahryar   1001:Arabian Nights
Shakespeare   3:thrice; 9:nine men's morris
Shakespearean sonnet   14:sonnet
shamash   8:Hanukkah
shangha   3:aum
Shan T'ao   7:Seven Sages of the Bamboo Grove
Shantung Province   7:Seven Sages of the Bamboo Grove
Shan-yang   7:Seven Sages of the Bamboo Grove
Sharjah   7:United Arab Emirates
Shechem   6:Cities of Refuge
Shell Oil   7:Seven Sisters, 2
Shemites   6:Hyksos
Shepherd-Kings   6:Hyksos
Shichi Fukujin:7
shilling:$\frac{1}{8}$
shilling   2:twopence
shin:21
shin   22:major arcana
Shing-bya-can   5:Five Great Kings
Shinn, Everett   8:Eight
Shirer, William L.   8:Murrow's Boys
shiva:7
Shoreditch   5:"five farthings"
short six:6
Shoshoni   4:Serrano
shotgun   10:ten-gauge shotgun; 12:twelve-gauge shotgun
show (someone) a clean (*or* fair) pair of heels:2
Shrewsbury Two:2

shrike   9:Enneoctonus, nine-killer;
   40:forty-spot
Shrove Sunday   50:quinquagesima
Siberian Seven:7
sibyl:10
Sibylline Books   2:duumvir;
   9:Amalthea; 10:decemvir, sibyl;
   15:quindecemvir
Sicilies, the Two:2
sick, visit the   7:corporal works of
   mercy
sickness   12:house
sickness, house of:6
Siebengebirge:7
Siege Perilous   150:Round Table
sight, second   2:deuteroscopy
sigma:18
sign:12
silk   2:dupion, 2
silver   7:alchemy
silver wedding anniversary
   25:wedding
Simeon   10:Lost Tribes of Israel
Simeon's prophecy   7:Sorrows of
   Mary
Simmons, E. E.   10:Ten
Simonides   3:Iambic Poets; 9:Lyric
   Poets
Simon of Cyrene carrying the cross
   14:stations of the cross
simulcast   2:binaural
sin:22
Sinai, Mt.   10:Decalogue
Sindbad the Sailor   3:Cyclops
"Sindbad the Sailor"   1001:Arabian
   Nights
**"Sindbad the Sailor, Seven
   Voyages of":7**
"Sing a song of sixpence,":6
single:1
single   1:only, unique
'single'   1:mono-
single-cut file:1
single file:1
single space:0
singleton:1
singular:1
sinister base   ¼:grand quarter
sinister base point   9:point
sinister chief   ¼:grand quarter
sinister chief point   9:point

sinners, admonish   7:spiritual
   works of mercy
sins, seven deadly:7
sins, seven mortal   7:sins, seven
   deadly
Siren:3
SIS   6:M. I. 6
Siva   3:Trimurti, trisula
six:6
'six'   6:hexa-, sex-
Six, les:6
Six Acts:6
sixain:6
Six Articles:6
Six Companies:6
Six Counties:6
six-day bicycle race:6
Six-Day War:6
Six Dynasties:6
sixfold:6
six-gun   6:six-shooter
six hundred   600:sexcentenary
**Six Idlers of the Bamboo
   Streams:6**
Six Islands:6
Six Months' War:6
Six Nations:6
six of one:6
six-pack:6
sixpence:¹/16; :6
sixpence   6:"I've got sixpence"
sixpenny:6
sixpenny nail:6
**Six Schools of Hindu
   Philosophy:6**
six-shooter:6
sixte:6
sixteen:16
Sixteen:16
sixteen   16:hexadecimal,
   sextadecimal
Sixteen Kingdoms:16
sixteenmo   16:sexto-decimo
sixteenpenny nail:16
sixteenth-note:¹/16
sixth cranial nerve   6:abducens
   nerve
Sixth-day:6
sixth day   6:Creation
sixth disease   6:roseola infantum
sixth hour:6

sixty   60:sexagenary
sixty-four-dollar question:64
sixty-fourmo:64
sixty-fourth note:$^1/_{64}$
sixty-four-thousand-dollar question
   64:sixty-four-dollar question
sixty-nine   69:soixante-neuf
sixtypenny nail:60
six-year molar:6
Sketch Book of Geoffrey Crayon,
   Gent.   20:Rip Van Winkle
Skirophorion:12
Skuld   3:Norn
"Slaughter on Tenth Avenue"
   10:Tenth Avenue
Sleepy   7:"Snow White and the
   Seven Dwarfs"
slip me five:5
Sloan, John   8:Eight
sloth   2:ai, unau; 3:three-toed
   sloth; 7:sins
slow-gait   5:five-gaited
small casino   2:little casino
small pica:11
small
   rhombicosidodecahedron:62
small rhombicosidodecahedron
   13:Archimedean solid
small rhombicuboctahedron:26
small rhombicuboctahedron
   13:Archimedean solid
small stellated dodecahedron:12
small stellated dodecahedron
   4:Kepler-Poinsot solid
smelling   7:senses
Smith, Howard K.   8:Murrow's
   Boys
Smith College   7:Seven Sisters, 1
smiths, patrons of   3:Cyclopes
Smyrna   7:Homer
snail   1:univalve
snake-eyes:2
Sneezy   7:"Snow White and the
   Seven Dwarfs"
Snowdon   7:Seven Wonders of
   Wales
"Snow White and the Seven
   Dwarfs":7
snub-cube:38
snub-cube   13:Archimedean solid
snub-dodecahedron:92

snub-
   dodecahedron   13:Archimedean
   solid
soccer team:11
Social Democratic Party
   4:Limehouse Four
soixante-neuf:69
sol:5
sol $^1/_{10}$:dinero
Soledad Brothers:3
Soledad Brothers   6:San Quentin
   Six
Solemn Procession   5:litany
solidus   3:triens, 2
Solomon   10:animals in paradise
Solon   7:Seven Sages of Ancient
   Greece, Seven Wise Men
soma   3:trichotomy
Son   3:Trinity, tritheism;
   4:tetratheism
song   3:Muses
song of degrees:15
sonnet:14
Sonoma County,
   California   33:Bear Party
Sons of Liberty   13:stripes
Sophocles   3:tritagonist; 13:Tragic
   Poets
Sorbonne   13:Louis Treize
sorrowful, comfort the   7:spiritual
   works of mercy
Sorrows of Mary:7
Sosicles   13:Tragic Poets
Sosiphanes   13:Tragic Poets
Sositheus   13:Tragic Poets
Souda   10:wives
soul   2:second death; 3:three, 2,
   trichotomy
soul and intelligence   3:aum
south   4:cardinal point
South Africa   10:Club of Ten
South Carolina   13:original
   colonies; 50:United States of
   America
South Dakota   50:United States of
   America
Southern Sea   4:four seas
Southern California, University of
   10:Pacific Ten
Southern Ch'i   6:Six Dynasties

sovereign or Chief of
State   21:salute
sovereign/subject
relationship   5:five social
relations
Soviet Union   3:Big Three, 2; 4:Big
Four, 2; 5:Five-Year Plan
space, tree of   3:Yggdrasil
spades   78:tarot
spades, ace of   3:matador
spades, two of   2:little casino,
manille
Spain   2:Hermanas; 3:Triple
Alliance, 2; 10:Ten Years' War;
12:European Economic
Community; 80:Eighty Years' War
Spain, patron saint of   7:Champions
of Christendom
span:2
Spanish Armada   15:fifteen decisive
battles
Sparta   7:Homer; 10:Trojan War;
30:Thirty Years' Truce, Thirty
Years' War; 500:agathoergi
Speaker of the House   19:salute
speech   2:double talk; 7:senses
speech, king of   5:Five Great Kings
speech, right   8:Noble Eightfold
Path
Speech day:4
Speedway   500:Indy 500
Spenserian sonnet   14:sonnet
spermatozoa   2:dispermy
sphere   ½:hemisphere
Spindletop   1,000,000: "Bet-a-
million" Gates
spirit   3:trichotomy
Spirit, first fruits   50:Pentecost, 2
Spirits of God   7:seven gifts of the
Holy Ghost
spiritual sword   2:Doctrine of the
Two Swords
spiritual works of mercy:7
splendor   10:Sephiroth
split fifty-fifty   50:go fifty-fifty
spondee   2:dispondee
Sporades   12:Dodecanese
'spouse'   ½:better half
Spurius Oppius
Cornicen   10:Wicked Ten
spy   2:double agent

square   2:ditesseral
square one, back to:1
Šri   14:Trisala
Stair, Earls of   9:Curse of Scotland
Stalin, Josef   3:Big Three, 2; 5:Five-
Year Plan
stamen   2:diandrous, didynamous;
6:hexandrous, hexastemonous;
7:heptandrous; 9:enneander;
10:decander
Standard Oil of California   7:Seven
Sisters, 2
Standard Oil of New
Jersey   7:Seven Sisters, 2
Standard & Poor's 500 Index:500
stand on one's own two feet:2
Stanford, Leland   4:Big Four
Stanford University   10:Pacific Ten
star   2:dyaster
Star   22:major arcana
star, five-pointed   5:pentacle
star, six-pointed   6:hexagram, 2
Star of David   6:hexagram, 4
stars:7
stars   6:Creation; 7:Pleiades
stations of the cross:14
statue of Zeus   7:Wonders of the
Ancient World
Stephen   2:unlucky number;
7:seven men of good repute
Stepney   5: "five farthings"
stere   ⅟₁₀:decistere; 10:decastere;
100:hectostere
Steropes   3:Cyclopes, Cyclops
Stesichorus   9:Lyric Poets
Stevenson, Robert
Louis   15: "Fifteen men on a Dead
Man's Chest"
Stevinus   10:decimal system
Steward of the Chiltern Hundreds
100:Chiltern Hundreds
Stewart, Doug   5:Vancouver Five
Sthenelus   12:Dodekathlos
stigma   2:distigmatic
stirrup cup   1:one for the road
"stitch in time . . . ":9
stiver   ⅛:doit
stone:14
Stonehenge   7:Seven Wonders of
the Middle Ages

Stones, Nine Precious   9:**Precious Stones**
stop on a dime:10
stopover   1:**one-night stand**, 1
'storm'   3:**Harpies**
strain-relaxation   3:**tridimensional theory**
Strasbourg   4:**Tetrapolitan**
strategic distribution of the army   1:**M. O.** 1
Strauss, Richard   6:**Six**; 7:**Dance of the Seven Veils**
strength   10:**Sephiroth**
Strength   22:**major arcana**
string of cash:500
stripes:13
stripping of the clothes   14:**stations of the cross**
strontium-90:90
Stuart, James Edward   15:**Fifteen**
Stuarts:88
Stuldreher, Harry   4:**Four Horsemen**
Sturgis, Frank A.   7:**Watergate Seven**
Stygian Lake   8:**Hell**
style   2:**digyn, distylous**; 10:**decagyn**; 23:**dodecagyn**
Stymphalian Birds   12:**Dodekathlos**
Styx   5:**rivers of Hell**
subduple ratio:2
substitutionary atonement   7:**Fundamentalism**
subtraction   4:**four species**, 1
succedent   12:**house**
suffering   4:**Noble Truths**
Suffolk   15:**Danelagh**
suffrage:19
suicides   8:**Hell**
Sui dynasty   6:**Six Dynasties**
Suleiman the Magnificent   9:**Worthies of London**
Summanus   9:**Etruscan gods**
summer solstice   21:**midsummer**
sun   3:**Agni**; 6:**Creation**; 7:**alchemy, stars**; 9:**Nine Heavens**; 14:**Trisala**
Sun   22:**major arcana**
Sunday:1
Sunday   1:**First-day, Sabbath**

Sundays in Lent   4:**Ember days**; 5:**Passion Sunday**; 6:**Palm Sunday**; 40:**Lent**
Sung dynasty   5:**Five Dynasties**
Sun I   4:**Four Masters of Anhwei**
Sunith   6:**Magi**
Sun King   14:**Louis Quatorze, Louis XIV**
Sun Yat-sen   5:**Five-Power Constitution**
Superabilis   12:**Knights of the Round Table**, 1
superlative:9
'supply exhausted'   86:**'all gone'**
Supreme Court:9
Supreme Perfection   4:**Noble Truths**
Supreme Truth   14:**Tantras**
Surya   3:**Agni**
'Suspended Ones'   7:**Seven Odes**
Sussex   7:**Associated Counties, heptarchy**, 2
Sutra   3:**Tripitaka**
Sutter's Mill   49:**Forty-Niners**
Svetambara   14:**Trisala**
swastika   4:**tetraskele**
Sweden   3:**Triple Alliance**, 1; 7:**Seven Years' War, Seven Years' War of the North**; 21:**Twenty-one Years' War**
sweet sixteen:16
Swift   5:**Big Five Packers**
Swinburne   6:**sestina**
swine   6:**Lucifera**
syllable   1:**monosyllable**; 2:**dissyllable**; 3:**trisyllabic**; 4:**proceleusmatic, quadrisyllabic, quaternion, tetrasyllable**; 5:**pentasyllabic, quinquesyllabic**; 6:**sexisyllabic**; 7:**heptasyllabic, septisyllable**; 8:**octosyllable**; 9:**enneasyllabic**; 10:**decasyllabic**; 11:**hendecasyllable**; 12:**dodecasyllable**
syllogism   1:**enthymeme of the first order**; 2:**enthymeme of the second order**; 4:**four species**, 2
Syme   12:**Dodecanese**
Symmachus   6:**Hexapla**
symmetry   2:**disymmetry, dyad**, 4
syphilis   606:**salvarsan**
Syracuse   15:**fifteen decisive battles**

Syria   6: Six-Day War
Syria, Roman province of
   10: Decapolis
systyle: 2
Szechuan Province   5: Five Pecks of
   Rice

T

T: 160
T̄: 160,000
T, t: 16; : 19; : 20
Tailleferre, Germaine   6: Six
tailor   9: ninth part of a man
Taiwan   5: Five-Power Constitution
"Take Me Out to the Ballgame"
   7: seventh-inning stretch
take-out double: 2
taking down from the cross
   7: Sorrows of Mary
Tammuz   4: Fast of the Fourth
   Month
T'ang dynasty   5: Five Dynasties
Tantras, the Four Great: 14
Tao Mien   6: Six Idlers of the
   Bamboo Streams
Tarbell, Edmund Charles   10: Ten
tarok   22: major arcana; 78: tarot
tarot: 78
Tarquin the Proud   9: Amalthea;
   10: sibyl
Tasso   1500: Cinquecentesti
tasting   7: senses
tau: 19
Taurus   ⅓: trigon; 7: Pleiades;
   12: sign
tav: 23
tav   22: major arcana
tax   1/10: tithe
Taygete   7: Pleiades
Taylor, Brent   5: Vancouver Five
Taylor, Roger   8: New York Eight
tea   6: Six Dynasties
Tebeth: 4; : 10
Tebeth   10: Fast of the Tenth Month
teeth: 20; : 32
"Tell me this riddle while I
   count eight": 8
Temperance   22: major arcana

temperance   7: virtues; 9: fruits of
   the Holy Spirit
temple of Artemis   7: Wonders of
   the Ancient World
Temple University   5: Big Five, 2
Ten: 10
ten: 10
ten   10: decimal system;
   100: honors; 150: honors,
   quaternary number
'ten'   10: deca-
Ten American Painters   10: Ten
ten-bore shotgun   10: ten-gauge
   shotgun
Ten Commandments
   10: Decalogue;
   50: Pentecost, 1
ten-dollar word: 10
ten-fifteen: 10
ten-fifty-five: 10
ten-fifty-seven: 10
ten-fifty-two: 10
ten-four: 10
ten-fourteen: 10
ten-gallon hat: 10
ten-gauge shotgun: 10
Tengyeling   4: Four Lings
Ten Hours Act: 10
Ten Kingdoms   5: Five Dynasties
"Ten Little Indians": 10
Ten Little Niggers   10: "Ten Little
   Indians"
Tennessee   50: United States of
   America
Tennyson   600: Light Brigade
"ten o'clock scholar, . . . ": 10
tenpenny nail: 10
tenpins: 10
Ten-pound Act: 10
Ten Provinces: 10
ten-seventy: 10
ten-seventy-eight: 10
tenspot: 10
tentacle   10: decapod
ten-ten: 10
'tenth'   1/10: deci-
Tenth Avenue: 10
tenth cranial nerve   10: vagus nerve
ten-thirty-four: 10
ten-thirty-one: 10
ten-thirty-three: 10

ten-thirty-two:10
Tenth Month, Fast of the   10:Fast of
   the Tenth Month
Ten Thousand:10,000
Ten Thousand Immortals:10,000
Ten Tribes of Israel   10:Lost Tribes
   of Israel
Ten Years' War:10
terce   3:third hour; 7:Canonical
   Hours
tercentenary:300
tercentennial:300
terdiurnal:3
tern:3
ternary:3
Terpander   4:Lesbian Poets
Terpsichore   9:Muses
terra mercurialis   3:earths
terra pinguis   3:earths
terra vitrea   3:earths
tertian:3
Tertiaries   3:Third Order of St.
   Francis
Tertiary:3
tertium quid:3
terza rima:3
tessaradecad:14
tessarescaedecahedron:14
tête-à-tête:2
tête-à-tête   2:love seat
teth:9
teth   22:major arcana
Tethis   12:Titan
tetrabiblion   4:quadripartite, 2
tetrachloride:4
tetrachord:4
tetractomy:4
tetractys number   10:decade, 2
tetracycline:4
tetrad:4
tetradactyl:4
tetradactyle   4:tetradactyl
tetradactylous   4:tetradactyl
tetradecapod:14
tetradecapodous   14:tetradecapod
tetradrachm:4
tetragammadion:4
tetragammation   4:tetregammadion
tetragon:4
tetragonal trisoctahedron
   24:icositetrahedron

tetragram:4
Tetragrammaton:4
tetrahedron:4
tetrahedron   5:Platonic solid
tetrahexahedron:24
tetrakishexahedron
   24:tetrahexahedron
tetralemma:4
tetralogy:4
tetrameter:4
tetramorph:4
tetrapharmacon:4
tetrapharmacum   4:tetrapharmacon
tetraphyllous:4
tetrapod   4:quadruped
tetrapolis:4
Tetrapolitan:4
Tetrapolitan Confession
   4:Tetrapolitan
tetrapteran:4
tetraptych:4
tetrarch:4
tetrarchate:4
tetrarchy   4:tetrarchate
tetrascele   4:tetraskele
tetraskele:4
tetraskelion   4:tetraskele
tetrastigm:4
tetrastyle:4
tetrasyllable:4
tetratheism:4
tetravalent:4
Teutonic legal system
   100:centenarius
Texaco   7:Seven Sisters, 2
Texas   1:Lone Star State; 50:United
   States of America
Thales   7:Seven Sages of Ancient
   Greece, Seven Wise Men
Thalia:23
Thalia   3:Graces; 9:Muses
thanatos   8:theta
thanksgiving   5:prayer
Thebes   7:Seven Against Thebes
Theia   12:Titan
Themis   3:Fates; 12:Titan
Theocritus   7:Pleiad, 1
Theodotion   6:Hexapla
theology   10:tripos
Theophrastus   5:Philosophers
The Raven   4:tetrameter

Theses, 95:95
Theseus   7:seven youths
Thespis   1:protagonist
theta:8
Thiard, Ponthus de   7:Pleiad, 3
thickness   3:tridimensional;
    4:dimension
Thiery   12:Paladins of
    Charlemagne
think twice:2
Third Avenue:3
third base:3
third class:3
third class mail:3
third cranial nerve   3:oculomotor
    nerve
Third-day:3
third day   6:Creation
third degree:3
third-degree burn:3
Third Department:3
third estate:3
third eyelid   3:nictitating
    membrane
third hour:3
third house:3
Third Law of Thermodynamics:3
third man:3
Third Order of St. Francis:3
third party:3
third person:3
Third Philippic:3
third position:3
third rail:3
third-rate:3
third reading:3
Third Reich:3
Third Republic:3
Third Section of . . . Chancery in
    Russia   3:Third Department
third sex:3
Third World:3
thirsty, give drink to
    the   7:corporal works of mercy
thirteen:13
Thirteen Years' War:13
thirty:30
"Thirty days hath September
    . . .":30
Thirty-nine Articles:39
thirtypenny nail:30

thirty-second degree Mason:32
thirty-six righteous men:36
thirty-third degree Mason:33
Thirty Tyrants:30
Thirty Years' Truce:30
Thirty Years' War:30
Thomas, Michelle   8:Grand Jury
    Eight
Thomond   5:Five Bloods
Thompson, Tom   7:Group of
    Seven, 2
Thor:3
Thor   12:Asir
thoracic   3:trisplanchnic
Thoth   3:Hermes Trismegistus
thought, right   8:Noble Eightfold
    Path
thousand:1000
'thousand'   1/1000:milli-; 1000:kilo-,
    milli-
Thousand and One Nights
    1001:*Arabian Nights*
thousand days:1000
Thousand Guineas   1000:Guineas
thousand-legged table:1000
thousand-legger:1000
'thousandth'   1/1000:milli-;
    1000:milli-
Thrasybulos   30:Thirty Tyrants
thread of life   3:Fates
three:3
'three'   3:tri-
three-a-cat:3
"three bags full . . .,":3
three-banded armadillo:3
"Three blind mice,":3
three-cent piece:3/100
three-color Ming ware:3
three-day fever   3:phlebotomous
    fever
Three Days' Battle:3
three-dimensional:3
three-dog night:3
three-dollar bill   3:queer as a three-
    dollar bill
Three Emperors, Battle of the:3
Three Emperors' Alliance   3:League
    of the Three Emperors
three F's:3
Threefold Collection   3:Tripitaka
threefold ministry:3

Threefold Refuge:3
three hours:3
three hours' agony   3:three hours
three hours' service   3:three hours
Three Jewels   3:Threefold Refuge
Three Kings (of Cologne)   3:Magi
three-legged mare:3
three-legged stool   3:three-legged
   mare
three-line pica:36
"Three Little Pigs":3
three L's:3
"Three men in a tub . . . ":3
Three Mile Island:3
three-mile limit:3
three-million bill:3,000,000
'three-mothered'   3:Agni
Three Musketeers:3
three-o'-cat   3:three-a-cat
three old cat   3:three-a-cat
three on a match:3
threepenny nail:3
Threepenny Opera:3
three percent consolidated annuities
   3:consolidated threes
three-point landing:3
three-quarter:3/4
three-quarter face   3/4:three-quarter
three-quarter time:3/4
three-ring circus:3
three R's:3
three score and ten:70
three sheets in the wind:3
threesome:3
three-square file:3
Three Stooges:3
three-time loser:3
three-toed sloth:3
Three Treasures   3:Threefold
   Refuge
three trees:3
Three Vedas   7:sacred books
three-way switch:3
"Three wise men of Gotham":3
thrice:3
thrones   9:angels
thunderbolts of Zeus   3:Cyclopes
Thursday:5
Thursday   5:Fifth-day
Thymbria   6:Troy
Tibet, Lhasa convents   4:Four Lings

Tibetans   16:Sixteen Kingdoms
Tiburtine Sibyl   10:sibyl
Tiedemann   3:Hakatist
tierce:3
tiercet:3
Tigris   4:rivers of Eden
Tilos   12:Dodecanese
time   4:dimension
time, tree of   3:Yggdrasil
Timon   7:seven men of good repute
tin   7:alchemy
Tindall, Connie   10:Wilmington
   Ten
Tinia   9:Etruscan gods
Tiphereth   10:Sephiroth
Tisiphone   3:Furies
Titan:12
Titans   3:Cyclopes
tithe:$^1/_{10}$
tithing $^1/_{10}$:Saladin's tenth;
   10:decennary^2, 2, 3
Titian   1500:Cinquecentesti
Titus Antonius Merenda   10:Wicked
   Ten
toe   2:didactyl; 3:tridactyl;
   4:quadridigitate, quadrisulcate,
   tetradactyl; 5:pentadactyl;
   6:hexadactylous, sexdigitate
'to go':97
Tomlinson, Eric   2:Shrewsbury
   Two
Tomyris   9:Nine Worthy Women
tongue   2:diglossia, diploglossate
tooth   1:unicuspid; 2:dicynodont,
   dilambdodont, diodont,
   diphyodont, diprotodont;
   3:tritubercular;
   4:quadricuspidate; 6:six-year
   molar; 10:decemdentate. (See also
   teeth in Index.)
topography   1:M. O. 1
topology   4:four-color map
   problem
top ten:10
Tor   12:Knights of the Round
   Table, 1
tortoise   10:avatar
to the nines:9
Touche, Ross   8:Big Eight, 2
Tours   15:fifteen decisive battles
Tower   22:major arcana

315

towns, association of   4:tetrapolis
trade, freedom of   14:Fourteen
  Points
tragedy   9:Muses
Tragic Pleiades   13:Tragic Poets
Tragic Poets:13
Trailokyavyago   8:Dhyanibodhi-
  sattva
Trajan   5:Adoptive Emperors
transcontinental railroad, first in
  U. S.   4:Big Four
transportation   14:fourteen
  penn'orth of it
transubstantiation   6:Six Articles
Transylvania   15:Fifteen Years' War
trapezohedron   24:icositetrahedron
Trapezus   10,000:Ten Thousand
Trath Treroit   12:Arthur
treasure   7:Shichi Fukujin
Treasure Island   15:"Fifteen men"
Treasures, Three   3:Threefold
  Refuge
Treaty of Prague   7:Seven Weeks'
  War
Treaty of the Eighteen
  Articles:18
Treaty of the Eighteen Articles
  24:Treaty of the Twenty-four
  Articles
Treaty of the Twenty-four
  Articles:24
Treaty of Versailles   10:Big Ten, 1
Treaty of Vienna   7:Seven Weeks'
  War
Trebellius Pollio   6:Augustae
  Historiae Scriptores
treble fitché:3
treble-tree:3
trecento:1300
tredecillion:10⁴²; :10⁷⁸
tree of knowledge   3:Yggdrasil
tree of life   3:Yggdrasil
Tree of Life   10:Sephiroth
tree of space   3:Yggdrasil
tree of time   3:Yggdrasil
trefoil:3
Trelawney   7:seven bishops
trench fever:5
trénise   4:quadrille, 2
trental:30
trente-et-quarante   30:rouge-et-noir

tri-:3
triacontahedral:30
triaconter:30
triact   3:triactinal
triactinal:3
triactine   3:triactinal
triad:3
triad   3:Pythagorean letter,
  Pythagorean triangle, three, 2
triakisicosahedron:60
triakisoctahedron:24
triakistetrahedron:12
trialogue:3
triangle:3
triangular file   3:three-square file
Triangulum:3
trianthous:3
triapsidal:3
triarchy:3
Trias:3
Triassic:3
triathlon:3
triaxial:3
tribasic:3
tribrach:3
tribrachial:3
tribracteate:3
tricennial:30
tricentenary   300:tercentenary
tricentquinquagenary:350
triceps:3
tricerion:3
trichotomy:3
trichroism:3
trichromatic:3
triclinium:3
tricolor:3
triconsonantal:3
triconsonantic   3:triconsonantal
tricorn:3
tricycle:3
tridactyl:3
tridactyle   3:tridactyl
tridactylous   3:tridactyl
trident of Poseidon   3:Cyclopes
tridigitate   3:tridactyl
tridimensional theory:3
triduum:3
triennial:3
Triennial Act:3
triens:3

trierarch:3
trifacial:3
trifid:3
trifold:3
trifoliate:3
trifoliated   3:trifoliate
trifoliolate   3:trifoliate
triforium:3
triform   3:triformed
triformed:3
triformous   3:triformed
trifurcate:3
trifurcated   3:trifurcate
trigamist:3
trigeminal nerve:5
trigintal   30:trental
triglot:3
triglyph:3
trigon: ⅓
trigoneutic:3
trigonometry:3
trigrammatic:3
trigrammic   3:trigrammatic
trigraph:3
trihedral:3
trilateral:3
trilemma:3
trilingual:3
triliteral:3
trilith:3
trilithon   3:trilith
trillion:10¹²; :10¹⁸
trilogy:3
trimaran:3
trimensual:3
trimester:3
trimeter:3
trimorphism:3
Trimurti:3
trinary:3
trine   ⅓:trigon, 2
trine aspersion   3:trine immersion
trine immersion:3
Trinity:3
Trinity   2:Filioque; 3:three, 1;
    7:church year, Fundamentalism
Trinity, Hindu   3:Trimurti
Trinity Sunday   7:church year
trinomial:3
trinominal   3:trinomial
trio:3

trioctile:135
trionym:3
triphthong:3
Tripitaka:3
Tri Pitikes   7:sacred books
triple:3
Triple Alliance:3
triple alliance   3:dreibund
triple bogey:3
triple-decker:3
triple double:3
Triple Entente:3
triple Hecate   3:triformed, 2
triple play:3
triple rime   3:feminine rime
triple space:2
triplet:3
tripod:3
tripos:3; :10
triratna   3:aum
trirectangular:3
trireme:3
trisacramentarian:3
Trisala, 14 auspicious dreams
    of:14
trisect:3
triskaidekaphobia:13
triskele   3:triskelion
triskelion:3
trispast:3
trispaston   3:trispast
trisplanchnic:3
Tristram   12:Knights of the Round
    Table, 1
trisul   3:trisula
trisula:3
trisyllabic:3
trisyllabical   3:trisyllabic
Trita   3:Agni
tritagonist:3
tritheism:3
tritium:3
tritocere:3
triton   3:tritium
tritubercular:3
triumvirate:3
Triumvirate:3
triumviri capitales:3
Triunity   3:tricerion
trivalent:3
triverbial:3

trivet:3
Trivia:3
trivium:3
trivium   7:seven arts
trochaic tripody   9:enneasemic
trochee:2
trochlear nerve:4
troika:3
Trojan   6:Troy
Trojan Sibyl   10:sibyl
Trojan War:10
trot   5:five-gaited
'trouble'   8:behind the eightball
Troup, Bob   66:Route 66
Troy:6
Troy   10:Trojan War
Truce, Thirty Years'   30:Thirty
   Years' Truce
Trucial States   7:United Arab
   Emirates
Trumbo, Dalton   10:Hollywood
   Ten
trumps, seven of   3:matador
trumps, two of   3:matador
truncated cube:14
truncated cube   13:Archimedean
   solid; 14:tessarescaedecahedron
truncated dodecahedron:32
truncated dodecahedron
   13:Archimedean solid
truncated icosahedron:32
truncated icosahedron
   13:Archimedean solid
truncated octahedron:14
truncated octahedron
   13:Archimedean solid;
   14:tessarescaedecahedron
truncated tetrahedron:8
truncated tetrahedron
   13:Archimedean solid
Truth   7:*Seven Lamps of Architecture*
Tryambaka   3:Agni
Trystan   12:Knights of the Round
   Table, 2
tsade   22:major arcana
Tsamo Ling   4:Four Lings
Tsecholing   4:Four Lings
Tsemchong Ling   4:Four Lings
tsien   500:string of cash
Tsui Tsungchih   8:Eight Immortals
   of the Wine Cup

Tuesday:3
Tuesday   3:Third-day
Turner   7:seven bishops
Turner, Anne Shephard
   10:Wilmington Ten
Turquine   12:Knights of the Round
   Table, 1
Tuscan   8:classic orders
Tuscarora   6:Six Nations
tusk   2:dicynodont
Twachtman, John Henry   10:Ten
Twelfth:12
Twelfth Avenue:12
Twelfth-cake:12
twelfth cranial
   nerve   12:hypoglossal nerve
Twelfth Night   12:Epiphany
Twelfth-night   12:Twelfth-cake
'twelve'   12:dodeca-
Twelve, quorum of   12:quorum of
   Twelve
Twelve Apostles   70:seventy
   disciples
twelve-bore shotgun   12:twelve-
   gauge shotgun
"Twelve Days of Christmas":12
twelve-gauge shotgun:12
Twelve Labors of Hercules
   12:Dodekathlos
twelvepenny nail:12
Twelve Sporades   12:Dodecanese
Twelve Tables:12
Twelve Tribes of Israel:12
Twelve Tribes of Israel   12:Ahijah
Twelve Wise Masters
   12:Meistersingers
Twentieth Century Limited:20
twenty   20:icosian
twenty   20:vintiner
twenty-eight:28
twenty-first rule:21
Twenty-five Articles:25
Twenty-four Proprietors:24
twenty-mule team:20
twenty-one   15:quince
Twenty-one Years' War:21
twenty questions:20
"twenty-seven different
   wigs":27
*Twenty Thousand Leagues
   Under the Sea*:20,000

twenty-three skidoo:23
twenty-twenty hindsight:20
twenty-twenty vision:20
twenty-vertexed dodecahedron
    5:Platonic solid
twibill:2
'twice'   2:bi-
'twice-born'   2:Duija
twicetold:2
twill:2
twin:2
twin   2:binary, didymous, 2, gemel
twine:2
Twins:2
Twins   2:Gemini; 12:sign
two:2
'two'   2:bi-, di-
two, rule of   2:rule of two
two-a-cat:2
two-alarm fire:2
two-bagger   2:double, 11
two-base hit   2:double, 11
two-bit:2
two-by-four:2
two can play at that game:2
two-cent piece:$\frac{1}{50}$
two cents:2
two cents plain:2
two-dimensional:2
two-dog night:2
two-faced:2
twofer:2
two-fisted:2
'twofold'   2:di-, bi-
'two-fold'   2:duplo-
two heads are better than one:2
two-legged mare:2
two-line brevier:16
two-line English:28
two-line pica:24
two-line small pica:22
two minds, be of:2
two-o'-cat   2:two-a-cat
two old cat   2:two-a-cat
twopence:2
twopenny:2
twopenny damn, not worth a:2
twopenny-halfpenny:2$\frac{1}{2}$
twopenny nail:2
two-platoon system:2

two's company, three's a
    crowd:2
two shakes of a lamb's tail, in:2
"two sides to every question":2
Two Thousand Guineas
    2000:Guineas
two-time (someone):2
"Two-Ton" Tony:2
two-way mirror:2
Two Years Before the Mast:2
Tydeus   7:Seven Against Thebes
Tyler, Wat   9:Worthies of London
Tyndareos   2:Twins
Tyndaridae   2:Twins
Tynwald, Court of   24:House of
    Keys
Tyr   12:Asir
Tyrone   6:Six Counties

U

U, u:5; :20; :21
ugly man:3
Ugra   8:Agni
Uller   3:Yggdrasil
Ullur   12:Asir
Ulster   5:Five Bloods
umbrella   11:number eleven
Umm al Qaiwain   7:United Arab
    Emirates
Umm Habiba   10:wives
unagitated   5:Dhyanibuddha
unau:2
'uncoordinated'   2:(have) two left
    feet
undecennial:11
'undecided'   2:two minds
undecillion:10^{36}; :10^{66}
understanding   7:senses, seven gifts
    of the Holy Ghost
'understood'   10:ten-four
'unenthusiastic' $\frac{1}{2}$:half-baked, 1
"Unfriendly Ten"   10:Hollywood
    Ten
'ungainly'   2:(have) two left feet
unicorn:1
unicorn   1:monocerous
unicuspid:1
unicycle:1
unidactyl:1

unidactyle   1: unidactyl
unidactylous   1: unidactyl
unidigitate   1: unidactyl
unilateral: 1
uninomial   1: uninominal
uninominal: 1
uniocular: 1
Union of Soviet Socialist Republics
   3: Big Three; 5: Big Five, 1
Union Pacific Railroad   4: Big Four
unipara: 1
uniparous: 1
uniped: 1
unique: 1
unit: 1
United Arab Emirates: 7
United Arab Emirates   13: OPEC
United Arab Republic   6: Six-Day
   War
United Kingdom   12: European
   Economic Community
United States   3: Big Three, 1, 2;
   4: Big Four, 2,   Four-Power
   Pacific Treaty; 5: Big Five,
   1;   7: Group of Seven, 1;   10: Big
   Ten, 1
United States of America: 50
unity: 1
unity   1: one, 2, oneness
univalent: 1
univalve: 1
universal monarch   14: Trisala
unlucky   4: Devil's four-poster;
   13: thirteen, triskaidekaphobia,
   unlucky number
unlucky number: 2; : 13
unlucky letter   8: theta
unused   1: first-hand
upanayana   2: Duija
upside-down year: 1961
upsilon: 20
Urania   9: Muses
uranium-238: 238
Uranus   3: Cyclopes; 9: planet;
   12: Titan
Urbs Septicollis   7: Rome
Urdur   3: Norn
'urination'   1: number one
Ursa Major   7: muni, septentrional
Utah   50: United States of America
Uther Pendragon   150: Round Table

V

V: 5
V, v: 17; : 21; : 22
vague year: 365
vagus nerve: 10
Vairocana   5: Dhyanibuddha,
   Herukabuddha
Vaisesika   6: Six Schools of Hindu
   Philosophy
Vajraheruka   5: Herukabuddha
Vajrapani   5: Manusibuddha;
   8: Dhyanibodhisattva
valence   1: univalent; 2: divalent;
   3: trivalent; 4: tetravalent;
   5: quinquevalent; 6: hexavalent,
   sexivalent; 7: heptavalent,
   septivalent
Valentinism   8: ogdoad, 2
Valhalla: 800
Vali   12: Asir
Valkyr: 9
Valmy   15: fifteen decisive battles
Valois, Marguerite   14: Henry IV
Vamana   10: avatar
van   5: cinquain
Vancouver Five: 5
Vanity   6: Lucifera
Vanyume   4: Serrano
Varaha   10: avatar
Varley, Frederick   7: Group of
   Seven, 2
Varnefrid   7: Pleiad, 2
Varro   9: Amalthea
Varuna   3: aum
vase   14: Trisala
vase, Greek   2: diota
Vashchenko, Augustina   7: Siberian
   Seven
Vashchenko, Lidiya   7: Siberian
   Seven
Vashchenko, Lilya   7: Siberian
   Seven
Vashchenko, Lyuba   7: Siberian
   Seven
Vashchenko, Pyotr   7: Siberian
   Seven
Vassar   7: Seven Sisters, 1
Vathek   5: Alkoremmi

vav:6
vav 4:Tetragrammaton; 22:major
  arcana
Veadar:13
Vedanta 6:Six Schools of Hindu
  Philosophy
Vedius 9:Etruscan gods
Vegetabilia:3
Vegetabilia 3:Animalia, Primalia
vegetable kingdom 3:kingdom
vegetation 6:Creation
Venezuela 13:OPEC
Ventôse:6
Venus 7:alchemy, stars; 9:Nine
  Heavens, planet
Verdandi 3:Norn
Vereen, Willi Earl 10:Wilmington
  Ten
Vergil 3:Harpies;
  100:Hecatoncheires
Vermont 50:United States of
  America
Verne, Jules 20,000:*Twenty
  Thousand Leagues Under the Sea*
Verona 4:quadrilateral, 3
Veronica wiping the face of Jesus
  14:stations of the cross
Versailles 14:Louis Quatorze
Versailles, Treaty of 10:Big Ten, 1
verse 2:distich, dyad, 3; 3:tiercet,
  trimeter; 4:quadrilateral;
  5:pentameter; 6:hexapody,
  sextain, sixain; 7:heptameter,
  heptapody; 8:octameter;
  10:decalet, decastich, dizain;
  11:hendecasyllable; 12:douzain;
  14:quatorzain
versicles 5:litany
vespers 7:Canonical Hours
vessel 2:bireme, catamaran,
  double-decker, double-ender;
  3:tern, 2, trimaran, triple-decker,
  trireme; 4:four-boater,
  quadrireme; 5:five-boater,
  quinquereme; 30:triaconter;
  50:penteconter
vessel, fishing 2:dogger
Vesta:4
Vesta 6:Vestal Virgins; 9:Sabines
Vestal Virgins:6
vestibulocochlear nerve:8

vice-chancellor:3
vicennial:20
Vice President of the United States
  19:salute
victory 10:Sephiroth
Vidar 12:Asir
Vietnam 17:Seventeenth Parallel
view, right 8:Noble Eightfold Path
vigesimal:20
vigesimal system:20
vigintillion:10^{63}; :10^{120}
'vigorous' 2:two-fisted
Villanova University 5:Big Five, 2
Viminal 7:Rome
Vinaya 3:Tripitaka
vingt-et-un 15:quince
vintiner:20
violence 8:Hell
violet 3:secondary color,
  trichromatic
violin, miniature 3:kit
Virgil (See *Vergil* in Index)
Virgin Birth 5:Fundamentalism
"Virginia" 10:Wicked Ten
Virginia 13:original colonies;
  50:United States of America;
  100:hundred, 2
Virgo 1/3:trigon; 2:double-bodied;
  12:sign
virtue 5:Five Blessings
Virtue, Festival of 5:Sans-
  culottides
virtue, king of 5:Five Great Kings
virtues:7
virtues 9:angels
Vishnu 3:Trimurti; 10:avatar
vision, double 2:diplopia
Visitation 7:Joys of Mary
visit the imprisoned 7:corporal
  works of mercy
visit the sick, widows, and orphans
  7:corporal works of mercy
Visvapani 5:Manusibuddha
Vitruvius $1\frac{1}{2}$:picnostyle; 2:systyle;
  $2\frac{1}{4}$:enstyle; 3:diastyle; 4:areostyle
Vixen 8:Santa's reindeer
vowel 1:monophthong;
  2:diphthong; 3:triphthong
Vulcan 9:Etruscan gods
Vulcatius Gallicanus 6:Augustae
  Historiae Scriptores

Vulgate   3:Kings

W

W, w:18; :22; :23
Wagner, Richard   6:Six
Wales, patron saint of   1:St. David's
 day; 7:Champions of Christendom
walk   5:five-gaited
Walküre   9:Valkyr
waltz   2:deux-temps
waltz time   ¾:three-quarter time
Walworth, Sir William   9:Worthies
 of London
Wang Chih-jui   4:Four Masters of
 Anhwei
Wang Jung   7:Seven Sages of the
 Bamboo Grove
Wang Meng   4:Four Masters of the
 Yüan Dynasty
wapentake   100:hundred, 1
war   4:Four Horsemen of the
 Apocalypse
War Between the States   7:Seven
 Days' Battles
ward eight:8
Wareham, Roger   8:New York Eight
Wareham, Wanda   8:Grand Jury
 Eight
War of the First Coalition:1
War of the Second Coalition:2
War of the Third Coalition:3
Warren, Dennis   2:Shrewsbury
 Two
Washington   50:United States of
 America
Washington, George:1
Washington, University
 of   10:Pacific Ten
Washington State University
 10:Pacific Ten
watch:4
'water':81
water   4:elemental quality; 5:Five
 Agents; 100:centigrade
Water-bearer   12:sign
Watergate Seven:7
Waterloo   15:fifteen decisive
 battles; 100:Hundred Days
waters   6:Creation

watery trigon   ⅓:trigon
wave   10:decuman, 2, 4
wax   4:tetrapharmacon
wealth   5:Five Blessings; 7:Shichi
 Fukujin
wear two hats:2
Weber, Wilhelm   12:Meistersingers
Webster, Daniel   3:Great
 Triumvirate
wedding:25; :50; :75
Wednesday:4
Wednesday   4:Fourth-day
week:52
week   7:hebdomad, 2, sennight
weekly:7
weekly   7:hebdomadal
weeping, river of   5:rivers of Hell
Weill, Kurt   3:*Threepenny Opera*
Weiner, Lee   7:Chicago Seven
Weir, J. Alden   10:Ten
Wellesley   7:Seven Sisters, 1
Wellington   15:fifteen decisive
 battles
Wells Fargo
 robbery   13:Macheteros
Wells Fargo
 Thirteen   13:Macheteros
Wenderoth, Joseph   6:Harrisburg
 Six
Wesleyan University   3:Little Three
Wessex   7:heptarchy, 2
west   4:cardinal point
Western Sea   4:four seas
West Germany   7:Group of Seven,
 1; 12:European Economic
 Community
West Jersey   24:Twenty-four
 Proprietors
West Riding   ⅓:riding
West Virginia   50:United States of
 America
wet   4:secondary
whacks, forty   40:Borden, Lizzie
whale   10:animals in paradise
What   5:five W's
"What this country needs is a
 good five-cent cigar":5
Wheel   22:major arcana
wheelbarrow   6:Six Dynasties
When   5:five W's
Where   5:five W's

whiffle-tree   3:treble-tree
whip with six strings   6:Six Articles
whisky   40:forty-rod lightning
whist   2:double dummy; 3:dummy
White   7:seven bishops
White, Sir Thomas   9:Worthies of
   London
white bull   14:Trisala
white elephant   14:Trisala
White Friars   4:four orders
white lion   14:Trisala
white moon   14:Trisala
white post:144
Whitsunday:7
Whitsunday   4:Ember Days;
   50:Pentecost, 2
Who   5:five W's
Why   5:five W's
Wicked Ten:10
widows, visit the   7:corporal works
   of mercy
'wife'   ½:better half
wife   1:monogynous
wigs, twenty-seven different
   27:"twenty-seven different wigs"
Wilde, Oscar   7:Dance of the Seven
   Veils
William I   3:League of the Three
   Emperors
William II   2:unlucky number;
   24:Treaty of the Twenty-four
   Articles
William and Mary   2:diarchy
William of Normandy   15:fifteen
   decisive battles
William of Prussia   2:Louis
   Napoleon
Williams, Shirley   4:Limehouse
   Four
Williams College   3:Little Three
Wilmington, North Carolina
   10:Wilmington Ten
Wilmington Ten:10
Wilmot proviso   3,000,000:three-
   million bill
Wilson   5:Big Five Packers
Wilson, Woodrow   3:Big Three, 1;
   4:Four Ends, Four Principles;
   5:Five Particulars, "What this
   country needs . . ."; 14:Fourteen
   Points

Wimmin's Fire
   Brigade   5:Vancouver Five
Winchelsea   5:Cinque Ports
Wine Cup, Immortals of
   the   8:Eight Immortals of the
   Wine Cup
wine of one ear:1
wing   2:dipteral; 4:quadripennate,
   tetrapteran; 6:hexapterous
winged white horse   10:avatar
Winner, Septimus   10:"Ten Little
   Indians"
winter   3:Yggdrasil
winter solstice   21:midwinter
Wisconsin   50:United States of
   America
Wisconsin, University of   10:Big
   Ten, 2
wisdom   7:seven gifts of the Holy
   Ghost; 10:Sephiroth
wise men   7:Seven Wise Men
wise men of the east   3:Magi
'wishy-washy'   ½:half-baked, 2
witches' sabbath   13:devil's dozen
witnesses   9:possession
wives:10
Wojciechowicz, Alex   7:seven
   blocks of granite
wolf   6:Lucifera
Wolkenburg   7:Siebengebirge
womb   2:didelphian
Wonders of the Ancient World:7
wood   5:Five Agents
word:1
word   2:dissyllable, doublet, 1,
   female rime, feminine rime;
   3:trisyllabic; 4:quadrisyllabic,
   quaternion, 2, tetrasyllable;
   5:quinqueliteral, quinquesyllabic;
   6:hexasyllabic; 7:septisyllable;
   8:octosyllable; 9:enneasyllabic;
   10:decasyllabic
word, taboo   4:four-letter word
World   22:major arcana
Worthies of London, Nine:9
'worthless'   2:twopenny damn;
   6:sixpenny
Wrath   6:Lucifera
wrath   7:sins
Wrexham steeple   7:Seven
   Wonders of Wales

Wright, Joe   10: Wilmington Ten
wrongs, bear patiently   7: spiritual
 works of mercy
Wu   6: Six Dynasties
Wu Chen   4: Four Masters of the
 Yüan Dynasty
Wundt   3: tridimensional theory
Wynkyn de Worde   15: points of a
 good horse
Wyoming   50: United States of
 America

X

X: 10
X, x: 19; :23; :24
Xenophon   5: Philosophers;
 30: Thirty Tyrants; 10,000: Ten
 Thousand
xi: 14
XV: 15

Y

Y: 3: Pythagorean letter; 150
Y, y: 20; :24; :25
Yale   3: Big Three, 3
Yale University   8: Ivy League
year of confusion: 707
yellow   3: primary color; 7: perfect
 color
Yesod   10: Sephiroth
Yggdrasil: 3
Yggdrasil   3: Norn
YHVH   4: Tetragrammaton
YHWH   7: names of God
yod: 10
yod   4: Tetragrammaton; 22: major
 arcana
Yoga   6: Six Schools of Hindu
 Philosophy; 14: Tantras
yoke-toed   2: zygodactyl
York Rite   10: Knight Templar
Yorkshire   ⅓: riding
Young, Alexander   7: Group of
 Seven, 2

Young Co., Arthur   8: Big Eight, 2
"Young Pretender"   2: unlucky
 number; 45: forty-five
yüan   5: Five Power Constitution
Yüan Chi   7: Seven Sages of the
 Bamboo Grove
Yüan dynasty   4: Four Masters of
 the Yüan Dynasty
Yukon Territory   10: Ten Provinces

Z

Z: 2000
Z, z: 21; :25; :26
Zacynthus   7: Ionian Islands
Zadkiel   7: menorah
Zahorsky's disease   6: roseola
 infantum
Zainab   10: wives
zayin: 7
zayin   22: major arcana
Zeba'ot   7: names of God
Zebulon   10: Lost Tribes of Israel
Zendavesta   7: sacred books
Zeppo   3: Marx Brothers
Zero: 0
zero-dimensional: 0
zero in on: 0
Zeroth Law of
 Thermodynamics: 0
zeta: 6
Zeus   2: Twins; 3: Cyclopes, Fates,
 Golden Apples of the Hesperides;
 9: Deucalion, Muses; 12: Olympians
Zeus, Cretan   2: double ax
Zeus, statue of   7: Wonders of the
 Ancient World
Zimri   6: Magi
zodiac   2: feminine sign;
 12: dodecatemory; 36: face
zodiac, divisions of   12: sign
Zoroaster   4: Eloquence
Zuñi Indians   7: Cibola, Seven
 Golden Cities of
zwanziger: 20
Zwingli, Ulrich   4: Tetrapolitan
zygodactyl: 2